A Traveler's Guide to Mars

The Mysterious Landscapes of the Red Planet

William K. Hartmann

WORKMAN PUBLISHING • NEW YORK

Featuring images from NASA Mars missions, including Mars Global Surveyor images, with thanks to Malin Space Science Systems and Caltech's Jet Propulsion Laboratory; and with additional image preparation by Daniel Berman, Gil Esquerdo, Kathleen Komarek, Kunegunda Belle, David Crown, Ron Miller, and the author, all through the Planetary Science Institute.

Notice: Twenty-five percent of the royalty proceeds from this book go to support Mars research and public outreach at the Planetary Science Institute, Tucson, Arizona.

Library of Congress Cataloging-in-Publication Data
Hartmann, William K.
A traveler's guide to Mars / by William K. Hartmann.
p. cm.
ISBN 0-7611-2606-6
1. Mars (Planet) I. Title

QB641.H335 2003
919.9'2304—dc21 2003041149

Design: Janet Vicario with Jarrod Dyer

WORKMAN PUBLISHING
708 Broadway
New York, NY 10003

Printed in Thailand

First printing June 2003
10 9 8 7 6 5 4 3 2 1

ACKNOWLEDGMENTS

Spacecraft photos of Mars in this book are from various NASA missions. Public-domain photos from the Mariner and Viking missions are courtesy of NASA and Caltech's Jet Propulsion Lab (JPL). Images from NASA's Mars Pathfinder mission were received through JPL and processed at the University of Arizona and made available through the lander imaging team leader, Peter Smith. Images from NASA's recent Mars Global Surveyor (MGS) mission were received through JPL and processed by the builders of the spacecraft camera at Malin Space Science Systems in San Diego, with additional processing at the Planetary Science Institute (PSI) in Tucson; they are used with the kind cooperation of MGS imaging team leader Michael Malin. The roughly 100,000 released MGS images can be seen on the Malin Space Science Systems Web site at www.msss.com.

Radar maps of Mars were prepared from the Mars Orbiter laser altimeter (MOLA) on the MGS spacecraft and made available through the MOLA team and team leader Maria Zuber. A differently processed version of the MOLA altimetry map was also kindly made available by Dr. Nick Hoffman and his colleagues at the Victorian Institute of Earth and Planetary Sciences in Victoria, Australia. Additional pictures from NASA's ongoing Mars Odyssey mission have been taken from daily public releases by the Mars Odyssey imaging team.

Photographs in this book are identified according to the missions that produced them. In the case of the Mars Global Surveyor mission, we have simply used the official mission catalog designations that begin with the letters AB, SPO, M, or E, which stand for various phases of the MGS mission. Further geological and topographic mapping of Mars is supported by NASA through the U.S. Geological Survey astrogeology branch in Flagstaff, Arizona, and through Lowell Observatory, also in Flagstaff; sample maps are individually credited.

Earth-based telescopic photos of Mars are from Lowell Observatory and the Lunar and Planetary Laboratory of the University of Arizona and are individually credited. Hubble Space Telescope images of Mars are released through the Space Telescope Science Institute in Baltimore.

Additional images used as chapter openers were constructed digitally by Ron Miller of Black Cat Studio in King George, Virginia, using Mars imagery wrapped onto a globe and reprojected by Gil Esquerdo at PSI.

Additional digital and painted conceptual renditions of Mars by Ron Miller, Chesley Bonestell, and the author are credited individually.

The author thanks all of the above and editor Richard Rosen, designer Janet Vicario, and the great team at Workman for collegial support and for their work—helping to make Mars exploration an exciting and beautiful adventure for armchair explorers the world over.

CONTENTS

PART IV HESPERIAN MARS: A TIME OF TRANSITION 197

PART V INTERLUDE: ROCKS FROM MARS 255

PART VI AMAZONIAN MARS: THE RED PLANET TODAY 273

PART VII WHERE DO WE COME FROM, WHERE ARE WE GOING? 409

FOREWORD

Why Mars?

For many years, I've been a part of the ambitious program of lunar and planetary exploration in the former Soviet Union, and now in Russia. I've lived through the many vicissitudes of our programs and those of the United States. Looking back, one can appreciate the importance of the first space flights to Mars and Venus in the 1960s and '70s because they not only brought us new knowledge about our closest planetary neighbor, but also paved the road to future cooperative programs. That period was the peak of the Cold War, when the very existence of space exploration was driven by political motivations, yet Mars—the planet of war—inspired the first Agreement on Cooperation in Space between NASA and the Soviet Academy of Sciences, signed in Moscow in 1971 (see page 181). Based on recent breakthroughs in exploration throughout the solar system, the first Martian exploration efforts may not seem too impressive, but their contribution to the origins of later, more ambitious projects is difficult to overestimate.

Negative results about potential Martian life, returned by the Viking landers in the '70s, resulted in a temporary loss of interest in Mars for nearly a decade, but they inspired deep thought. Time was required to comprehend the results and come to the realization that the data were still not sufficient to rule out extinct or even extant life on Mars. We recognize now that those questions are intimately related to the many underlying problems of the planet's evolution.

A new phase of resumed interest in Mars began in 1988 with the Russian mission to Phobos, one of the two Martian satellites. The spacecraft did reach orbit around Mars but was only partially successful; some good observations of Mars and Phobos were obtained, although we were unable to carry out the planned close approach to the enigmatic small moon. The new surge of Martian interest was further clouded in the 1990s by the losses of the American Mars Observer mission in 1992 and the Russian Mars 96 mission in 1996.

The loss of Mars 96 was disastrous for the Russian planetary program. It not only prevented the planned deployment of a network of small stations and devices to penetrate the Martian surface that were intended to operate jointly with an orbiter, but also negatively affected international cooperation. About 20 countries had participated in the development of the Mars 96 scientific payload. The loss also resulted in the cancellation of even more ambitious planned missions—conceived as follow-ups to Mars 96—to land a rover and launch a balloon into the atmosphere of Mars. Coupled with the tough economic situation in our country after the disintegration of the Soviet Union, the failure of Mars 96 led to a dramatic shrinking of the Russian space budget and placed severe constraints on the whole human blueprint of planetary exploration.

Meantime, NASA's approach after the Mars Observer loss was to start a new "faster-cheaper-better" program with smaller and cheaper spacecraft. It is debatable how advantageous this and other similar programs were, especially after the failure of two missions to Mars (a Mars climate orbiter and a Mars polar lander) at the turn of the twentieth century. NASA's efforts eventually resulted, however, in three successful missions (the Pathfinder lander, and the Mars Global Surveyor and Mars Odyssey orbiters) that greatly contributed to our understanding of the planet.

Today, the scientific community generally supports a very ambitious NASA program of Martian study with missions planned for every launch window through 2012. Similar interest in Mars is exhibited by the European (ESA) and Japanese (NASDA) space agencies. In addition, there is a mission planned by the Russian Space Agency to return samples from the Martian moon Phobos and also to monitor Mars from a small station left behind on the surface of Phobos after the launch of the sample return rocket.

All of this leads to the real question: Why Mars? What are we going to learn from the detailed investigation of this planet, which involves investment of substantial intellectual and financial resources? What is the implication of studying Mars within the general framework of solar system origin/evolution, and how do we view the planet in terms of further development of our civilization?

The first questions we need to address about Mars are how it evolved and, in particular, why it followed a different path of evolution from that of its neighbor Earth. What caused the dramatic change from a presumably more benign original climate to catastrophic drought conditions early in its history (roughly 3.5 billion years ago)? We have found numerous riverbeds, meandering stream systems, geological landforms associated with ancient lakes, and underground ice, but where did the water disappear to? What kinds of mineral deposits precipitated from these waters? Where are the contemporary reservoirs of water, and what were the processes of interaction between the rocks, the water, and the atmosphere? This list of questions is far from complete, but it focuses on the principal mysteries of Martian geology.

A second set of questions relates to the nature and evolution of potential Martian biology. We can assume that life on Mars originated in a way that's similar to the beginnings of biology on Earth and at about the same time (roughly 3.8 billion years ago) when both planets are thought to have been alike, with plenty of running water. Springs and volcanic vents in an ancient ocean might have served as life's first harbors. Another possibility is deep-seated microbial life on Mars—essentially independent of solar energy and photosynthesis for its primary energy supply, and even independent of the surface conditions, because its energy may come from chemical sources (perhaps from fluids migrating upward from deeper levels). To investigate such mysteries, new methods need to be employed, specifically laboratory studies of samples collected from different sites and depths on Mars and delivered to Earth.

Another kind of evidence for ancient Martian life-forms came from a particular rock described in this book: a meteorite blasted off Mars and found in Antarctica. It triggered its own wave of studies because some scientists think it contains signs of ancient life, but the reality of its supposed "nanobacteria" remains controversial. Nonetheless, the proven concept of natural transport of rocks between solar system bodies, by means of collisional processes, makes for plausible and fascinating scenarios: a potential for the seeds of life to have been exchanged between the nearby planets.

We can think of Mars as an extreme model of Earth's geological evolution (balanced by Venus, an extreme at the opposite end of the climatic scale). It is also a great natural laboratory for studying biological evolution. Generally speaking, all the nearby planets (Mars, Earth, Venus, and Mercury) need to be studied together in the sense of "comparative planetology."

There is a final reason for interest in Mars: It is the most accessible and challenging goal for future human space flights. Getting to Mars has been a dream for over a century, ever since the American astronomer Percival Lowell thought he detected traces of a civilization there. Twentieth-century exploration discouraged those hopes, but Mars has never lost its attraction. Human flights to Mars will be plausible during the first half of this century—it's possible that our children or grandchildren will be the first astronauts to step on Mars. Although the capabilities of robotic missions have not yet been exhausted, the possibilities of human flight are being seriously discussed in space agencies, industry, and the scientific community—along with their political, financial, technological, medical, and social challenges.

Obviously, the first human mission to Mars should be not a single, short-lived "flagship" program similar to the Apollo missions, but rather a long-term enterprise, with the ultimate goal of a permanent human outpost, the baseline for future Mars settlement. To make it feasible and beneficial for all humanity, from the very first expeditions this enterprise should be an international undertaking involving the pooled efforts of many countries. Human flights beyond the Moon's orbit would open new frontiers in the evolution of our civilization—the chance for its expansion throughout the whole solar system. Provided that humankind avoids economic, social, and military catastrophe, I predict we will start this adventure in the twenty-first century.

Mikhail Yakovlevich Marov
Keldysh Institute of Applied Mathematics
Russian Academy of Sciences
Moscow, Russia

PREFACE

Humans have a fitful record of exploration. Spanish conquistadors abandoned the American Southwest for about 40 years after the monumental Coronado expedition of 1540, and no one returned to the North or South Poles until a generation after the pioneering forays of Scott, Amundsen, and Peary. Human interest in the Moon declined after the last landing in 1972, and we abandoned the surface of Mars for 21 years following the robotic Viking landings of 1976. Perhaps after each wave of exploration, it takes us a while to absorb the new reality and figure out why we want to go back.

Two decades after those first Martian landers came a rising tide of interest in our neighboring planetary island, not too far across the sea of space. In 1997, two space probes reached Mars. Pathfinder landed and deployed its small rover, Sojourner, to look at nearby rocks. The Mars Global Surveyor went into orbit around Mars and has since made more than 100,000 images and maps of the surface, much more detailed images than Viking was able to provide. In 2001, Mars Odyssey, a spacecraft named for Arthur C. Clarke's famous novel, began a new round of photography and sophisticated orbital measurements to study mineral composition, leading to the confirmation of near-surface buried ice deposits. *A Traveler's Guide to Mars* uses new data and some of the best images from these recent probes in order to take the reader to the most interesting natural wonders we've discovered on the red planet.

As this book appears in print, Mars is making one of its closest approaches to Earth in many years—only 34 million miles away! Such an approach makes Mars appear brighter in the evening sky and bigger in telescopic views than it has been for a generation. Amateur astronomers with modest backyard telescopes can see the polar ice caps and many of the larger dark markings and bright deserts mentioned here. In the first weeks of 2004, three more robotic landers will attempt to touch down in carefully chosen landing spots on windswept Mars; then they will send out rovers to look at different rocks, soils, and landscape formations. More Martian probes will follow, from several different countries, all working together toward possible human journeys to Mars, perhaps in the 2020s.

In this book, the findings from the Martian probes help tell the story of the planet's evolution. I start with an introductory section about the fundamental nature of Mars and the mysterious markings that were charted from Earth by generations of telescopic viewers. The book is then divided into three major sections, with interludes in between. In the first major section, we'll visit sites that characterize the primordial, or Noachian, conditions of the planet during its infancy, when conditions were strangely more Earth-like than today. Second, we'll visit places that characterize the transition of adolescent, or Hesperian, Mars toward its modern conditions. Third, we'll tour places that shed light on the modern, or Amazonian, landscapes and processes. Among other attractions along the way, we'll see giant impact craters, the solar system's largest volcano, and a grand canyon that dwarfs our own.

The interludes that separate these three sections deal with the first attempts to land instruments on the surface of Mars and with Martian rocks that tell us about the planet's environment. Finally, throughout the book, in a series of personal reminiscences called "My Martian Chronicles," I share some memorable experiences from my own interplanetary adventures.

Any quest to understand Mars sooner or later induces a cosmic perspective. Earth seems precious and cultural differences small when seen from a neighboring planet. From this perspective, I add some thoughts in part VII on the meaning of it all, the goals of future exploration, and the problems we face as a world community in getting human explorers to Mars to finish what our generation has started.

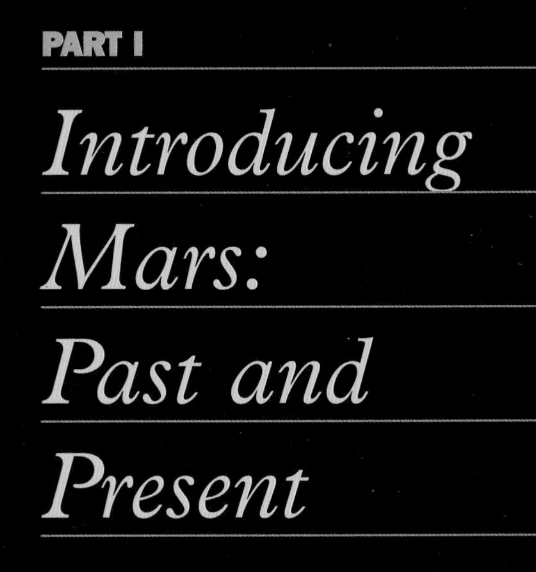

PART I

Introducing Mars: Past and Present

CHAPTER 1

THE HISTORY OF A MYSTERY

One hundred years ago, the mention of Mars called forth visions of unearthly blue-green vegetation, canals built by unknown civilizations, and malevolent invaders bent on colonizing our planet. In 1976, however, the first two probes on Mars—the Viking landers—showed none of these wonders, just fields of barren rocks. The science popularizer Carl Sagan, never quite willing to give up on the possibility of Martian life, remarked with memorable understatement that the Viking photos proved only that Mars isn't "teeming with life from pole to pole."

Our understanding of Mars has continued to grow since that time. The red planet, it turns out, is more intriguing than the geologically dead world pictured by many Viking-era researchers of the '70s. We've found evidence of geologically young lava flows and recent releases of underground liquid water onto the surface. We've found hints of ancient seas. Surprising amounts of the Martian surface have been buried by strata and then once again exposed. These discoveries lead to an exciting new mystery. Did ancient seas spawn life? Could erosive removal of sedimentary strata reveal ancient seafloor fossils? Could the sporadic aqueous and volcanic activity provide the heat and moisture needed for bacterial life even today? What about the claims that fossil Martian microbes have already been found in rocks that were blasted off

Mars by asteroid impacts and later crashed onto Earth as "Martian meteorites"? If such microbes did exist, what strange forms of DNA might they carry, and what might they tell us about the origins of life 4,000 million years ago on our own planet? Or, at the other extreme, has Mars managed to stay sterile despite its long history of aqueous activity?

This "traveler's guide" to Mars draws from the photographic mapping carried out by early orbiters, such as Mariner 9 and the two Viking craft, and from the 100,000 new photographs made by the Mars Global Surveyor spacecraft (MGS) since it entered orbit around Mars in 1997. We'll also review the discoveries made by Viking and Pathfinder landers on the surface of Mars. Armed with this information, plus data from other MGS instruments, the Hubble Space Telescope, and the Martian meteorites, we will explore alien landscapes where wind, water, and fire have shaped formations never before seen by humans (or any other sentient creatures—at least to our knowledge). Because planetary scientists and funding are both limited in numbers, most MGS photos have not been studied in detail, and readers of this book will be among the first human beings to study some of the ones chosen here.

But first, an introduction to the planet that, for all its alien features, is still the planet most like Earth.

A FIRST LOOK AT THE MARTIAN ENVIRONMENT

Mars is half the size of Earth but has roughly the same land area. Early science fiction portrayed Mars as totally alien and unfamiliar, but some aspects of the Martian surface would seem surprisingly recognizable to a human visitor. The Martian day is a bit longer than 24 hours—almost the same as Earth's. The sun rises in the east and sets in the west. The land is a cold but beautiful desert of sand, gravel, rocks, lava, dunes, and strata. Seasons of spring, summer, fall, and winter follow each other as on Earth, except that they add up to a year that is about twice as long as that on Earth, consisting of 669 Martian days.

Mars has virtually no surface liquid water because the temperatures are too cold and the air is too thin. A pan of water would evaporate in minutes if it didn't freeze first. To be more specific, the boiling temperature in the higher regions would be so low that the pan of water would bubble and sputter away. The water would be boiling, even though it would be cold. This relates to a familiar phenomenon on Earth: Water boils at lower temperatures on mountains, where the air pressure is lower, than at sea level. In the

WHAT TO WEAR: A LOOK AT MARTIAN WEATHER

On Mars, typical daily air temperatures range from about –87° C (–125° F) at night to a "balmy" –25° C (–13° F) in the afternoon. The soil and rocks absorb sunlight and become much warmer than the air; summer afternoon soil temperatures can rise to 10° C (50° F) or higher. However, soils in the morning and evening, as well as soils just below the surface, are usually much colder, with temperatures of –70° C (–94° F).

The atmosphere of Mars is very thin, almost pure carbon dioxide, with an air pressure typically just less than 1 percent of that on Earth's surface. This is still much less than the pressure of the thin air at 35,000 feet, where commercial jets cruise; it is more similar to the pressure encountered by a high-flying spy jet, 110,000 feet above Earth.

To wander among the dusty hills of Mars, you'd need a space suit similar to that worn by Apollo astronauts on the Moon. Because the soil and rocks of Mars can be much colder than those on the daylit Moon, on which Apollo astronauts landed, Martian visitors would need boots and gloves that are especially insulated.

A particular hazard facing the space-suited explorer is dust. Apollo astronauts had problems with the fine lunar dust; on Mars, this could be worse because occasional strong winds can blow the dust into suit joints, oxygen regulators, and vehicle parts. Local dust storms may strongly reduce visibility and cause blinding "brownout" conditions like arctic whiteouts, during which visibility drops to a yard or two, destroying all sense of direction. (In my novel *Mars Underground*, I conjured an encounter between an astronaut and a Martian dust devil, whose whirling winds blast a space suit with sand grains, with dangerous consequences.)

lower regions of Mars, the air pressure is great enough to prevent the water from boiling, though it would evaporate rapidly.

Despite the harsh conditions, the trackless landscape is much more inviting than that of the Moon. A visitor on Mars is greeted by vistas of rocks and hills, sand dunes and lava flows—strangely attractive in their awesome desolation. The sky is not black but bright pinkish tan in color, due to the fine reddish dust carried aloft by Martian winds. Thin clouds occasionally form overhead, especially at dawn and dusk. The wind stirs up eddies of blowing dust. Dust devils sometimes wander the landscape, leaving ghostly tracks in the otherwise pristine surface. Mysterious dry riverbeds testify to a history of ancient water flow, even though the surface is now dry as a bone. Stratified hills reveal bedded layers of ancient rock and sediments. In our own visits, later in this book, we will find tongue-like flow features on slopes hinting at glacial activity, and dry hillsides where gullies seem to have been carved by geologically recent outbursts of water. The stars are brilliant at night after the glow of hazy sunsets fade, and the constellations are the same as the ones we see from Earth, with one exception: a blue-glowing "evening star" with a faint companion "star" is sometimes prominent for an hour or so after dusk.

This stranger to our skies is the Earth-Moon system, 50 million miles away.

MYTHIC MARS

For two hundred generations, Mars teased humanity with its mysteries. To the ancients, it was distinguished by its odd color, a firelike reddish orange, and our Mediterranean forebears named it after their god of war. This red beacon in the sky was not a world but a god, a celestial force, a ruler of fates. In those days, many more people believed in astrology, the idea that planetary positions affect our destinies. This is just one of many superstitions, along with palm reading and tea-leaf reading, that share the foolish underlying theory that visual patterns found in nature can

A drawing of Mars by the Dutch observer Christiaan Huygens on November 28, 1659, is one of the first to show recognizable markings of Mars. This is believed to show the dark triangle now known as Syrtis Major.

This Huygens drawing from August 13, 1672, may be the first to show one of the polar ice caps of Mars—noted as the bright spot at the bottom.

be used to explain and predict human events. As a result of such widespread beliefs, the main motivation for observing Mars until the 1600s was not what we'd call science but rather the desire to cast horoscopes.

A more naturalistic view of Mars evolved during the dramatic Copernican revolution that started in the 1600s and continues today. It's a story of changing conceptions, slowly zeroing in on the real Mars. The developing conceptions of Mars form what I call the story of mythic Mars. This is a saga of dreams and theories, a tale of scientists' gradual progress toward truth, expressing in any given decade our best estimates of what Mars *might* be. We've been on a four-century quest to learn what wonders and secrets are hidden on the red planet. Today, we are living in the closing chapters of this saga. The story of mythic Mars will end only when humans first set foot on the red planet and the myths are replaced by gritty experience.

FIRST TELESCOPIC VIEWS

Until the early 1600s, no human gazing at Mars could see more than a point of ocher light. Then Dutch spectacle-makers invented the telescope, which the Italian naturalist Galileo improved and turned on the heavens around 1610. After 10,000 years of questions about supernatural celestial beacons, Galileo needed only a few days to prove that the Moon was not a supernatural orb but a *world,* like Earth, with mountains and plains. In the case of Mars, Galileo could see little but a small ruddy disk; but by the mid-1600s, telescopes improved and other observers began to record dusky markings on the red disk. In 1659, the Dutch observer Christiaan Huygens accurately drew the most obvious feature, a dark triangle now known as Syrtis Major. By tracking this marking, he correctly deduced that Mars turns once in about 24 hours, its day being just slightly longer than ours. By the 1670s, Huygens and his French contemporary Giovanni Cassini began to detect the bright white caps at the poles, as well as additional dusky markings.

To the generation after Galileo, Huygens, and Cassini, in the late 1600s, these dry facts had deep philosophic implications. For the first time in all of human history, people began to realize that the lights in the skies were not supernatural entities, but other worlds. Mars was revealed as another spinning globe, possibly quite like Earth! During that generation, humanity's eyes were opened and the cosmos was seen anew.

The Reverend W. R. Dawes, in England, made fine observations of Mars, including this 1865 drawing of the dark triangle of Syrtis Major. The sketch shows Syrtis Major tapering to a narrow streak curving to the north and west, and gives a sense of the streaky character of other markings.

That was only 350 years ago. Of the thousand generations since Neanderthals disappeared, only the last fifteen or twenty generations have understood the real facts of life—our relationship with the cosmos! The Renaissance, the voyages of Magellan, and the new science of astronomy had already proven that Earth was neither a flat land centered around Jerusalem nor the center of the universe, surrounded by seven crystalline spheres holding the sun, planets, and heavens—as construed by most medieval thinkers. Theologians and laypeople alike were rudely forced by direct observations to abandon the cherished myth that Earth was the imperial capital of the cosmos. Instead, the sun was the center of our planetary system, and our little world was just one of several orbiting around the sun. We were part of the universe, not its landlords.

Were the other worlds in the sky really like ours? Cassini pointed out that if Earth were observed from another planet, it, too, would display markings, bright polar caps, and rotational changes similar to what he saw on Mars. So why not an Earth-like Mars? Perhaps all these worlds were ordinary geographic places, with landscapes like Earth's. The universe might be full of alternate Earths.

A PLURALITY OF WORLDS

This new idea was called the hypothesis of the *plurality of worlds*. It electrified thinkers. Huygens and others proposed that other planets might have plants and animals. If there were alien creatures, what was their relation to God and the universe? Europe was already undergoing a convulsion dealing with the discovery of the "new world" of America and the West Indies across the sea. In the 1500s, there had been debates about whether the "Indie-ans" were humans or some kind of lesser, soulless people who could rightfully be enslaved, according to Aristotle's dictum that some persons were meant for servitude. To their credit, the scholars and the Pope settled on the official position that the Indians were fellow sentient humans with souls, even though the conquistadors, in practice, produced an exploitive political reality that was much less exalted. In 1600, a mystical and rather contentious scholar named Giordano Bruno had been burned at the stake by the Church for daring to hypothesize that God's realm might include many worlds comparable to Earth, with souls on each. In 1632, the Church's Office of the Inquisition had arrested Galileo and later threatened him with torture for teaching that the Earth was not the center of the universe, but one of several planets moving around the sun. Those struggles and debates set the stage for human civilization's dawning planetary awareness. In the medieval view, we were masters of the universe, with dominion over all creation. But if we were only part of the great sweep of things . . . what then? What would be our relationship to the hypothetical alien species of organisms on other new worlds in the sky?

As for Mars, a still clearer appreciation of reality became possible toward the end of the 1700s, when the German musician William Herschel moved to England, studied astronomy, and built large, improved telescopes. With these, he reaffirmed that the polar bright spots seemed to be arctic ice fields that expanded in winter and shrank in summer. He saw moving bright patches that he correctly interpreted as clouds, like those in Earth's skies. He noted also that Mars has not only a 24-hour day similar to Earth's (actually, it's 24.6 hours) but also an axial tilt like ours. This is why Mars has seasonal patterns of summer and winter that resemble ours, though they are stretched out over the long year of 669 Martian days. Many observers began to think that the dark Martian markings might be seas or oceans. Herschel was appointed to be England's court astronomer, and his observations strengthened the idea of an Earth-like Mars and the concept of the plurality of Earth-like worlds.

The idea of a plurality of worlds was on the right track, but overzealous. Observations later proved that all planets are indeed worlds, but most of them are not very Earth-like. In the inner solar system, our neighbor Venus, named after the Roman goddess of love, is sometimes called our sister planet because it's the nearest planet to Earth and similar in size. But with its thick carbon dioxide atmosphere, its greenhouse-effect temperature averaging 900° F, and its air pressure being 90 times greater than ours, it's hardly Earth-like! Airless Mercury is much more Moon-like than Earth-like. The other planets are in the outer solar system; they are far away, cold, rich in ices, and laced with poisonous, hydrogen-rich compounds such as methane.

Yet Mars has undeniably Earth-like features, and so by the 1800s the red planet had acquired a reputation as the one place in the solar system that might turn out to be another Earth, a habitable world.

THE NAMING OF NAMES

In the mid-1800s, several observers made maps of Mars to show features they observed with the powerful new telescopes of that period. Many still assumed that the dark grayish areas were seas and that the bright orangish areas were deserts. As reviewed by the Mars historian William Sheehan, their maps did not agree perfectly with each other and had a confusing variety of names, many

In 1884, 20 years after Dawes's drawing, the French observer E. L. Trouvelot also showed Syrtis Major narrowing into a north-pointing streak with a more gradual curve to the west.

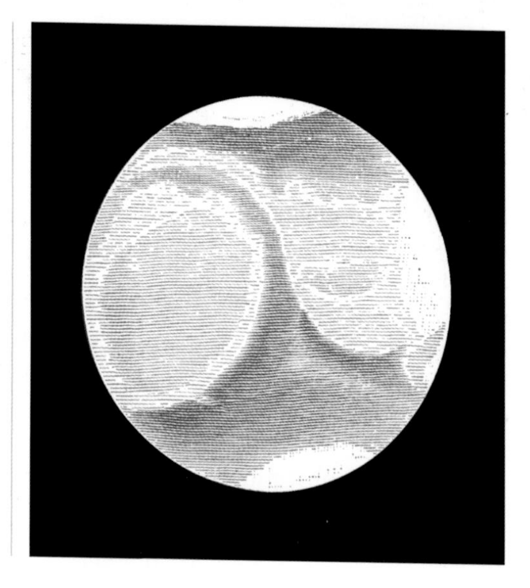

associated with the observer who charted them: Dawes Sea, Cassini Land, and Secchi's Continent. A breakthrough came when the Italian observer Giovanni Schiaparelli compiled a more reliable map during the close approach of Mars in 1877. He was an accomplished astronomer, the man who first figured out that meteor showers are caused by debris from specific comets. Schiaparelli invented a more complete and rather romantic system of names, drawing from Mediterranean, Near Eastern, and biblical sources of classical antiquity. For his Mars map, he followed the old idea that dark areas were associated with bodies of water, bright areas with lands. (This system was also used for convenience on the Moon, though everyone realized that the dark-colored lunar "seas" were not oceans but plains.) Schiaparelli's system created wonderfully euphonious and evocative names, such as Mare Sirenum (the Sea of Sirens), Solis Lacus (Lake of the Sun), Atlantis, Arabia, Eden, and Utopia. As Arthur C. Clarke remarked in his 1951 novel, *The Sands of Mars*, "Even to look at those words on the map was to set the blood pounding." The International Astronomical Union adopted this system in 1958 (see the first fold-out map in the front of the book).

This drawing by Giovanni Schiaparelli in 1888 shows the same side of Mars as Trouvelot's sketch and confirms Syrtis Major narrowing into a north-pointing curve, but shows how Schiaparelli perceived the streaks as more sharp-edged lines and added a whole network of fine linear features he called canali.

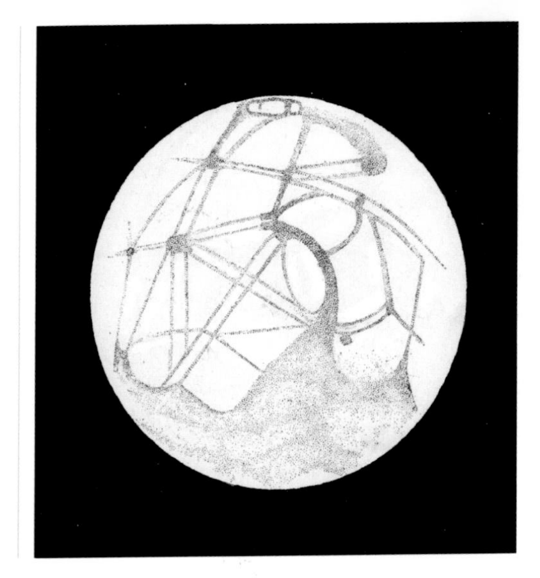

CANALS ON MARS?

At about the time Schiaparelli began making his new nomenclature of Mars, keen-eyed observers with good telescopes were noticing that the dark markings often tapered into fine streaks, as shown in the sketches by the English observer W. R. Dawes in 1865 and the French observer E. L. Trouvelot in 1884. Some observers saw these streaks as narrow straight lines; others saw them as cruder alignments of wispy patches. Schiaparelli was one of those who saw them as linear features, as shown by his sketch of 1888. The claim of the straight-line observers was that the better the telescope and the better the observing conditions, the more clearly the streaks could be seen as narrow lines. Carrying on the tradition of naming dark markings for bodies of water, Schiaparelli gave them the fateful Italian name **canali**—channels, or canals (see the Glossary at the back of the book for further definition of boldface terms). Schiaparelli did not mean artificial canals, but merely narrow waterways, fitting his larger geographic system. However, this was soon to change, as the term *canal* began to take on a more literal connotation.

The first step was a change in the explanation of the dark markings. Put yourself in the shoes of the observers in that era as they tracked the Martian dark patches from season to season. Many markings faded and became more colorless in the Martian winter; then, come spring, the contrasts and color often increased. From the time of the northern hemisphere's summer solstice through its early summer, Syrtis Major in particular appeared to become strikingly dark and bluish or greenish gray. These seasonal changes couldn't be denied, and they long tantalized astronomers. They seemed to advertise an awesome fact, put into words by Schiaparelli, one of the first to document the sweeping changes in the dark areas. He wrote to the French observer Flammarion: "The planet is not a bare uninhabitable desert of rocks. It lives!" One can well understand why these observers abandoned the idea that the dark markings were seas and adopted the hypothesis that they were wide tracts of vegetation.

THE MARS OF PERCIVAL LOWELL

In 1894, a wealthy Boston astronomer and adventurer, Percival Lowell, built a new observatory in Flagstaff, Arizona, and dedicated it to learning more about enigmatic Mars. Marveling at the excellent sky conditions at his site, he charted Mars night after

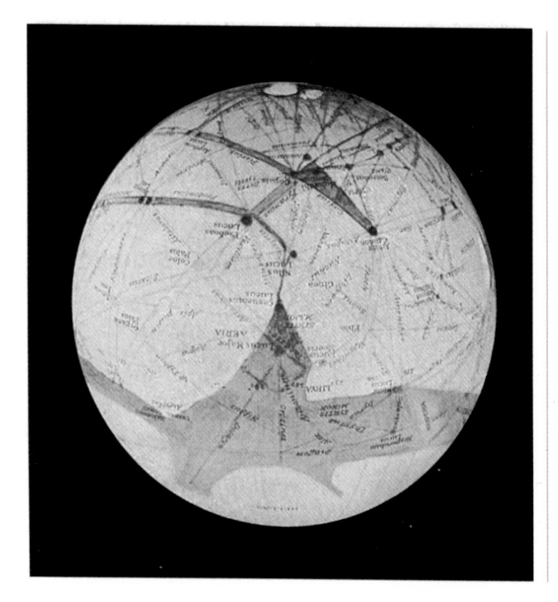

A globe of Mars, made by Percival Lowell from observations in the 1890s with his telescope on "Mars Hill" in Flagstaff, Arizona. Lowell perceived Mars as laced with straight-line canals that crossed both the lighter and darker areas, including Syrtis Major, near the middle of the disk. Note Lowell's labels are oriented with south at top, as seen in telescopes.

The planet Mars as it might have appeared to close-up spacecraft if Percival Lowell had been right. The network of canals was considered to consist of 100-km-scale strips of vegetation bordering the Martian civilization's waterways. This view shows the Syrtis Major side of the planet, based on a globe made by Lowell and preserved at Lowell Observatory in Arizona. (Painting by author.)

night with various assistants. According to Lowell, Schiaparelli's *canali,* if seen under the best conditions, were straight lines that connected end points with "wonderful directness." What could these lines be? Lowell soon devised a new theory that turned two worlds upside down!

Most astronomers of Lowell's day assumed that the planets were hot when they formed (as we now know to be true from modern measurements). Lowell correctly surmised that if the planets formed when hot, a small planet would cool off faster than a large one and its geological activity would decline faster. Also, a small planet would have less gravity than a big one to hold the molecules

of its air from escaping into space. Therefore, Lowell pictured Mars as a cooling, dying, drying planet, with declining atmospheric pressure and declining supplies of water. So far, so good—this general picture is still upheld today.

In his widely read book *Mars as the Abode of Life*, Lowell next made a leap by invoking the then-new scientific revelations of Charles Darwin. If simple life-forms had appeared on ancient

What Lowell thought we would find on Mars. This aerial photo shows a small stream winding across the Arizona desert, bordered by a wide band of dark vegetation. Although he could never have seen this view, which was taken from a modern jet, Lowell proposed that canals were similar strips of vegetation cultivated by Martians along their irrigation canals. (Photo by author.)

Earth and evolved, why not on Mars? Darwin showed how competition and natural selection helped life to evolve from simpler forms to more complex forms, eventually producing intelligent creatures. In the same way, argued Lowell, why couldn't creatures evolve to fit the colder conditions on Mars? Mars's variable dark patches, changing with the seasons, seemed to prove that at least vegetation had evolved and thrived.

Lowell wasn't through. The canals were straight lines, he said, and straight lines don't appear in nature; therefore, they must be artificial. In other words, intelligent Martians must exist who had built them! And why would they do that? The answer was clear. On cold, drying Mars, intelligent life would live in the warmest regions, at the equator, but would need the water that melted off the polar ice caps each spring to farm the Martian vegetation. The previously bland term *canals* now took on new meaning as artificial constructions. The Martians had built a canal system to bring the water from the poles to their equatorial cities! Lowell also observed dark spots on Mars, some of which were real. He called them oases and ascribed them to points at which canals intersected, creating locales with abundant water.

Actually, admitted Lowell, plausible canals would be too narrow to see from Earth; the astronomers' *canali* were really the strips of vegetated land, hundreds of miles wide, that bordered the canals. The theory fit the times. The French had tried to build a mighty canal across Panama, and the Americans completed it. Great canals were considered the highest marvel of planetary engineering.

The new research on Mars electrified intellectuals. Tennyson wrote a poem about the view of Earth as seen from Mars, and

H. G. Wells penned *The War of the Worlds*, inventing the science-fiction staple of Martian armies invading Earth. In a plot twist with twentieth-century resonance, Earth's armies try ineffectually to counterattack the Martians; the Martians are eventually defeated not by our best technology, but rather by succumbing to lowly terrestrial bacteria, against which they have no immunity.

The surface of Mars as perceived in Victorian times, based on the theories of Percival Lowell. Intersections of canals marked Martian cities with pumping stations that moved water from the polar ice caps to the warm equator, where most Martian civilization was located. Strips of vegetation lined the canals and made them visible from Earth. The balcony scene on the left is based on a famous illustration by American painter William Leigh, published in Cosmopolitan *magazine around 1910, to illustrate Lowell's theories. (Digital image by Ron Miller.)*

Surface of Mars as visualized in the 1940s and '50s. Lowell's theory of civilizations had been abandoned, but astronomers correctly interpreted the bright areas as deserts, and many of them suspected the dark areas were masses of simple vegetation such as mosses. The air was known to be thinner than on Earth, and the sky was thus interpreted as deep blue with some low, blowing dust. (Painting by Chesley Bonestell, ca. 1948; courtesy Bonestell Archives and Ron Miller.)

MARS GROWS LESS HABITABLE

Lowell's view was a wonderful theory. It did everything a good theory should do: It drew on all the latest science, and it tied all the loose ends together into a believable picture. But it was all wrong.

One issue in question was whether the canals were truly straight lines. Lowell and his supporters said they were as straight and narrow as lines on a steel engraving, but other astronomers saw them as patchy streaks. Furthermore, in the early decades of the twentieth century, astronomers' newly improved spectrometers yielded data suggesting that the Martian air was much thinner and the conditions more harsh than Lowell thought. By the early 1900s, the British astronomer Edward Maunder was characterizing the best days on Mars as winter on top of a 20,000-foot mountain on Spitsbergen Island in the Arctic Ocean. By the mid-twentieth century, the best photos with the biggest telescopes could show details of Mars only about 200 km (130 miles) across.

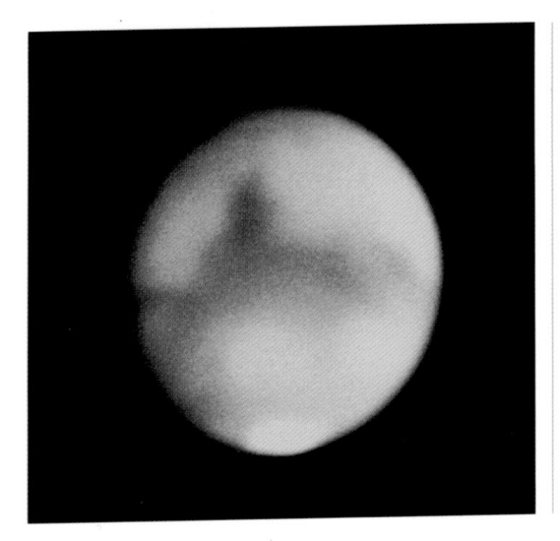

The best telescopic photography of Mars in the early 1960s was restricted by the limitations of the film of that period. Exposures of half a second or more were blurred by atmospheric shimmering. They showed the shifting dark markings, polar caps, and other major features, but could not reveal geological detail. (University of Arizona, Lunar and Planetary Laboratory.)

They showed no Lowellian canal network but were too fuzzy to allow any geologic detail to be discerned. Still, vegetation remained the best explanation of the changeable dark markings.

Then came a report in the late '50s that chlorophyll had been detected from the spectrum of Mars! Was simple Martian vegetation a proven reality? Alas, within a year or two, this information was shown to be a misinterpretation of the spectroscopic data, and we were back at square one.

As new instruments came on-line, Mars kept growing more hostile. By the 1960s, scientists knew from telescopic and spectroscopic studies that Mars's atmosphere was carbon dioxide, with less than 10 percent of the pressure of Earth's atmosphere. Researchers began to downgrade their view of what kinds of vegetation could exist on Mars, but many scientists still felt that the red planet might harbor very simple forms, such as algae or lichens. What *were* the dark markings? What was the Martian surface really like? A new approach was needed.

THE FIRST SPACE PROBES

In the summer of 1964, the United States sent the Mariner 4 probe to zip past Mars at close range and snap a few close-up photos. Would it spot artificial canals or lost cities or forests near rivers and oases? The blurry images showed not a single canal or

MEASURING MARS

In this book we will give dimensions of most features in terms of the metric system, because all Martian geographic science is done in this system, used by virtually all nations on Earth, except the United States. Even the English have abandoned the so-called English system of cumbersome miles, yards, feet, acres, pounds, ounces, quarts, et cetera. The metric system is much easier to use because all units are related by multiples of ten. If you are still uncomfortable with the metric system, remember that meters and yards are virtually the same, and that a kilometer (1,000 meters) is about a thousand yards or two-thirds of a mile. Here and there I'll give equivalents in miles to help those readers unused to metric measures.

straight line, but rather old, eroded impact craters like those on the Moon. While cautioning that their photos reflected only a tiny part of the planet, Mariner 4 scientists concluded that Mars is a world geologically as dead as the Moon, but with a little air to blow the dust around. The bottom dropped out of the market in Mars-life speculations.

Two more Mariner space probes, numbered 6 and 7, flew past Mars in 1969 and returned more pictures. This time, the images were a combination of full-disk images plus postage-stamp close-ups of small areas. Lowell's canals were still nowhere to be found. In their place were vague, discontinuous streaky markings. And no sign of Martians. As astronomer Carl Sagan quipped, Lowell's canals proved that there was intelligence involved, but at which end of the telescope? Apparently, Lowell's eye-brain combination was particularly tuned to see alignments in patchy features. He was demoted by many to the status of an overenthusiastic crank.

The Mariner 6 and 7 pictures showed more geologic variety than Mariner 4 had seen. Some bright plains seemed to have fewer craters, which implied cosmetic resurfacing processes that removed impact scars of old age. Other localized regions were unfamiliar, higgledy-piggledy morasses of hills and valleys; these were called **chaotic terrain**. Still, the pictures fortified the new scientific

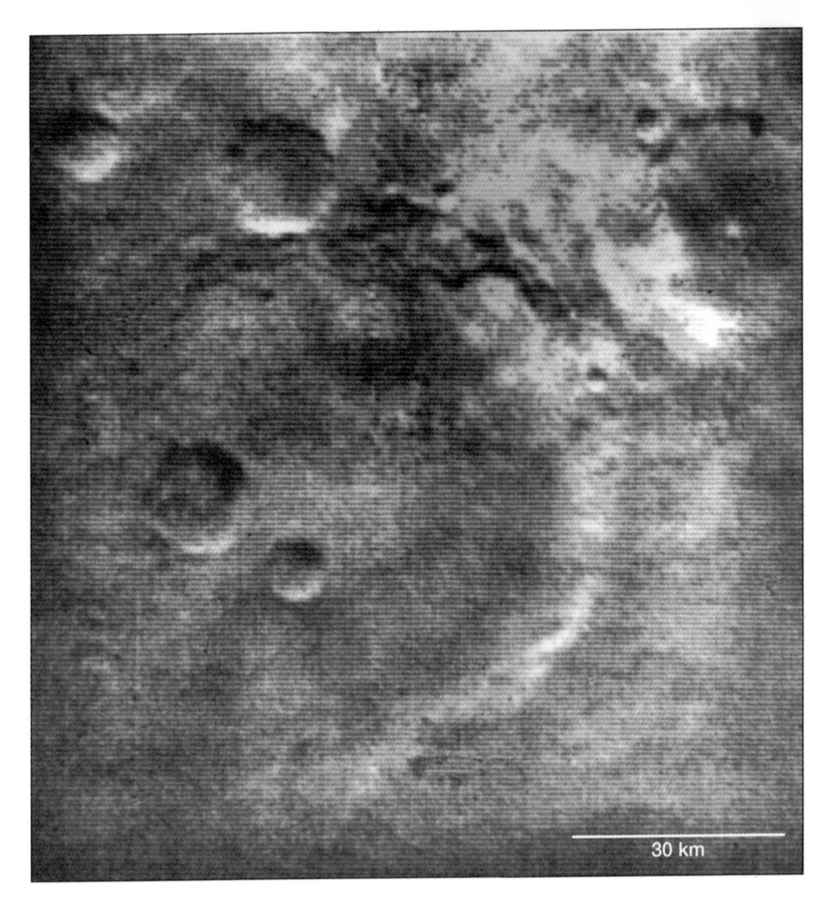

The best of the 1965 Mariner 4 close-up photos of Mars showed dimly visible craters, suggesting that Mars was a dead, Moon-like world. (NASA and Jet Propulsion Laboratory of Caltech.)

opinion that because Mars had impact craters, it must be dead as the Moon, where, as was said among astronomers, "Nothing ever happens."

By 1969, the year humans first reached the Moon, humanity seemed to have been forced to accept a picture of a dry, lifeless, hostile Mars.

ON SECOND THOUGHT

Scientific opinion, it turned out, had moved in the right direction but had overshot the mark. By a strange quirk of fate, Mariners 4, 6, and 7 had managed to photograph some of the least interesting parts of Mars—and completely missed the most exciting features.

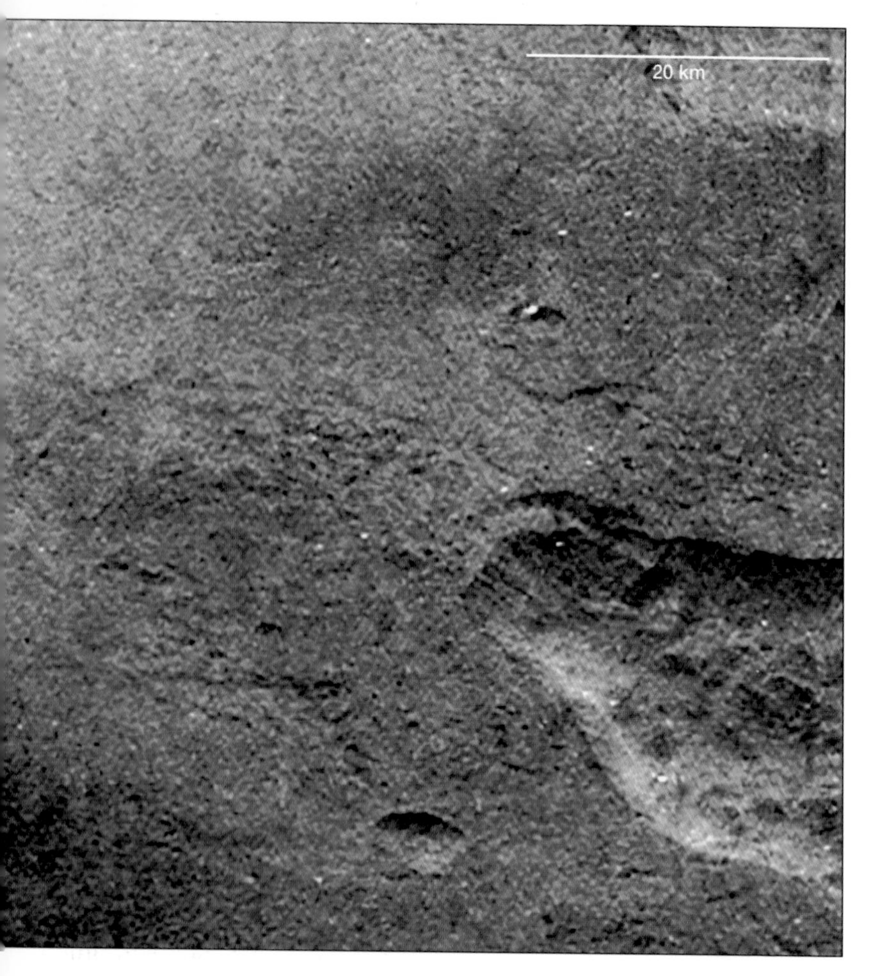

Mariner 6 and 7 views of Mars in 1969 revealed not only cratered terrain, but tantalizing glimpses of other unfamiliar terrain types. This Mariner 6 view through thin Martian haze shows some "normal" cratered plains and a hint of "chaotic terrain," seen at the right edge of the frame. (The chaotic terrain was later found to be part of the collapsed ground near the east end of the Valles Marineris canyon system, longitude 26W, latitude 13S; NASA and Jet Propulsion Laboratory of Caltech.)

With the next Mars mission, Mariner 9, the pendulum swung dramatically back toward a more interesting Martian world. Mariner 9 arrived in 1971. It was a new type of mission: Instead of zipping by the planet and snapping only a few pictures, it was the first probe to go into orbit around Mars, and it mapped the whole planet during its

Mariner 9 revealed that the chaotic terrain glimpsed by Mariner 6 and 7 is actually collapsed ground associated with release of water. This shows an area where collapse of overlying terrain (as in the lower right) caused the release of water that carved channels leading northward into a region of additional collapse in the upper left. (Mariner 9; 36W, 4N; NASA and Jet Propulsion Laboratory of Caltech.)

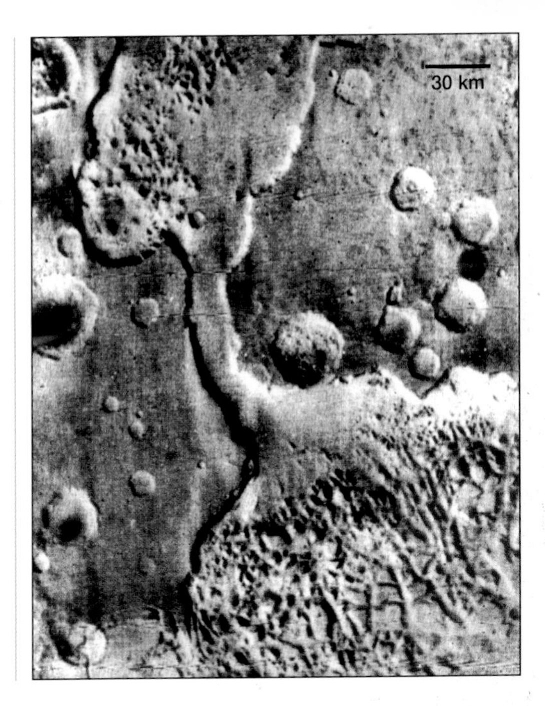

many months of operation. The first days of mapping were a washout because Mariner 9 arrived at the height of a dust storm that obscured almost the entire planet. As the dust settled, immense volcanic mountains were revealed, nearly three times as tall as Mount Everest's 29,000 feet above sea level! The air cleared, revealing a fabulous new planet, with far more geological complexity and youthful vigor than anyone had dreamed. Roughly a third of the planet indeed seemed old and geologically dead, cratered by aeons of asteroid impacts—the kind of terrain picked up by the earlier Mariners. Another third consisted of lava-covered plains and the tall volcanoes, and these displayed few impact craters—meaning that they were much younger than the ancient, cratered provinces. Other scattered areas, comprising another third of the planet, showed complex and exotic features, including polar ice fields lying atop stratified layers of sediment. A canyon near the equator was so big that our Grand Canyon would match only one of its tributary valleys. The so-called chaotic terrain testified to complicated processes of collapse and drainage, producing jumbled badlands of hills and ravines.

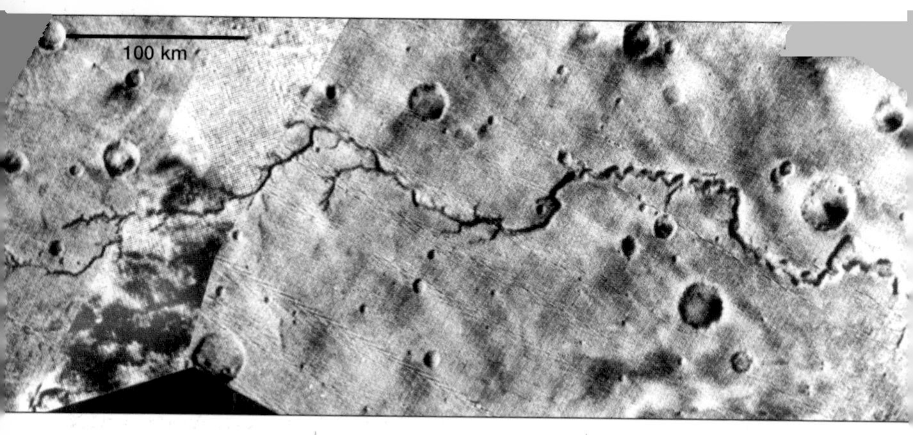

The most astonishing discovery of the Mariner 9 orbiter, which mapped kilometer-scale structures all over the planet in 1972, was that Mars is laced with ancient dry riverbeds, or channels. (Mariner 9 mosaic of Nirgal Vallis; 42W 28S; NASA and Jet Propulsion Laboratory of Caltech.)

One important mystery was solved right away. The dark markings were revealed to be not vegetation, but patches of windblown dust that could change shape as windstorms built up and subsided, capriciously leaving deposits of dust that depended on wind-flow conditions. Seasonal wind patterns explained the markings' seasonal variability. Instruments aboard Mariner 9 confirmed that the planet had no liquid water and that, given the surface conditions, liquid water would either freeze or simply evaporate rapidly into the thin air.

Amid this treasure trove of new data, the biggest advance of Mariner 9's pioneering mapping was a stunning surprise. Mars—the dry, frozen, desert planet—was laced with dry riverbeds! They appeared in the images as winding valleys, broadening in the downhill direction, and often had short tributary valleys. Some had delta deposits of sediment, where they emptied out of the uplands onto the plains. In short, they had all the features of riverbeds carved by flowing water on Earth. They looked like dry arroyos in Arizona or Utah. These features were named **channels** —not to be confused with canals. Their distribution bears little relation to the old maps of canal systems. Most are in other areas.

Most of them are found in the older uplands, not in the young, uncratered plains, implying that most of them are very old features. Most have a scattering of impact craters superimposed on them, confirming that they are old, dating from the first half or third of the planet's history. Yet, as we will see, recent work has revealed surprising evidence that the channels are only part of the story of water flow on Mars.

Mariner 9's discovery of ancient riverbeds on seemingly dry Mars changed all conceptions of the planet forever. Early Mars must have been much more Earth-like than modern Mars. Once again, we are faced with many of the old questions. Has the Martian climate changed radically from a more halcyon past? If so, why? Did Mars once have oceans as well as rivers? Could life have evolved in those ancient seas, as it did on Earth?

TOPOGRAPHY: AN ADDITIONAL SYSTEM OF NAMES

After space probes began mapping the topographic features of Mars in the 1960s and '70s, a new problem arose. Because the classical dark and bright markings are associated primarily with blowing dust, they drape almost randomly over the underlying topography. Depending on prevailing winds, one volcanic plain might show the dark coloration of fresh lavas, while another similar plain might be mantled by bright dust. Worse yet, a dark patch might appear for no rhyme or reason in the middle of a nondescript plain or might drape across a boundary from a rough upland into a smooth basin. In short, the classical dark and bright markings have little better than a random correlation with topography.

A new set of names was needed to describe the newly discovered, underlying landforms. These names were decided primarily by the Mariner 9 team and U.S. Geological Survey scientists in 1972, then adopted by the International Astronomical Union in 1973. For better or worse—probably worse—the mappers chose mostly Latin terms for topographic features, so the new Martian names can be opaque to modern readers. Some of the more important terms, in order of usefulness, are:

planitia (pl.: *planitiae*)	Low plain
mons (pl.: *montes*)	Mountain
vallis (pl.: *valles*)	Valley, usually a dry river channel
tholus	Small, domed mountain, usually a volcano
patera	Shallow, scalloped crater, usually a volcanic caldera
fossa (pl.: *fossae*)	Long, narrow valley
labyrinthus	Network of intersecting valleys
vastitas	Vast wide lowland
planum	Plateau or high plain
rupes	Scarp, or cliff
rima	Narrow fissure

To summarize, Schiaparelli's names are still used to designate the dark and bright regions visible from Earth, but additional names have been given to the underlying three-dimensional landforms. (For reference, see the "classic" fold-out map, located in the front of the book.) Thus, the bright patch known as Hellas is associated with Hellas Planitia (the plain of Hellas). A stubby dark "canal" named Coprates was revealed as the great canyon and renamed Valles Marineris (the valley system discovered by Mariner 9). A variable bright spot that Schiaparelli named Nix Olympica (Snows of Olympus) was revealed as the largest volcano in the solar system and is now named Olympus Mons (Mount Olympus). In this book, I'll emphasize the new topographic names but will also use the old regional names as appropriate for certain broad light or dark regions.

THE NEW TOPOGRAPHIC MAP OF MARS

Between 1997 and 2002, the Mars Global Surveyor (MGS) spacecraft not only obtained 100,000 photos of Mars (of which this book reproduces some of the best), but also completed an extraordinary new topographic map that we will see here. This map was made by the Mars Orbiter Laser Altimeter (MOLA) instrument. To construct the map, the laser in the orbiter reflected a flash of light off the surface of Mars and measured the time it took for the signal to return, using the principle of radar to measure the distance of the spacecraft from the tops of mountains and the bottoms of craters. These distances were then converted to altitudes, relative to the average surface. The system worked with amazing precision, measuring relative altitudes to an accuracy of within several feet, as verified by repeated orbital passes. The laser altimeter measured Martian altitudes one point at a time, but after several years it built up so many points that a detailed map emerged, like a newspaper photo made up of a million dots. The resulting map exceeds in clarity the earlier maps that had been constructed from Viking-era photos. I will use various versions of it to illustrate regional geography of the red planet.

In later chapters, I'll describe the most interesting of the strange new landscapes being revealed on Mars. But to set them in the context of geologic time, as well as in terms of Martian geographic space, I need to describe one more system of terms—the one used to describe the ancient history of the red planet.

THE MARTIAN STRATIGRAPHIC TIMESCALE: RELATIVE AGES

As planetary geologists began to piece together the history of Mars from orbital photos and topographic maps, they were keenly aware of how the pioneering European geologists, tromping around the hills and cliffsides of the British Isles and Europe around 1800, pieced together the history of planet Earth. The first step was to recognize layers, or strata, in the soils and rocks, and the second step was to classify these layers according to a time sequence. As the nineteenth-century geological pioneers studied their maps of strata layers, they recognized a threefold division of Paleozoic, Mesozoic, and Cenozoic layers—old life, middle life, and recent life—based on the fossil life-forms in the rocks. (The suffix *-zoic* relates to zoology, or life-forms.) Once the system was set up, geologists could see that surfaces in some areas were much older than surfaces in other areas. In the same way, twentieth-century planetary geologists recognized differences in geological strata on Mars. The older the surface, the more meteorite hits it had taken and the more craters it had. Thus, Martian surfaces range from highly cratered ancient upland surfaces to sparsely cratered recent lava flows and sand dunes.

But how old was "recent"? What are the absolute ages of features on Mars? The planets themselves are 4,500 million years old; no surfaces can be older than that. In setting up the Martian geologic system, the first step was to assign relative ages, and the second step was to try to figure out the absolute ages corresponding to each unit. For example, do the youngest lava flows date from ancient volcanoes 3,000 million years ago, or did they form "only yesterday" at 3 million years ago?

A Martian map combining the classic dark and light markings with the topography discovered by space probes in the 1970s. The markings correlate only poorly with the topography. (NASA.)

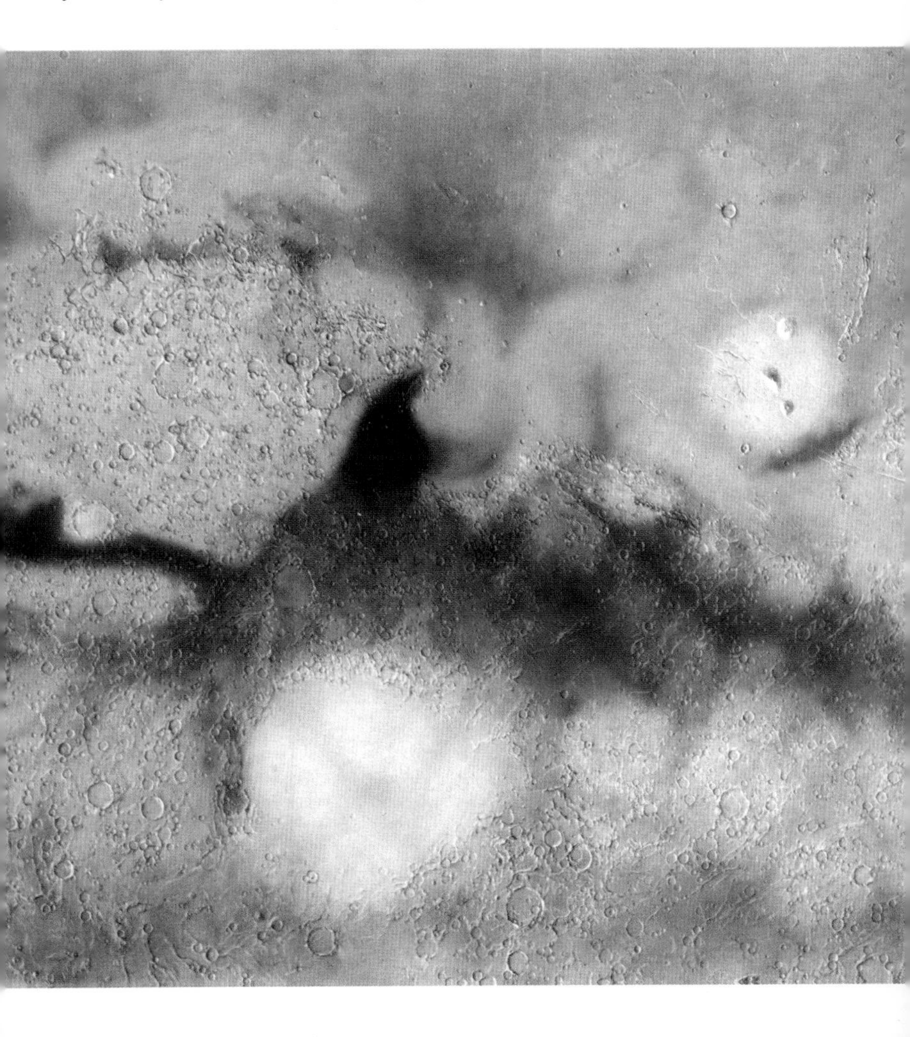

TELLING PLANETARY TIME:
A MILLION YEARS IS BUT A DAY

The solar system itself is middle-aged, and has another 4 or 5 billion years to go before the sun runs out of its hydrogen fuel (so go ahead and make those plans for next year's vacation). If you think of the present planetary age as analogous to, say, a thirty-year-old human, then 1 million years is like a couple of days in the history of a planet—a good unit for measuring planetary time. For 1 million years, we'll use the acronym MY. In this book, we need to think like geologists and get used to the idea that 2,000 MY is about halfway back in geological time, 10 MY ago is "recent," and 1 MY ago is like yesterday—essentially the present.

To carry out the first step, the U.S. Geological Survey astrogeology program in Flagstaff, Arizona, set up a system in which crater numbers were an important index for recognizing the different layers. Through work carried out partly at our Planetary Science Institute in Tucson, we have been able to tie the numbers of craters to absolute ages. Here are some rules of thumb that let us judge age from the appearance of a surface on Mars, the Moon, or Earth.

- Surfaces crowded with craters nearly rim to rim tend to be older than 3,500 MY.
- Surfaces with craters scattered here and there might be 1,000 to 3,000 MY old.
- Surfaces where it is hard to find a crater tend to be less than about 500 MY old.

On Earth, for example, we have very few impact craters, because Earth's surface is so active, geologically speaking, that most large land expanses are less than a few hundred MY old. An example of an older terrestrial surface is the central part of Canada, and careful aerial surveys starting in 1960 have revealed many ancient impact scars, tens of kilometers across, mostly eroded by glaciers into circular lakes. These are fossil craters, pale remnants of their former majesty.

Combining all these principles, U.S. Geological Survey scientists such as David Scott, Michael Carr, and Kenneth Tanaka, working with colleagues at other institutions, used the crater counts and observations of overlapping Martian strata to build up a mapping system that Tanaka formalized in 1986. It's fitting that much of this work was done in Flagstaff, just across town from the observatory at which Percival claimed to map canals on Mars a century earlier. The karma of Mars hangs thick in the piney hills of that town, a mecca for Mars buffs, being not only the home of Lowell Observatory and the Geological Survey astrogeology branch but also a jumping-off point for Mars-like terrain such as Meteor Crater, Sunset Crater National Monument, the Painted Desert, the mesas of Monument Valley, and the Grand Canyon.

Laboring over their maps in this Martian power spot, Tanaka and the U.S. Geological Survey gave the Martian time system three broad divisions, or eras.

- *Noachian Era.* The Noachian era probably lasted from the beginning of Mars, 4,500 MY ago, to roughly 3,500 MY ago, plus or minus 100 MY, according to recent estimates. It was an era of active erosion, intense volcanic activity, and possible lakes or even oceans. Early intense impact cratering was winding down. The atmosphere was probably denser, at least in the early Noachian. There are many signs of liquid water on the surface, at least sporadically, along with massive transport and deposition of sediments by such water. The climate was probably significantly different, but it is unclear whether it was much warmer or colder, on average. This era was named after the ancient uplands of Noachis, in Mars's southern hemisphere.

- *Hesperian Era.* The Hesperian era probably lasted from about 3,500 MY ago to perhaps 2,500 to 2,000 MY ago, with an uncertainty of around 500 MY. It was an era of transition to Mars's modern, drier, dustier conditions. Much Noachian water may have frozen as massive underground ice deposits, or the depth to such frozen layers grew deeper. River-forming activity continued but declined. Possible sporadic local melting of the underground ice may have produced "breakouts" of water, the collapse of overlying terrain, and massive localized floods. Most river channels probably had their last water flows in this era. This era was named after middle-aged plains of Hesperia, not far from Noachis in the southern hemisphere.

- *Amazonian Era.* The Amazonian era began roughly 2,500 or 2,000 MY ago (plus or minus 500 MY) and continues to the present. It is an era of relatively modern conditions and lower rates of geologic activity. Volcanism and impact cratering continue at lower level. Mars in the Amazonian era has been mostly dry and dusty, but moisture (from melted ice?) percolates among Martian rocks, and water is occasionally released onto the surface. This era was named after very young lava-covered plains called Amazonis Planitia, in the northern hemisphere.

GETTING ROCK-HARD DATES

As mentioned above, the dates cited are based on counts of impact craters. But are they correct? Converting crater numbers to ages involves a chain of assumptions that began to be developed in the '60s and '70s. But a breakthough in understanding Martian history came in the 1980s, when certain strange meteorites were found not to be fragments of asteroids, like most of the thousands of rocks that fall out of space onto Earth, but rather chunks knocked off of Mars. The proof that they came from Mars? They contain gas that has the same exact composition of the Martian atmosphere.

As we will see in more detail in part V, nearly all non-Martian meteorites formed 4,500 MY ago, when the planets themselves formed, but these particular meteorites attracted attention because they were only a few hundred MY old. We can date their formation by using different groups of radioactive isotopes. All but one of the first two dozen are chunks of lava, or **igneous rock**, that formed between 170 and 1,300 MY ago. (Igneous rocks are crystallized directly from molten material, as opposed to rocks formed by chemical deposition of sediments, et cetera.) The 170-MY-old lava sample formed within the last 5 percent of the planet's history, and these results beautifully confirm the crater-count estimates that Mars has had geologically recent volcanic activity.

This, in turn, may mean that future visitors to Mars will be able to find geothermally active sites and perhaps tap into steam sources to obtain water and geothermal power.

To summarize, the combination of crater counts on known provinces and randomly sampled rocks blasted off of unknown provinces sketches a picture of the 4,500 MY evolution of Mars. In the planet's early days, until about 3,500 MY, the Noachian era experienced a different climate, probably with more water. In a transitional period of the Hesperian era, this gave way to the drier, frozen climate that Mars experiences still today in the Amazonian era. Yet even in the Amazonian era, a certain level of geologic activity is producing young lava flows and other geological wonders that await us in the coming chapters.

CHAPTER 2

TERRA TYRRHENA

The Secret of the Winds

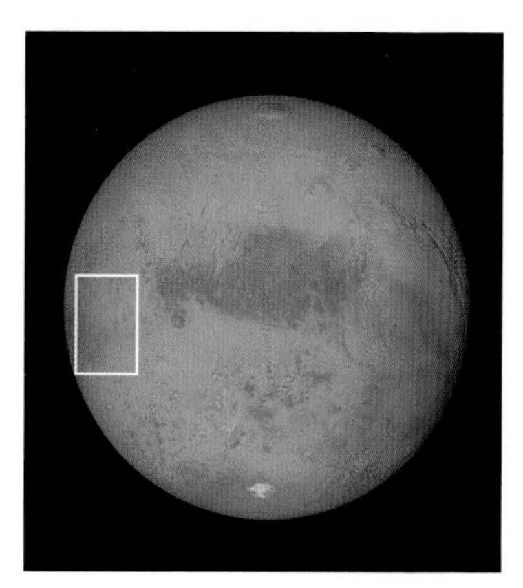

The Martian mystery that endured longest—from the 1600s to the 1970s—was the question of what caused the changeable dark markings. So our first visit is to one of the southern dusky areas of which photographs by Mariner 9 disproved the Lowellian-era ideas about vegetation and proved that the markings are related to windblown dust. This area was named Mare Tyrrhena by the Italian observer Schiaparelli in 1877. The name refers to the Tyrrhenian Sea, between Italy and Sicily, and dates from the time when a few astronomers thought the dark areas might be seas. Modern spacecraft mapping shows an old, cratered plain, necessitating a name change from Tyrrhenian Sea to Terra Tyrrhena, or Tyrrhenian Land.

The first orbital mapping photos by Mariner 9 in 1971 and 1972 showed extraordinary fields of streaks in this region. Individual fields of streaks spread over areas of hundreds of kilometers. In some areas, the streaks are lighter than the background, but often they are darker. Close examination showed that each streak emanates from a topographic feature. Usually this is a crater, but sometimes it is an isolated hill or cliff. These observations show that the streaks are **wind tails**, features created when prevailing winds blow across topographic irregularities such as craters. The irregularities disturb the wind flow, initiating turbulence that can either

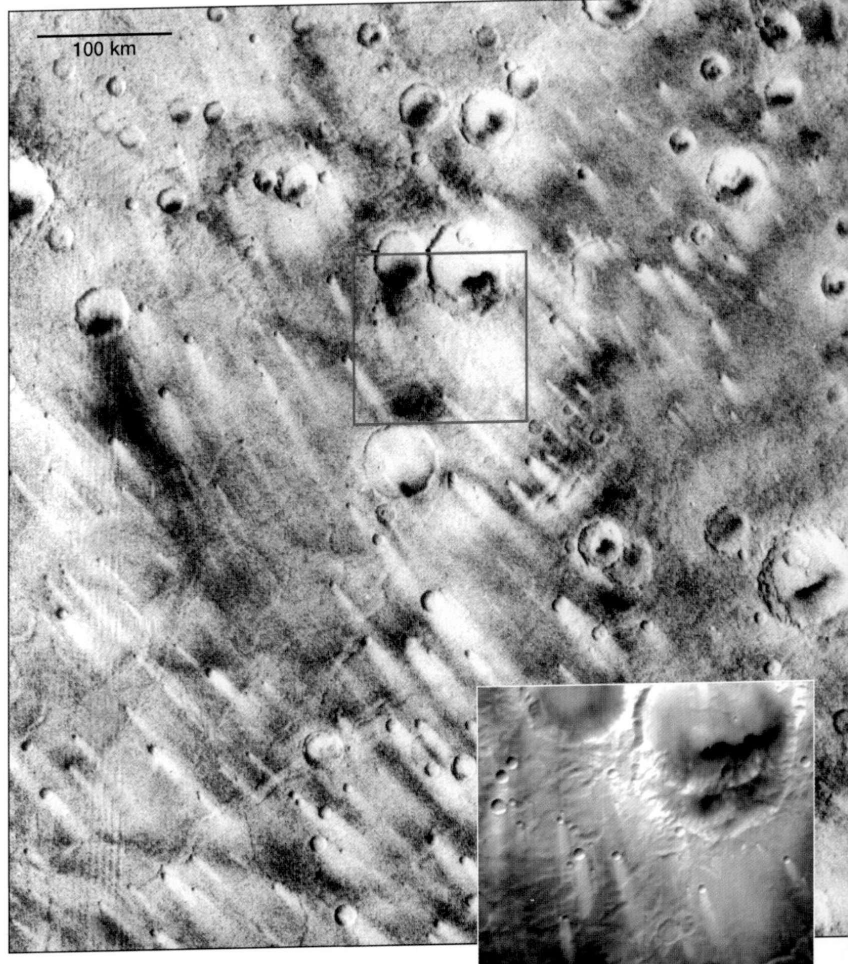

Streaks across the region of Terra Tyrrhena show the direction of prevailing winds that have left dust deposits on the leeward side of every crater rim in the area. This was one of the photos by Mariner 9 in 1972 that first indicated the relation of the bright and dark markings to windblown dust. Most of this area is heavily cratered and therefore very ancient, classified as Noachian uplands. Box shows location of next image. (Longitude 242W, latitude 17S.)

Inset: *Part of the region shown in the 1972 Mariner 9 photo was rephotographed at larger scale by Mars Global Surveyor in 2000. The pair of large craters in the top half can be found in the upper center of the Mariner 9 frame. Bright "wind tails" are still visible, but detailed patterns have changed in the intervening 28 years. For example, the dark patch in the left of the two craters has disappeared. (241W 16S, M18-01153.)*

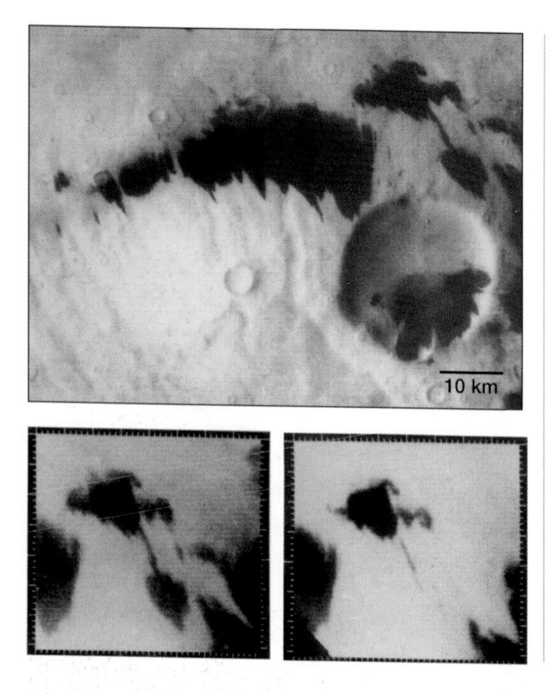

Changes in the dark markings of Mars over a period of months during the Mariner 9 mission supported the idea that they were due to windblown dust. Left: *View of cratered region with ragged markings inside and outside craters.* Below: *Close-up of spade-shaped marking in upper right shows complete disappearance of one dark area, indicating that it was a thin, surface deposit. (NASA, Stanford University.)*

scour loose dust from the surface or cause dust carried by the wind to fall out and leave deposits. On Mars, the brighter material tends to be finer, weathered dust, and the darker material tends to be coarser sand and gravel, as shown by spectroscopic and other studies, although these correlations may not be universal.

Detailed studies during the Mariner 9 and Viking missions of the '70s revealed cases in which individual dark patches actually changed shape on a scale of kilometers after windstorms. Subsequent observations on the surface of Mars, made by the Viking landers in 1976, revealed dunes and other signs of wind action and at the same time revealed a landscape with sterile soil devoid of vegetation. Taken together, these observations established once and for all that the dark, variable markings are not vegetation but instead are due to the capricious actions of the wind, which sorts light dust from darker gravels as it whistles over Martian topography.

The parallelism and prominence of the streaks over large areas, such as Terra Tyrrhena, reveal patterns of Martian prevailing winds. These operate in the same way that strong prevailing winds dominate certain areas on Earth. Under the global circulation pattern, prevailing winds tend to make stronger streaks in some areas than others. Comparison of spacecraft photos during 30 years of Mars exploration show cases where winds of one season have made dark streaks in one direction, but winds in a different season have added streaks of dust lighter than the background in a different direction! Some streaks are associated with specific dust storms that swept over the area in certain years.

A kilometer-wide impact crater shows a patch of light-toned dunes on the crater floor and a SE-trending tail, partly covered by more recent dunes (right center). (M20-01660.)

MGS pictures can show features the size of a bus, and if you were to pick the

An aerial view of wind streaks in the coastal desert of Peru (above, left) proves that both Earth and Mars have wind tails. Each house-size hillock has disturbed the prevailing winds, resulting in the creation of a streak on the leeward side.

Sand dunes in northwest Sonora, Mexico (above, right), show wind-tail effects as seen from ground level. Each bush has a streak of fine, light-colored dust deposited on its leeward side. (Both photos by author.)

darkest and most dramatic streaks, you might expect to see distinctive textures of material inside the streaks, at telephoto-scale resolution. But the most detailed images of streaks generally show nothing unusual—just a darker or lighter landscape. The dust that makes the ephemeral streaks must therefore be below the 10-meter scale resolution of the photos. It is a fine powder or sand, draping the topography in a dusty layer that can blow away with the next contrary wind.

This same process operates on Earth. As represented by the photo of the Peruvian desert, many desert areas show streaks caused by strong prevailing winds. Furthermore, clever experiments by Arizona State University geologist Ron Greeley, who used the NASA wind tunnel facilities, confirmed that wind streaks can develop behind craters due to turbulence under Martian conditions. Greeley made small model craters and blew dust across them, then observed wind tails of dunes piling up on the leeward sides, exactly as in the Martian interpretation.

The wind effects are hard to predict. In some regions, the wind may drop dunes on a leeward wall or directly in the center of the crater, but in other regions crosscurrents may press the dunes into a triangular patch against the craters' northwest walls, or in a square in the east or a bar on the south. It is often remarkable how all craters in a given region will have the same odd shape of deposits on their floors, because the prevailing wind acted the same on each circular depression. Leafing through this book, you will see photographs showing many different forms of floor deposits.

As we'll see in the next chapter, some of the largest examples of wind streaks and dust deposition played a role in the canal controversy.

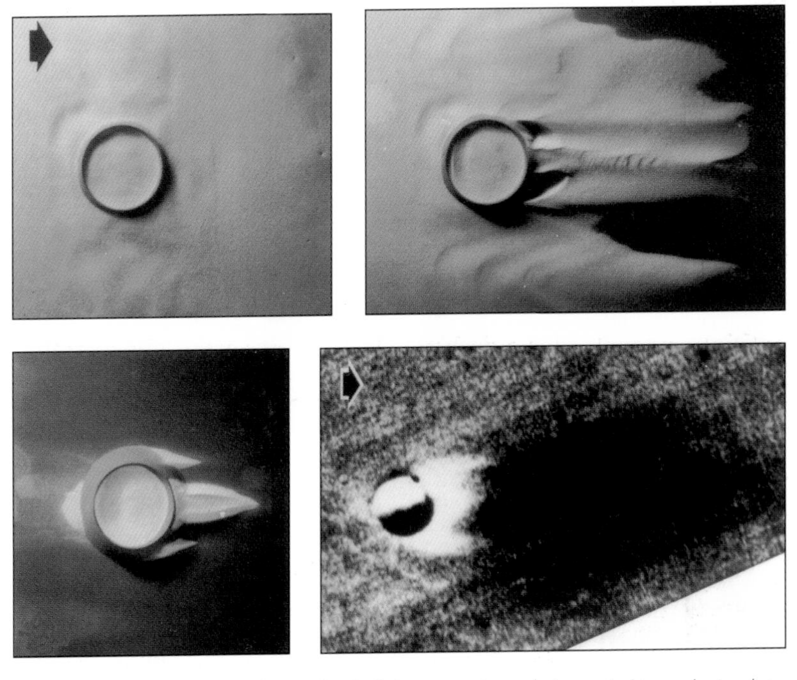

Wind-tunnel experiments by Ron Greeley, Arizona State University, used a model crater to show how wind flow can be disrupted by crater rim topography, leading to deposition of a wind tail of dunes on the leeward side of the crater. The first photo shows the crater model covered by dust, and the second and third photos show stages of formation of a wind tail as the dust is blown away in a wind tunnel, simulating Martian windstorms. The final photo shows an actual Martian crater with very similar wind-tail structure.

My Martian Chronicles

PART 1: HITCHING A RIDE ON MARINER 9

I've been lucky in planetary science. My big underlying piece of luck was to be included in the first generation, the first handful of students that tried to go into that then-nonexistent field.

I was in that generation described by Homer Hickam in his book *Rocket Boys,* later made into the film *October Sky,* which beautifully depicts the way it was for us. In grade school, I discovered a map of the Moon, showing mountains and plains that actually had names—the Moon was no longer just a light in the sky but a *place* where we could go. In high school, I built a telescope with my dad's help and knew I wanted to study planets. Grown-ups thought space travel was Buck Rogers foolishness, but we kids knew it was just a matter of time. In the early '50s, Wernher von Braun wrote detailed articles in one of the leading weekly magazines about how we could put artificial satellites in orbit around Earth and how we could fly to the Moon and to Mars. Looking back at one of these articles recently, I was amazed at the contrast between von Braun's articles aimed at kids—the exciting dream of engineering and exploration—and the vapid fluff aimed at kids in today's media. Von Braun taught us that we really could travel in space.

Launch of Mariner 9 from Kennedy Space Center, Florida. (NASA.)

I went off to college in 1957, and sure enough, that fall the Russians put the first satellite in orbit around Earth, shocking the West into convulsions of self-doubt and competitiveness. In 1961, just after the first human orbital spaceflight around Earth (again by a Russian, Yuri Gagarin), President Kennedy announced that we would try for the Moon within that decade—whereupon I went to graduate school at the University of Arizona with the man who might be called the first planetary scientist, the Dutch-American astronomer Gerard Kuiper. It was a golden age of science, NASA was throwing money at the Moon, and Kuiper had set up a major research program in lunar science, working with a superb collection of the best lunar photographs. By 1966, I had published some papers on impact craters and written my Ph.D. dissertation on how to use the number of craters on the lunar plains to estimate the lunar plains' age.

At about that time, the first Mars probe, Mariner 4, flew past Mars, and I published some comments on the significance of the craters it found. A few years later, a great new pair of Mars missions, Mariners 8 and 9, were being planned as the first probes to orbit Mars and map the entire surface with TV imagery. One day in 1970, on the eve of the missions, the telephone rang in my University of Arizona office, where I was now a somewhat naive assistant professor. On the other end of the phone was Bruce Murray, from Caltech—a scientist who had been a leader in the Mariners 4, 6, and 7 probes to Mars and who would go on to become the head of Caltech's Jet Propulsion Lab (JPL), which built all of the major American space missions of the '60s, '70s, and '80s. You need to understand—I was completely in awe of this guy. "Bill," says he, "we're looking for people to fill out the scientific teams for the new Mariner project. How would you like to be on the imaging team? We've got Carl Sagan and . . ."

He didn't have to twist my arm. As I say, I was lucky. There were only about a dozen young planetary scientists in the country at that time, and not much competition. All Murray had to do was pick up the phone to recruit the ones who were available. Today, perhaps a hundred young Ph.D.'s compete to get on every planetary mission. In order to participate, each one writes a proposal. A review committee is then assembled, referee reports are written, committees meet and debate, and only a few candidates are chosen, because NASA can fund only a few slots on each scientific team responsible for analyzing data from instruments sent into space. Sadly, many young Ph.D. researchers find they cannot get funding to work in science, and they drift off into the corporate world, working for computer firms, defense firms, or mining and oil companies.

In the golden days of the Mariner missions, invitations came out of the blue. The challenge for us today is to create a national and global society in which the best-trained young scientists get a similar chance to use their skills to advance the frontiers of long-term human knowledge—as opposed to being part of the brain drain into the global corporate efforts to make a buck off the latest cyberwhizbang, to build a smarter bomb, or to pump the last oil reserves to fuel the global frat party of twenty-first-century first-world consumers.

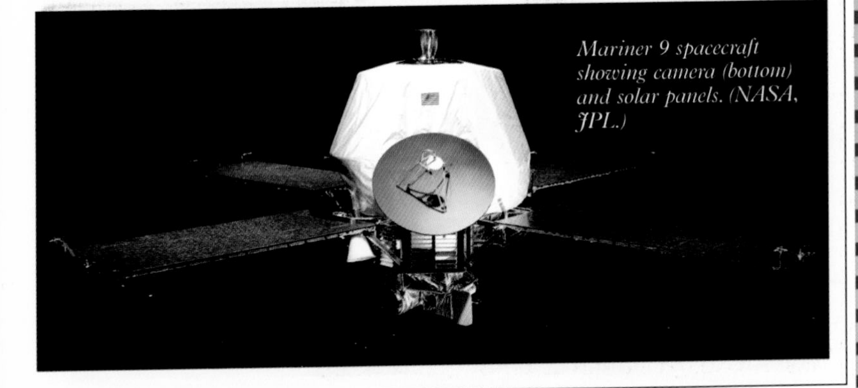

Mariner 9 spacecraft showing camera (bottom) and solar panels. (NASA, JPL.)

CHAPTER 3

THE "CANALS" OF XANTHE

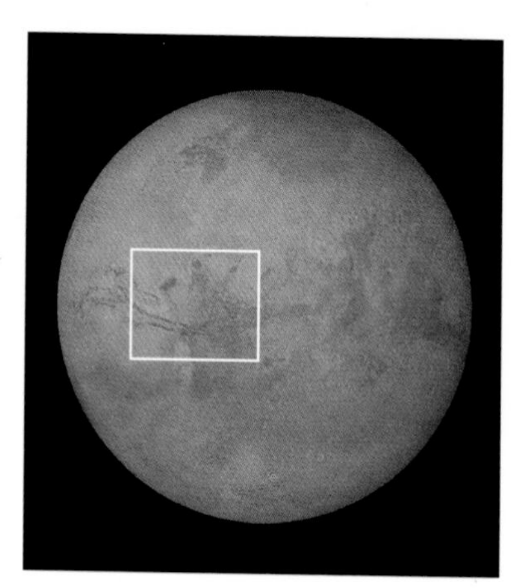

Mars is a desert world, and deserts spawn mirages. Percival Lowell's canal system may have been the biggest mirage of all. The greatest Martian riddle for 70 years, the canals were much ridiculed in the late twentieth century after the ragged, patchy, windblown nature of Martian dark markings became clear. Science historians considered Lowell and his followers to be deluded by their own wild imaginations when they drew narrow lines on Mars. But were Lowell and his followers really so crazy?

As hinted in the first chapter, the answer seems to be no. Proof that some of Schiaparelli's and Lowell's much-criticized canals were based on real features can be found in a Martian region known as Xanthe. This is a desert region a little north of the equator. The larger craters in this otherwise bright desert have dark patches of dust on their floors, and wind has blown long streaks of this dust from the craters, trailing to the southwest for hundreds of kilometers. These

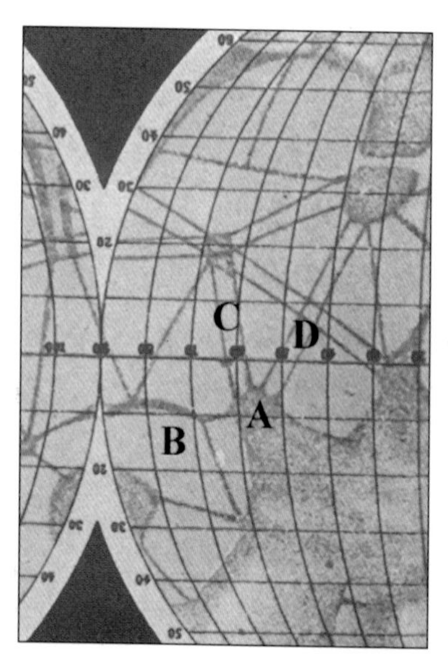

are areas where many earlier observers drew canals extending in this same direction. As shown in the figures below, there are specific matches between early observers' canals and the most prominent southwest-trending streaks.

Both Schiaparelli and Lowell often drew canals that intersected at the dark spots called oases, which probably correspond to the largest dark patches in craters. They also drew parallel double canals stretching across the desert like divided interstate highways. These probably correspond to glimpses of parallel streaks in areas crowded with wind tails.

The full answer to the riddle of Lowell's canals is a fourfold tangle of traits that relate to the human eye-brain complex. First, and

Progress in imagery of the region around Xanthe. Facing page: Schiaparelli's map of 1882–88, showing classical "canals" drawn in this area. Below left: Airbrush map from observations at Lowell Observatory, ca. 1971. Below right: Portion of global image from MGS spacecraft, 2000. Similar features can be seen in all images, revealing how crude streaks were transformed into linear canals. A = permanent dark area named Aurorae Sinus. B = Valles Marineris, the huge canyon system of Mars, often containing dark dust on its floor and seen as a canal. C and D = craters with dark wind tails, seen as canals. (Schiaparelli recorded these as double canals.)

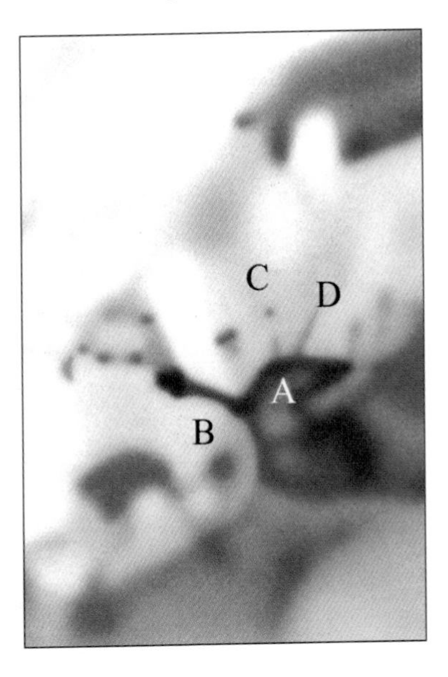

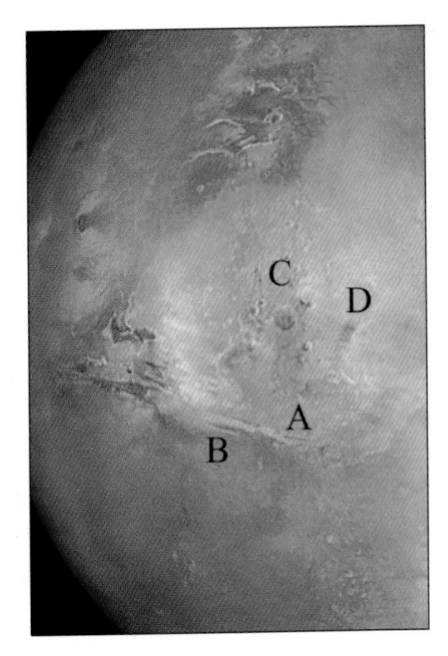

200m

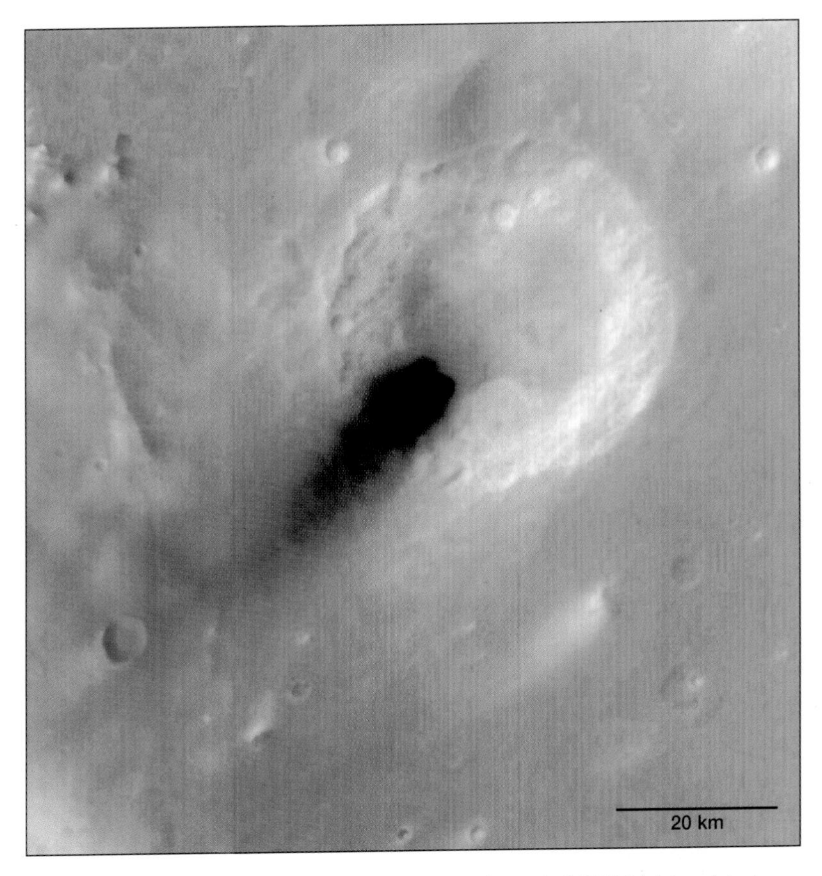

A beautiful desert impact crater (facing page) illustrates how fine, dark dunes have accumulated on the floor to create a dark patch. The whole area seems to be engulfed in smooth dust drifts. (56W 5N, M03-02480.)

The dark patch on the SW floor of a crater near 54W 7N (above) is the source of a dark tail extending to the SW. (M18-00410 context image mosaicked with M08-05762.)

most important, the wind has moved Martian dust in such a way that the dark markings really do tend to be streaky, especially in certain areas. Second, earthbound telescopes give tantalizingly imperfect views. As a telescope magnifies Mars, it also magnifies the natural shimmering of the air (visible in a more extreme case such as heat waves coming off a hot car roof). If you study Mars through a large telescope under normal conditions, you see mainly the larger dark markings, but in instantaneous moments of very sharp conditions you can catch fleeting glimpses of markings only 100 km

1 km

Detailed telephoto scan across the dark streak in previous frame shows the contrast between the streak and the surrounding plain, but aside from dunes in a few eroded craters (above), it does not directly reveal the cause of the streak, which is probably formed at the scale of fine dust particles. (M08-056761.)

(60 miles) wide, including the largest streaks. Then you must make a judgment as to what you've seen. Was it a straight line? Was it uniform in darkness? Or was it a patchy streak? This uncertainty is hardly an excuse for drawing a whole globe full of straight-line single and double canals, but it's clear that some observers glimpsed the crater-related streaks and spots, transforming them mentally into canals and oases.

This brings us to the issue of how the brain interprets what it sees and conceives that vision's meaning. Experiments show that if people are presented with a pattern of random patches, some observers will draw them fairly accurately, but others will perceive larger patterns and connections, even uniform straight lines. Lowell was one of these. He even drew linear markings on the virtually featureless white clouds of Venus!

Finally, and this is speculative, the "canals" may actually have been more prominent in the days of Lowell and Schiaparelli! Since the wind alters the dark and light dust patterns somewhat from year to year, it's possible that Lowell lived in a time in which the dark streaks were more prominent than usual.

At a finer scale, as seen in the MGS images, many of the smaller craters in Xanthe are partly draped in graceful drifts of dust. As the wind crosses the craters, its flow has been disturbed in such a way as to drop its dust load in the crater cavity, where the dune-forms can be seen clearly. This pattern is common in many windswept areas of Mars.

Dark dunes, deposited on an old, cratered background, on the equator at 55W, confirm the mobility of dark sand and dust in the Xanthe region. (M18-01659.)

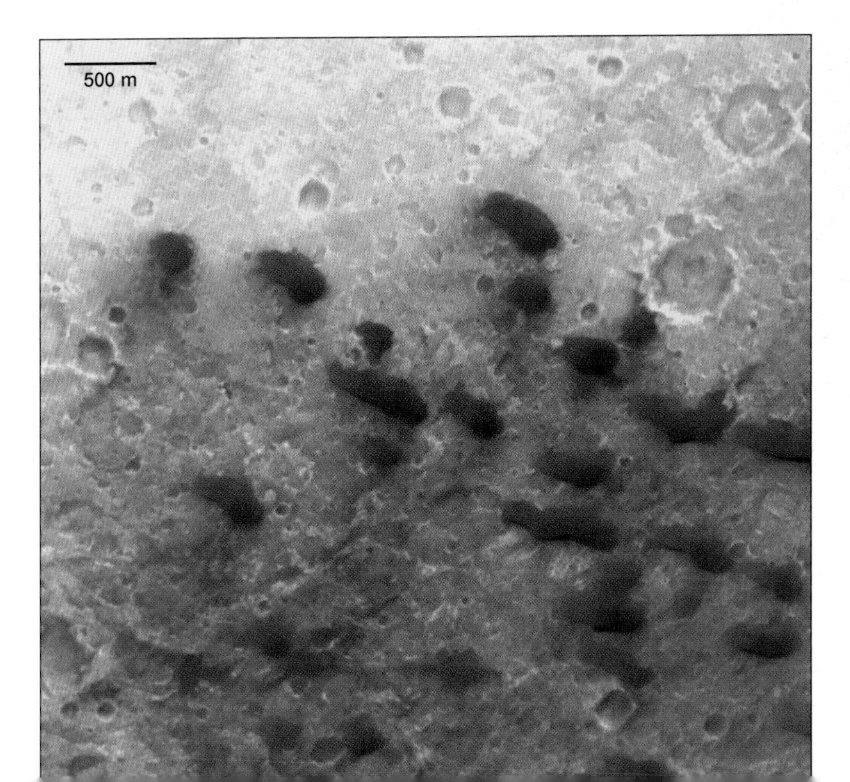

CHAPTER 4

SYRTIS MAJOR

The Mystery of Martian Soil

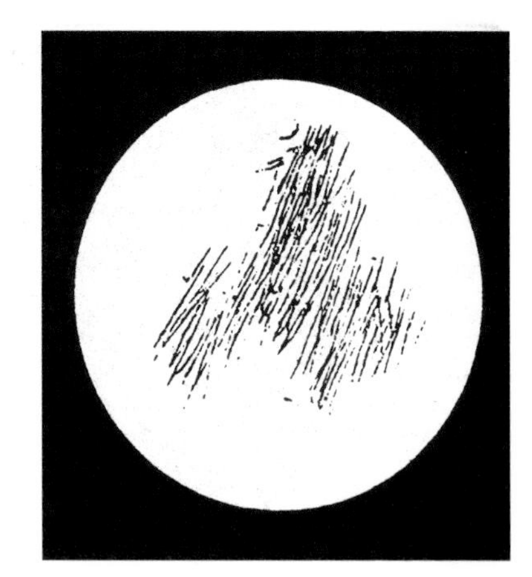

We seem to have solved the nineteenth-century question of the canals and the twentieth-century question of the nature of the dark markings, but our solution is incomplete because we've swept a mystery under the rug. If the markings of Mars are caused by the wind mobilizing deposits of bright and dark dust, then why hasn't the wind simply mixed all the dark and light stuff together, like stirring salt with pepper, to produce a uniformly bland, brown planet? Why have the major markings persisted, century after century? What's the basic difference between the dark stuff and the light stuff?

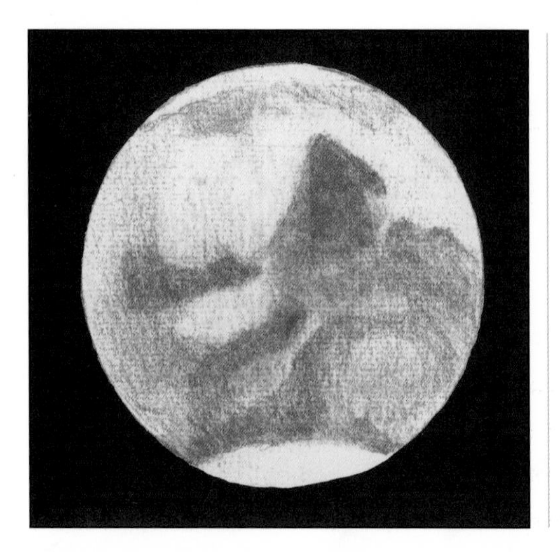

Drawing of Mars in July 1986 shows Syrtis Major as the darkest triangular feature. Compare with drawings and photos in Chapter 1. (Drawing by the author, using the University of Hawaii 21- and 88-inch telescopes at Mauna Kea Observatory, Hawaii.)

Hubble Space Telescope view of Mars shows much more detail of the Syrtis Major region than is possible from Earth-based observations. The Hubble telescope orbits in space and is thus immune to the blurring found in Earth-based views, which are due to the shimmering effects of Earth's atmosphere. This view dates from March 1997.

A great example of this problem lies in the most prominent of the dark markings: Syrtis Major, which was charted by the Dutch observer Christiaan Huygens more than three centuries ago. Why does Syrtis Major stay dark?

Syrtis Major mirrors the whole history of shifting ideas in Martian exploration. It was the first marking recorded from Earth, spied by Huygens in 1659 on November 28 (a date destined to appear again in the history of the planet, 305 years later, as the launch date of the first successful probe to Mars, Mariner 4). As shown in the first chapter, Huygens sketched a ragged dark triangle pointing north—about all he could see on Mars with his

A composite of wide-angle imagery from Mars Global Surveyor was used to make this view of the Syrtis Major side of Mars in June 2002.

crude telescope. When the next generation of telescopic observers began to map more dark patches on Mars, Huygens's dark triangle was so prominent that all the early observers could see it. Later observers timed when Syrtis Major faced Earth and then calculated all the way back to Huygens's timings of the same phenomenon, which allowed them to calculate the rotation period of Mars with extreme accuracy. For this reason, some astronomers nicknamed it the Hourglass Sea.

Until Schiaparelli's naming system was adopted, different observers had their own exotic names for Martian features. Around 1858, the Italian astronomer Angelo Secchi made the first

Color-contrast effects help show why the dark markings were observed to be bluish green. Both brownish gray blocks are the same pigment. However, when seen in a small patch against a bright orange background, as in the Martian deserts, the color appears slightly more bluish or greenish gray. If you use averted vision, the effect may be clearer.

serious attempts to represent the planet in color and showed Syrtis Major as blue-gray in color. He gave this patch, along with a nearby curved, tail-like extension, the striking name the Blue Scorpion. The purported bluish tones supported the hypothetical connection with seas, and Schiaparelli named the feature after the Gulf of Sirte in Libya.

Then came Lowell's generation, which abandoned the idea of open seas but construed the dark material as vegetation. Syrtis Major was now the thickest tract of alien, blue-tinted Martian mosses, bushes, or perhaps even forests! As a result of this theory, one of the main activities of Mars astronomers in the twentieth century was to chart the seasonal and year-to-year changes in the markings, in hopes that some clue would be found to confirm their nature. Syrtis Major was a prime candidate. Chapter 1 contains numerous images from different years, documenting the changes in the form of Syrtis Major. During much of the nineteenth century, the wedge of Syrtis Major sported a pigtail, the broad "canal" known as Nilosyrtis; it can be seen in the drawings in chapter 1, although it has since disappeared. A famous French observer of the early twentieth century, E. M. Antoniadi, found that the eastern side of Syrtis Major was streaked and narrow in the spring, but filled in and broadened in the autumn. By the 1950s, a large conspicuous patch—Nepenthes-Thoth—developed just to the northeast, but it too has faded. Late-twentieth-century drawings and photos show a stubbier triangle with no tail. All these are classic examples of the changes now explained by blowing dust.

DARK MYSTERIES EXPLAINED

If these markings are blowing dust, eroded from the gray-brown rocks of Mars, why did so many good observers report them to be bluish green? The question of color goes to the heart of the mysterious persistence of the dark markings. Is the Blue Scorpion really blue?

No. Telescopic observers such as the astronomer Tom McCord, starting in the 1960s and 1970s, mounted spectroscopes on large telescopes in order to isolate the dark areas and measure their color quantitatively. By doing so, they were able to detect subtle colorimetric signatures of different minerals and gases. McCord's pioneering measurements showed that the dark areas of Mars are not green or blue, but nearly neutral gray, with a warm brownish gray tint.

So why did visual observers call them bluish green? The "cool" tints perceived by the human eye were largely due to a color-contrast effect. If you make a neutral gray one-inch-size square (or patch of any shape) and put it on a neutral, white, or black background, it will look its true color. But if you place the same sample on a vivid orange background, it will seem to take on a bluish or greenish cast, because the brain registers it in contrast with the orange. Master painters have used this phenomenon to subtle effect for centuries to affect the brain's perceptions of paintings. This effect explains much of the fateful confusion over Martian colors that reigned for more than a century—"fateful" because it led to the myth of greenish vegetation.

An additional effect may be involved. Thin bluish white clouds often veil large regions of Mars—like a very thin, high, cirrus overcast on Earth. The bright orange deserts of Mars are too intensely colored to be affected by these thin veils, but the clouds may often impart their own bluish cast to dark, neutral-gray Syrtis Major, in the same way that a distant, dark mountain looks pale blue when seen through atmospheric haze.

Scientific measurements of the spectrum and thermal qualities of the darker and brighter regions, made from Earth-based telescopes and from spacecraft, together with soil studies from three landers, have greatly clarified the situation. The color of the dark regions comes not from water or vegetation, but from relatively unweathered grayish rocks and gravels. Generally speaking, the dominant rocky material is the same as the most common type of volcanic lava on Earth, a rock type known to geologists as **basalt**. Fresh basalt is typically a dark gray rock. When it is exposed to moisture, however, the iron minerals in the surface layers rust, weathering to a lighter, reddish brown color, just as an iron rake or hammer left outdoors will turn rusty red due to exposure to the elements. The rock itself can develop a reddish coating, and rock

Dark, coarse gravels have been separated from light-colored sand and dust by the wind action in this part of the Sonoran Desert near El Golfo in northwest Mexico. The background dune and the deposits on the downwind side of obstacles are formed by bright, wind-blown sand, but darker gravels dominate the foreground. Similar processes may help separate darker and lighter material on Mars. (Photo by author.)

particles can flake off to create a reddish dust. On Earth, basalts make up much of the sea floor **crust** (the main outer shell of Earth, typically around 5 kilometers thick under the ocean floor), as well as numerous lava flows found on land. Basalts on both Earth and Mars cover a range of compositions and are subdivided into different types, so we might better say that the rocks, gravels, and soils of the dark areas match the range of basaltic rocks found on Earth. To a casual observer, Martian basalts are almost indistinguishable from terrestrial basaltic lavas.

The basic message from this finding is that Mars probably went through an early geologic history similar to Earth's, forming a similar basaltic crust that makes up most of the rocks below Mars's shifting sands. Various surface and near-surface processes on both Earth and Mars—such as remelting of rocks, accumulation of wind-blown sediments, weathering, an alteration of rocks by water, et cetera—can produce local deposits of rocks different from fresh basalts, and we will look for those as signs of the climatic and geologic history of Mars.

As for the light-colored, orangish deserts of Mars, the scientific measurements reveal finer-grained soil and dust than found in the dark areas. In this reddish brown soil and dust, the iron minerals are rusted to a brighter, ruddier hue typical of rocky materials that have been weathered, especially by occasional exposure to moisture. Much of the dust is material flaked off the weathered coatings of basaltic rocks. The red rusted-iron colors are the same hues that can be seen in many desert areas of the American Southwest, Australia, or Africa.

THE SOURCES OF DUST AND THE TRUE NATURE OF THE BLUE SCORPION

Discovery of dark basalts and iron-rusted bright soils explains the *what* of Martian markings, but not the *why*. If the dark areas are simply darker gravels and the bright areas are powdery, rust-weathered dust and sand, then why haven't the Martian winds mixed the dark and light materials to produce a uniformly tan planet? Why haven't the dark rocks all been covered by wind-blown bright dust? The answers seem to lie in an ongoing contest between weathering of old rock, which produces light dust, and production of fresh, dark rock surfaces (for example, by eruption of lavas).

Syrtis Major holds a clue. To understand this clue, we need to point out that initial mapping of Syrtis Major was somewhat misleading. Early photos by Mariner 9 and Viking orbiters showed that central Syrtis Major contains a broad area relatively free of craters. The uncratered area stretched east to include a young, flat plain named Isidis Planitia after the Egyptian goddess Isis, queen of heaven. This plain occupied the floor of an old, circular impact basin. Viking images of Syrtis Major showed two large volcanic calderas near the middle: 61-km Nili Patera and 43-km Meroe Patera. Crude Viking data on altitudes showed that Syrtis Major rose out of the Isidis basin, so central Syrtis Major was initially mapped merely as a westward, upslope, lava-covered extension of the Isidis plain. It was given the modern spacecraft-era name of Syrtis Major Planum (Syrtis Major plateau).

But Syrtis Major is neither a plain nor a simple plateau. The precise altimetry from the MGS laser altimeter revealed the circular, sparsely cratered area making up most of Syrtis Major not as a *planum*, but as an enormous, low, conical mountain rising about 2 kilometers to the summit calderas. The base of this cone is about 1,200 km (750 miles) across, big enough to fill most of the United States east of the Mississippi, or to straddle most of Europe from London to Venice. This form of volcano is called a shield volcano, because its profile resembles the circular conical shields used by ancient warriors. The rise from the Isidis basin to Syrtis Planum is really the shallow east slope of the great cone.

This discovery offers a clue as to why Syrtis Major stays dark. The dark brownish gray color comes from the basaltic lava making up much of the volcano. As fast as this can be weathered to brighter red soil by wind, ice, moisture, or other weathering agents, the strong prevailing winds blow the soils away, exposing more fresh basalt. In modern geologic times, at least, Syrtis Major is a self-renewing dark area with changeable dusty windblown deposits trailing off in various directions, varying in shape somewhat from decade to decade at the whim of the wind.

Other areas of bright weathered dust—the so-called deserts of Mars—are thus the "sinks" in which weathered dust collects as it

Detail of map prepared from Viking observations (below), using the topographic data available at that time, shows the dark triangle of Syrtis Major superimposed on the smooth plain of Isidis Planitia (right) and a similar plain of Syrtis Major Planum (left center). (U.S. Geological Survey and Lowell Observatory, ca. 1973.)

Facing page: *Topographic map from MGS laser altimeter shows that the impact basin of Isidis Planitia is indeed a low plain (blue, right), but the region underlying Syrtis Major is a high volcanic cone (red, left) surmounted by a double summit caldera (yellow). The heavily cratered surroundings are Noachian in age. (MGS laser altimeter team map.)*

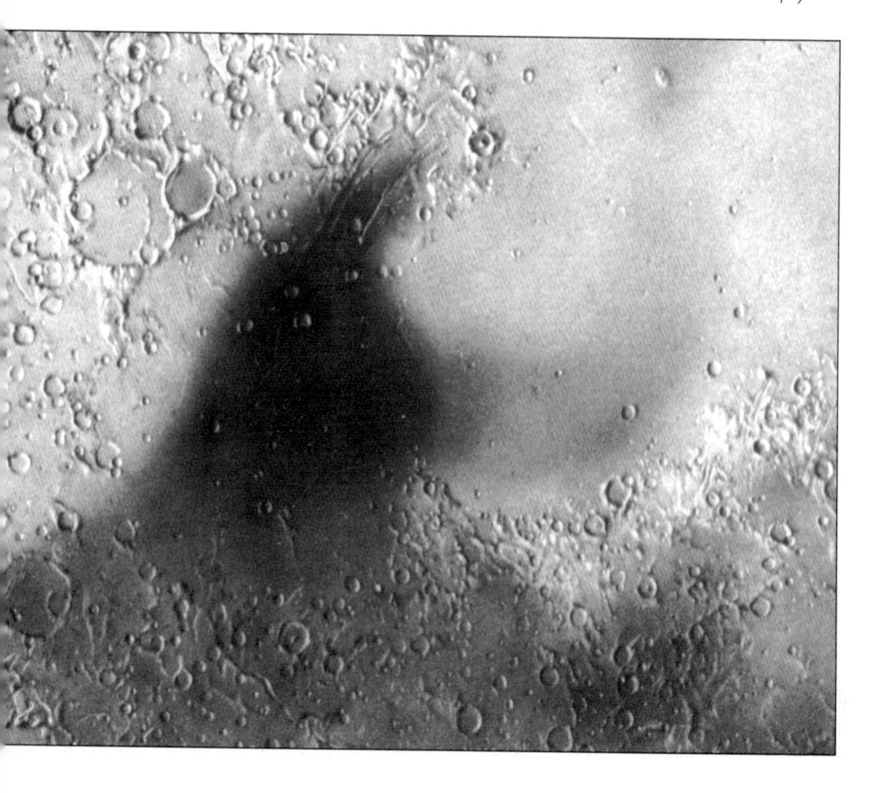

is wind-carried from the fresher rock sources' areas. The interplay between global wind patterns and these two types of areas would then give a gross explanation of the Martian markings.

COMPLICATIONS

So far so good, but we have to admit that the Syrtis Major solution does not explain everything. For example, not all volcanoes on Mars are dark, and not all dark regions are young volcanoes. Indeed, the tallest and grandest volcanic mountains on Mars, where young lava flows are visible, are bright; and spectroscopic-thermal measurements show that they are mantled by bright dust. Why aren't all fresh lavas dark? The answer again lies in the mobility of the dust. Martian global dust storms inject fine dust into the high Martian stratosphere, from which it slowly settles, mantling the entire planet with fine blankets of dust. According to the scenario we are imagining here, the wind in many regions sweeps this dust away from fresh lavas, creating dark regions like Syrtis Major, but

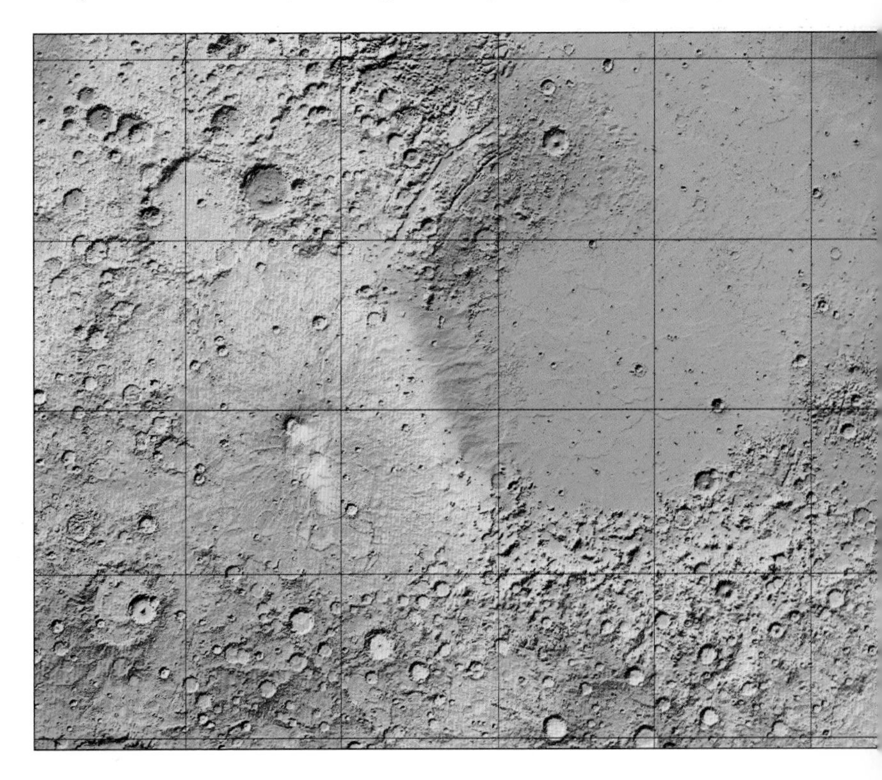

My Martian Chronicles:

PART 2: UNEARTHING MARS-LIKE ROCKS IN TUCSON

Parallel to work done by Tom McCord, mentioned on page 54, was work done by two friends who were fellow graduate students in Gerard Kuiper's lab at the University of Arizona.

Kuiper had been the preeminent astronomer to capitalize on infrared or "night vision" technology developed during World War II. He realized that the infrared wavelengths of light contained signatures (known as spectral absorption bands) of various types of rock-forming minerals and ices.

In the 1950s, he developed instruments to apply infrared technology. Alan Binder (who went on to head the 1999 Lunar Prospector mission that confirmed ice at the lunar poles) and Dale Cruikshank (who went on to discover a number of minerals and ices on the surfaces of satellites and asteroids in the outer solar system) were working in Kuiper's lab, building the next generation of spectrometers.

By 1960, there were already hints that the red color of Mars was due to rusting, or oxidation, of iron minerals in igneous or volcanic rocks on the planet. As described earlier in the text, iron minerals easily rust into Mars-red coatings. The process seems to work especially well in deserts, where sporadic rainfall moistens iron-bearing rocks, which rust as the water then evaporates off their surfaces. Tucson, where we lived, is in the middle of such a desert, and west of town is a range of igneous and volcanic hills, known as the Tucson Mountains, covered with brownish, weathered basaltic lavas and other igneous rocks. Inspired by the opportunity, Binder and Cruikshank simply drove out into the Tucson Mountains, and on hillsides near the famous Arizona-Sonora Desert Museum they collected some samples of these rocks. In the lab, they were among the first to show that the spectrum of sunlight reflected off these rocks was a good match to the spectrum of sunlight reflected off Mars. In a 1966 paper on this work, they concluded correctly that the reddish color of Mars was associated with various forms of rusted iron-bearing minerals, and gave a powerful demonstration that Mars was likely covered by igneous and volcanic rocks that had been affected by oxidation, possibly due to moisture at some time in the past.

For all of us, it was exciting to be there when powerful new instruments were coming on-line and when important new discoveries could be made by seizing an opportunity, picking up a few rocks, and following through on a clever idea.

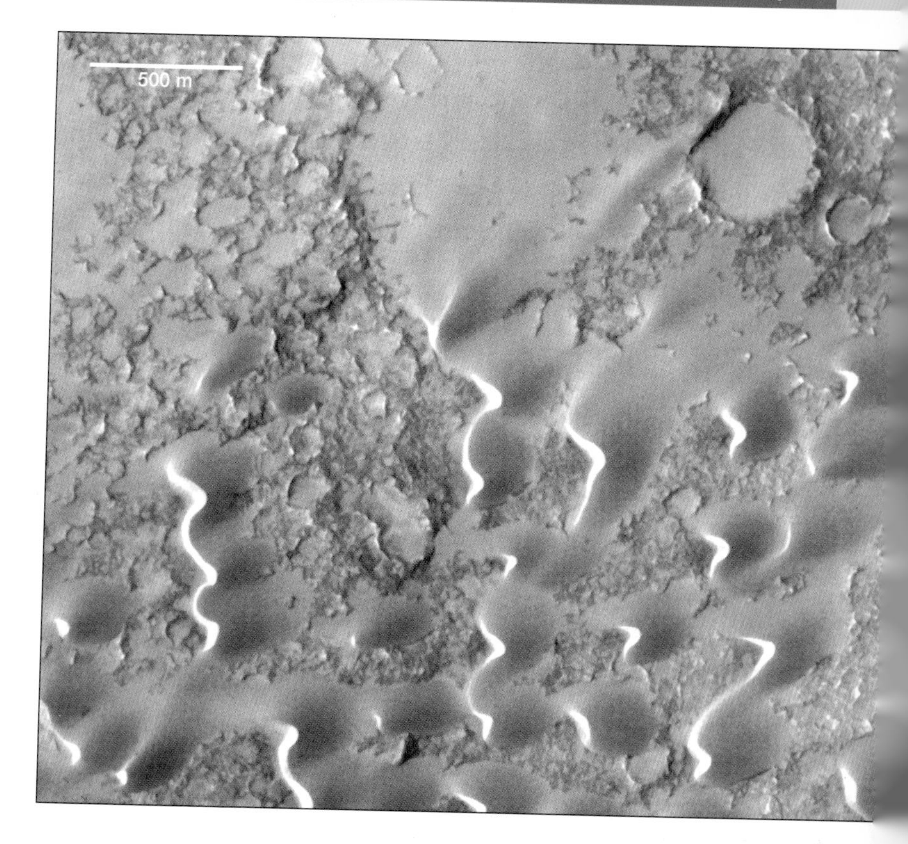

the extremely thin air atop the highest volcanoes may be less likely to sweep away a dust mantle.

Young dunes engulfing an old surface in the caldera of Nila Patera. A heavily cratered surface can be seen in the background. On top of this, the wind has arranged dunes of brighter-colored sand or dust. (293W 9N; M17-00435.)

Furthermore, even particle size differences in the same material can create local differences in tonality. Fine-grained dust tends to be lighter than coarse gravel derived from the same rocks, and so even a region with a single rock type may develop lighter-toned sand and dust that collects into light-toned dunes and dust drifts. Throughout this book, you will see images in which sand dunes and sheets of dust form patches that partially blanket older terrain. It seems impossible to predict where the winds will drop their load of fine material or whether a given deposit will be brighter or darker than the background.

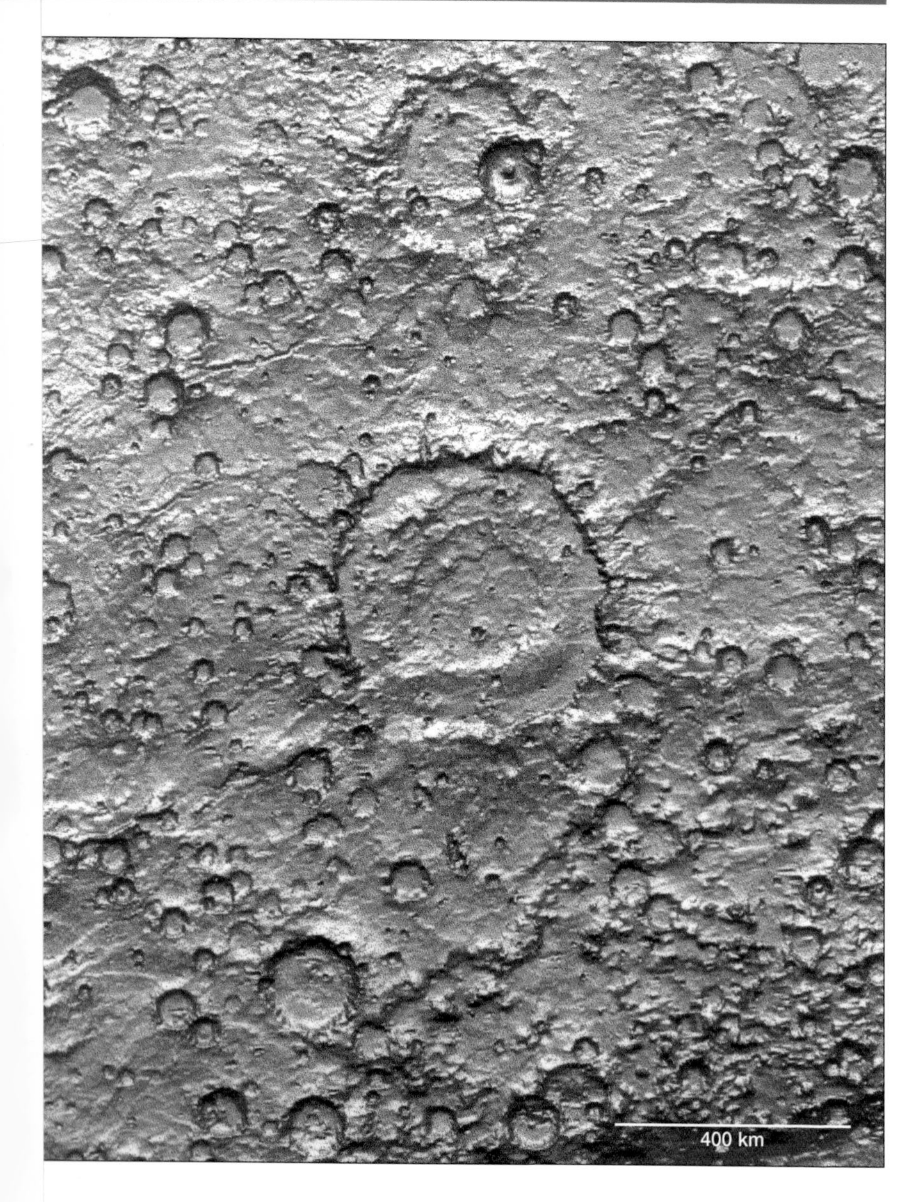

Huygens crater, as imaged by the MGS laser altimeter data. Huygens is the largest crater, in the center. Notice the double-ring structure, characteristic of large impact craters, and forming a transition to larger, multi-ringed impact basins. The blue (low) region at the bottom is the north edge of the Hellas basin. The laser altimeter reveals a faint, mountain-ridge arc (left middle to upper right), broken by the Huygens impact. This arc may be a remnant rim structure of a vast, earlier impact basin. (MGS MOLA team, processing of data courtesy Nick Hoffman, Victorian Institute of Earth and Planetary Science.)

into glassy material, rather than slow cooling, which allows time for large crystals to grow.

The complicated analysis of rock data types leads to an exciting possibility: The weathering and glassy compositions of the far northern basalts may have resulted when their lavas erupted through near-surface massive layers of underground ice or water. As we visit the high latitudes later, we'll look for more evidence to test this idea.

The Syrtis Major area has another noteworthy feature. Near the southwest corner of the dark triangle lies the giant crater Huygens, appropriately named after the first observer of Syrtis Major itself. With a diameter of 440 km, Huygens is one of the largest impact craters on Mars. The state of Ohio could just fit within its rim. The floor is partly filled in with windblown dark material.

Huygens has the distinction of being one of the few craters big enough to have been possibly spotted from Earth by telescopic observers. In the 1890s, the American observer Edward E. Barnard used the 36-inch-wide telescope at California's Lick Observatory to make visual studies of Mars, and he reported "a vast amount of detail," including various spots and patches, "irregular and broken up." In retrospect, Barnard has been recognized as one of the best Mars observers of all time, with one of the best telescopes ever used for extended visual observations of the red planet. Some later writers have claimed that when Barnard recorded a series of small spots on Mars, he was glimpsing individual craters. More likely, as we noted earlier, he was seeing the dark patches of dunes that are common on their floors. Conceivably, a dark deposit in Huygens crater was one of the spots fleetingly glimpsed by Barnard. Nonetheless, despite such tantalizing glimpses, neither Barnard nor any other observer was able to obtain real proof of craters on Mars, which is why the discovery of craters in the Mariner 4 photographs of 1965 came as such a surprise.

A MARS TOO FAR

This leads to a strange thought: If Mars had been just a little closer, or if our atmosphere had been just a little clearer, or our telescopes just a little better, astronomers would have recognized the craters of Mars sooner. In that case, by 1900 we already might have established the "Mariner 4 theory" that Mars was geologically and biologically dead—and a century of mythologic speculation about plants and canals and civilizations would have been avoided. But Mars was just far enough away to keep us guessing.

Paradoxically, then, if Mars had been more within range of our earlier instruments, we might know less about Mars today. If craters had been recognized, the sterile Mariner 4 vision of Mars as a dead, Moon-like world might have dominated the whole twentieth century, instead of lasting for only four years between Mariner 4 and Mariner 6. Mars's allure would have been lessened. Without the mythic and theoretical civilization of Lowell, we'd have lost the rich legacy of stories by H. G. Wells, Edgar Rice Burroughs, and Ray Bradbury. And without those, we might have lost the sense of wonder and curiosity that drove much of the early space program. Without "mythic Mars," the dream of space exploration might never have crested to produce the unprecedented fifty-year program of Martian exploration that our generation has been privileged to share.

So much for alternate realities.

At this point, we're armed with an understanding of the historical observations of Mars and a familiarity with a few Martian spots that solve the old mystery of the markings. Now we're ready to embark on a tour that will allow us to understand the evolution of the planet itself, from the oldest regions and their evidence of primeval conditions, to the youngest regions, with their astonishing evidence of modern-day geologic activity involving young lavas and liquid water.

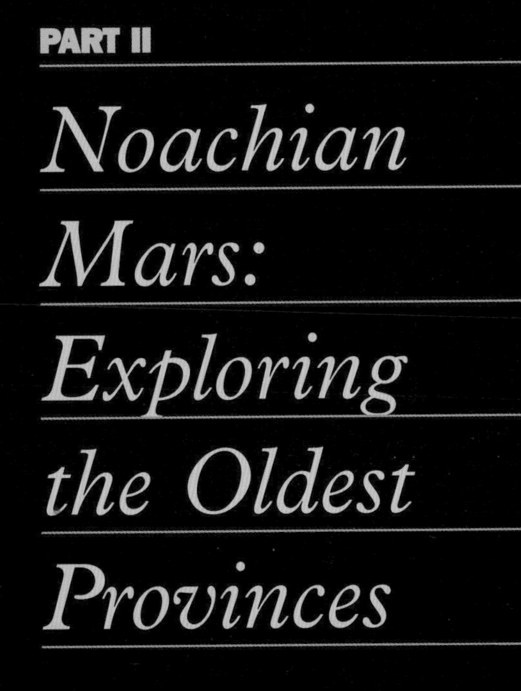

PART II

Noachian Mars: Exploring the Oldest Provinces

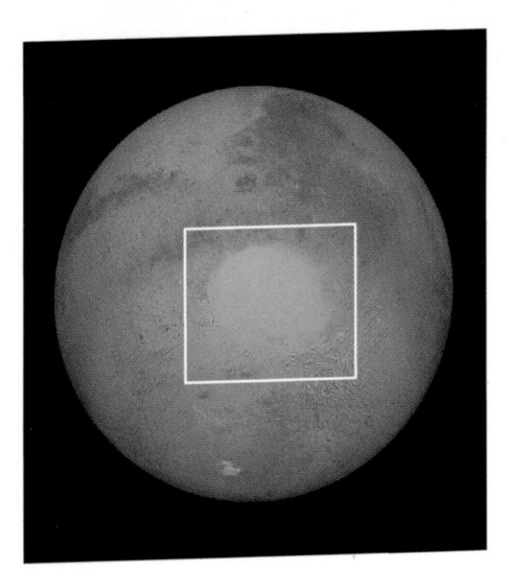

CHAPTER 5

HELLAS BASIN

A Giant Impact Scar

A traveler who wants to understand the entire history of Europe might start a tour with the Paleolithic caves of France, and in the same way we start our tour with the oldest regions of Mars, the southern uplands. They are heavily cratered, shoulder to shoulder with ancient impact craters. Remember, the more craters a planetary surface has accumulated, the older it must be. The saturation bombing of the southern highlands tells us that this area dates back to the earliest days of Mars, starting 4,500 MY ago to about 3,900 MY ago, when interplanetary space was still full of planetesimals, asteroid-like leftovers of solar system formations. These chunks of cosmic flotsam crashed into the newly born planets, creating the period known as the "intense early bombardment." Eventually, after about 600 MY, the supply of debris was swept up by the new-born planets and the rain of rocks from the sky dwindled to about the present rate on both Mars and Earth—many brick-size rocks fall every month and a massive chunk big enough to create a mile-wide crater falls once every 100,000 years or so.

The oldest Martian landforms, dating from the Noachian period, are areas of primordial, cratered crust, covering about a third of the planet. In the midst of the heavily cratered southern uplands, the pattern of random crater-pocking is interrupted by a huge, bright circular feature. This feature is easily visible in modest-size telescopes, including backyard 'scopes built by amateurs. For this prominent feature, Schiaparelli chose the name Hellas, after the ancient name for Greece.

DECIPHERING HELLAS

Mariner 9 revealed that the bright area is the dusty floor of a huge impact basin, about 1,700 km (or 1,100 miles) across—about the distance between New York City and Des Moines, Iowa—with a highly eroded rim structure and layered deposits covering part of the floor. The broad floor received the topographic name Hellas Planitia. According to the mapping by the MGS laser altimeter, the floor of Hellas Planitia lies some 8.2 km (27,000 feet) below the surrounding cratered uplands, making it the lowest region on the planet. The term *impact basin* refers to the giant form of an impact crater, typically larger than 500 km across and with a different structure than smaller impact features. The best-preserved ones have a bull's-eye rim pattern of concentric faulted cliffs and radial fractures, not unlike a bullet hole in plate glass, formed as the planetary crust responds to the overwhelming shock of the impact explosion. The full sequence of craters, from small to large, starts with simple bowl-shaped craters, then 20-km-scale craters with central mountains, then a broadening of the central mountain into a ring of peaks, and finally a full-fledged, 1,000-km-scale, multi-ring basin. Huygens, the 440-km crater we visited in the last chapter, is a rare transitional form in which the interior central ring of peaks is well developed.

The edge of the Hellas impact basin, which would normally be occupied by a high, mountainous rim, is instead occupied by a strange low zone of dissected or broken terrain, composed of rugged valleys and hills. This is particularly well shown by the topographic maps generated by the MGS laser altimeter. Here is another puzzle: What happened to the Hellas basin rim? Presumably, there was a high rim composed of explosion-ejected debris, sloping outward into the surrounding countryside, as seen in lunar multi-ring basins.

Global view of Mars, simulated from Mars Global Surveyor mapping images, reveals Hellas as a bright circular patch south of the dark area of Syrtis Major. The bright patch marks the sediment-covered floor of a large circular impact basin. The winter polar frost cap is prominent just south of Hellas, and transient frost sometimes covers part of the basin floor. (Malin Space Science Systems.)

Any regional water drainage, therefore, should have been *outward,* off the Hellas rim. Today, however, as shown in the photographs on page 76, parts of the region are traversed by ancient drainage channels, which wind hundreds of kilometers *into* the basin floor. How can the drainage have changed direction?

The whole pattern gives the impression that the entire rim structure, being composed of loose, rubbly debris, was some of the most easily eroded material on Mars, and that early periods of fluvial activity and wind erosion washed and blew it away. The overall drainage patterns testify to the fact that water was an active erosive agent on early Mars.

Map of the Hellas impact basin based on Mars Global Surveyor laser altimeter measurements of topography. Blue central area is depressed 5–8 km below average Mars surface. Green = 1–2 km depression; yellow = average elevation; and orange areas are highest at 1–2 km above the average. The original rim, which should be the highest area in the scene, would have occupied the outer edge of the green area, but has apparently been destroyed by erosion. The basin, its surviving rim structure, and the background cratered terrain date to the Noachian era. The floor is covered by Hesperian and Amazonian deposits, some of which were washed in along river channels such as those visible in the green zone at the east edge of the basin. (MGS MOLA team.)

THE SURPRISING FLOOR OF HELLAS

In the days of the first Mariner 9 photos of the Hellas basin, Mars seemed simpler, and many Mars researchers assumed that the bright floor of Hellas was simply a smooth, sand-filled depression—a massive sinkhole of dust. However, Viking orbiter images of the '70s revealed ridges, layers, and other complex structures. MGS's telephoto camera revealed regions of astonishingly complex layering. This layering suggests that Mars is more complicated than we thought and that Hellas has seen a complex geological history. Windblown or water-carried sedimentary layers were deposited in some areas during some eras, but they were partly eroded away in other places during other eras. Ancient winds and rivers seem to have brought in layer after layer of sediment. The layers seem to be weakly consolidated, like terrestrial desert gravel layers. The net result in some areas is a remarkable landscape of pancake-like strata, eroded into fantastic curvilinear forms.

DETECTING DUST, ROCKS, AND FROST

Maps of the thermal properties, derived from infrared detectors, confirm the complexity. Fine dust is a good insulator; it does not transmit its heat into the subsurface layers, so it heats up in the sun. This is why beach sand gets so hot. Bare rock is a poor insulator and a good transmitter, so it drains its heat into the rock's interior. So the surface of a dusty Martian area heats up fast in the day and cools off fast at night, because it hasn't stored up heat. Rock outcrops and bouldery surfaces take a long time to re-radiate their heat and cool off at night. Infrared thermal detectors have measured the cooling times in various parts of Mars and confirm that the floor of Hellas is not merely a basin full of fast-cooling dust, but an area of more solid, rocky material.

This portion of a 1997 view of Mars shows the Hellas basin partly filled with bluish white frost and haze. The transient frost tends to form under winter conditions but may rapidly "burn off."

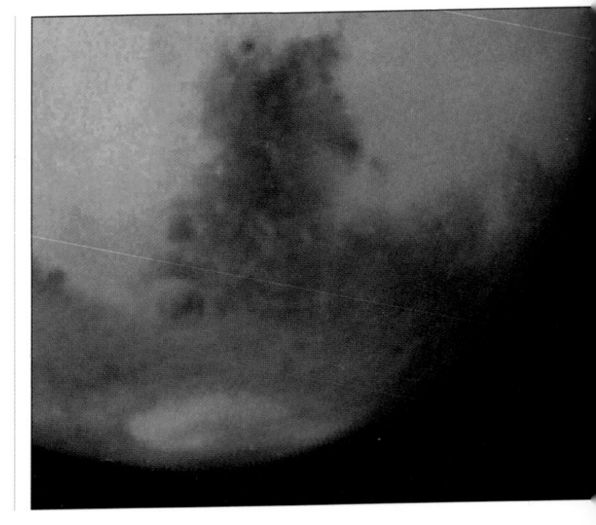

Morning frost patches on Earth. The two images at left, taken at McDonald Observatory in Texas, show frost covering the floor of a drainage basin at dawn, but burning off by 9 A.M. Similar behavior of frosts on Mars creates transient bright areas, including frost deposits on the floor of Hellas basin.

Layered sediments in the Hellas basin on a boundary between regions mapped as a "dissected floor unit" of Hesperian age, and a "channeled plains rim unit" of Amazonian age. This striking image reveals an area where sediments have been deposited in layers during Noachian and Hesperian times (by ancient water or wind action?), but are now being exposed by erosion, revealing old, pancake-stacked layers. (275W 41S, M20-00092.)

Because Hellas is at a high latitude, near the south pole, the climate is especially cold, even for Mars. Sometimes, Earth-based observations show the floor as unusually bright, coated with carbon dioxide frost deposits. This is one reason that Hellas Planitia appeared on early charts as a very bright circular patch—even brighter than the normal dust color. A typical Martian phenomenon, visible from Earth, occurs on cold mornings when Hellas and other similar areas are covered with bright frost deposits and/or bright haze, which then burns off by midday. The same thing can be seen on Earth, as shown above.

FOSSIL MAGNETISM

In 1998, the MGS made a striking discovery about Hellas that illuminated Mars's geologic evolution. The discovery involves Mars's magnetic field and requires some preliminary explanation. Basically, Mars has almost no magnetic field today. Russian and

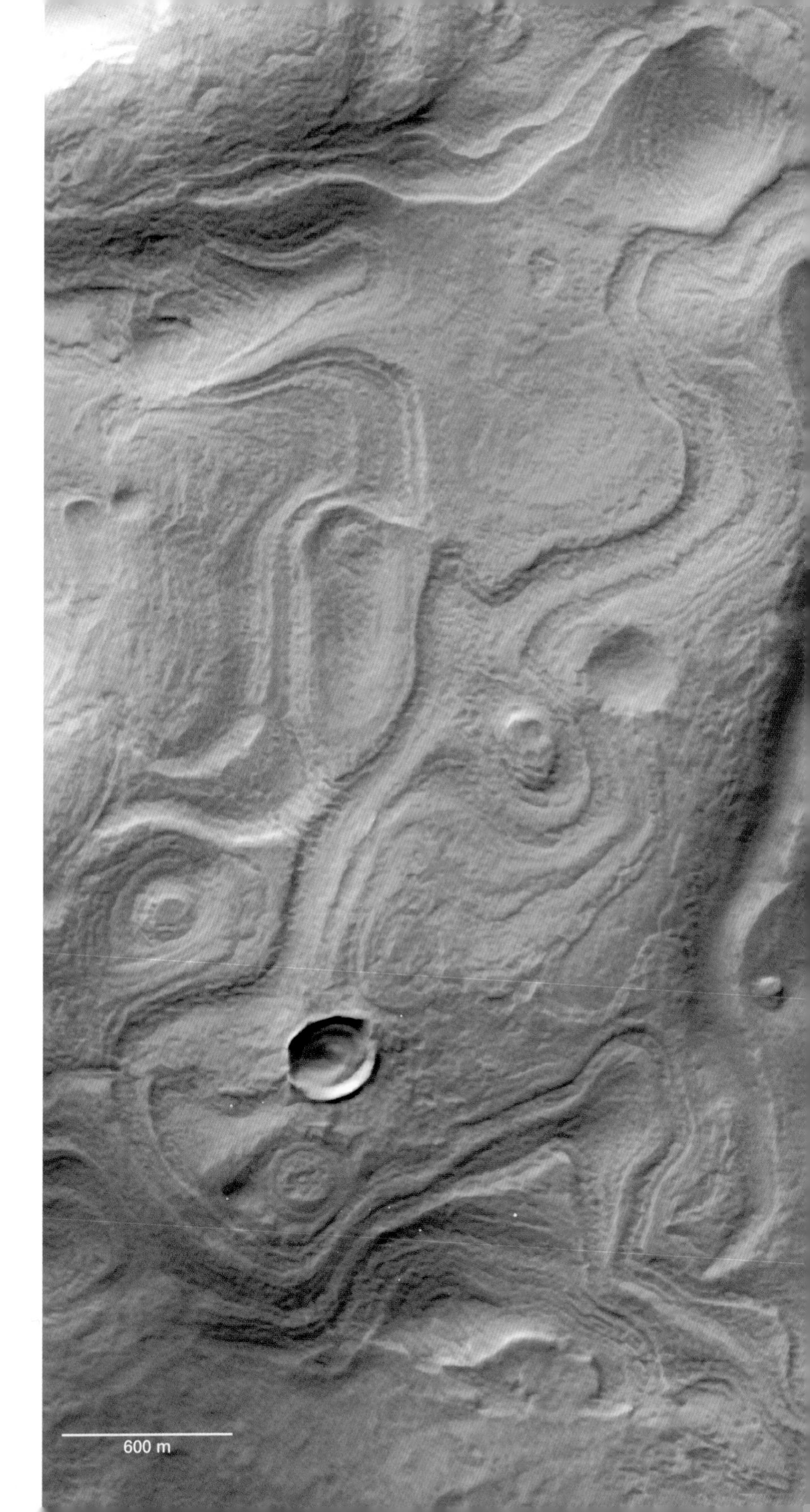
600 m

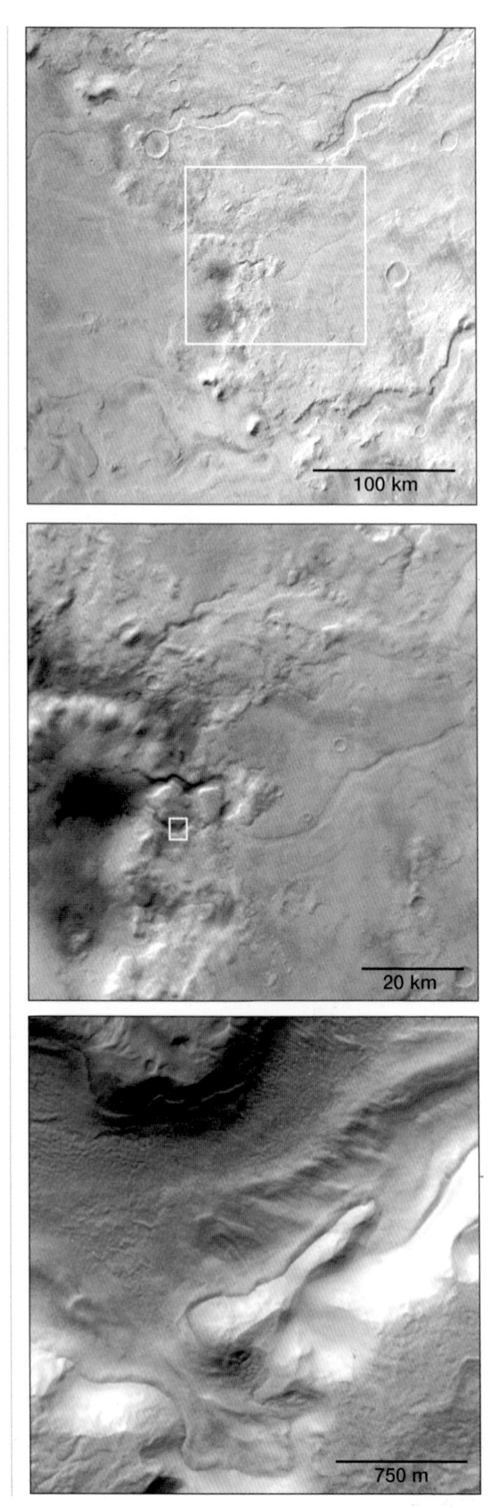

Regional image of the eastern rim of the Hellas basin (hills and cliffs crossing the image, upper left to lower middle). Ancient rivers drained across the eroded rim country (right) onto the basin floor (darker area, left). White box shows area of next image. (Viking mosaic.)

Closer view of the rim shows still smaller channels winding toward the basin floor. White box shows the area of the next telephoto image. (Mars Global Surveyor context image, M20-00093.)

Telephoto view of one of the channels shows that its floor is covered by a mysterious layer of dark material, looking as if a muddy slurry flowed down the channel and lapped up on its banks. The true nature of these deposits is unknown. This is part of the same frame that showed the pancake-stacked layers, located in an adjacent region to the north. (275W 41S, M20-00093.)

American space vehicles from the 1960s onward tried in vain to detect a Martian **dipole magnetic field**, meaning a field with a strong north magnetic pole and south magnetic pole, as on Earth. (Lack of a strong field means that a compass would be of little use in identifying "north" on Mars.) This tells us something important about the history of the planet. Since planetary magnetic fields are created by the sluggish flow of molten iron material in the central cores of planets, the lack of a modern field told scientists that Mars has no actively churning molten iron core. This in turn substantiates Lowell's idea that Mars, being smaller than Earth, cooled off faster. The absence of a magnetic field led many researchers in the 1960s and '70s to assume that the geologic engine of Mars had run down, and that the red planet now lacks any significant geologic activity.

MGS added a new wrinkle. Its sensitive instruments confirmed the lack of a strong dipole magnetic field, but it did discover that the old, southern upland crust was magnetized! How did this happen? If any molten magma cools and solidifies in a magnetic field, its iron mineral crystals will be frozen in a pattern of alignment with the field, like tiny magnets all aligned in the same direction. Thus, early Mars really did have a strong dipole magnetic field at the time the ancient crust was forming. This again fit the idea that early Mars was geologically active, with a molten core, but then cooled off to a point at which the iron core could no longer create a magnetic field.

Hellas comes into this story in the following way. As MGS mapped the crustal magnetic patterns, it discovered that the Hellas impact punched a hole in the initial magnetic pattern. Unlike the surrounding uplands, Hellas is not magnetized, which means we can set up a timeline. The Hellas impact is not too well dated, but lunar data indicate that all the giant basin-forming impacts occurred by 3,800 MY ago, and the number of impact craters superimposed on Hellas suggests that it, too, is old, probably forming before 3,700 or 3,800 MY ago. Mars must have had a magnetic field before then, but not afterward, since Hellas never reestablished its magnetism after the impact.

There is another clue about the early chronology. In chapter 25, we'll discuss an extraordinary handful of Martian rocks that have been found on Earth. One that dates back to 4,500 MY ago is magnetized. This is an amazing rock! It is a piece of the primordial crust of Mars—the crust that Hellas basin punched into. It seems to prove that iron cores formed very early, as planets melted from the heat of their own formation. This particular rock tells a story about early Mars; it shows some evidence of the occurrence of impact stresses and/or chemical disturbance 3,900 MY ago, but not enough to destroy the magnetization. Caltech researcher Benjamin Weiss, working with colleagues in Tennessee and Canada, showed that carbonate grains that formed in this rock during the disturbance 3,900 MY ago also picked up magnetization, so the magnetic field was still going at that time. Strong heating (not enough to melt the rock, but greater than about 575° C) destroys magnetization, so the rock was never heated above that after 3,900 MY ago. The overall evidence from this rock supports the Hellas scenario that the magnetic field started as the planet formed 4,500 MY ago, and magnetized the crust before 3,900 MY ago.

PLANETARY ARCHAEOLOGY

The Swedish-American planetary scientist Hannes Alfvén once remarked that planetary studies are like archaeology; we have to assemble the past from fragmentary clues. Putting our various clues together, our early timeline goes like this:

4,550 MY ago: Mars forms, along with the other planets and asteroids.

4,500 MY ago: Mars's crust forms as molten layers solidify. Mars is partly molten inside, forming an active molten iron core. Its magnetic field is turned "on."

4,500 to 3,900 MY ago: The magnetic field remains "on."

3,900 to 3,700 MY ago: As the interior core of Mars cools, the magnetic field weakens.

After 3,900 MY ago: Hellas basin-forming impact punches a hole in the crust and disrupts the pattern of magnetized rock. The magnetic field is now too weak to remagnetize the crust.

3,900 MY ago to approximately 3,000 MY ago: As the climate evolves, water flows, causing deposition of sediment layers on the Hellas basin floor, as well as enhanced erosion, dissection of rim structure, and the formation of some river channels.

Approximately 3,000 MY ago to present: Less fluvial activity occurs, and the magnetic field is now "off." Occasional impacts form craters on top of the eroded Hellas and dry river channels. Gradual erosion and exposure of sediment layers take place.

Some researchers take this scenario further. They hypothesize that the turnoff of the magnetic field stopped the deflection of charged solar particles around Mars (the same particles that cause auroras when they are not fully deflected around Earth by our magnetic field), and that this, in turn, allowed the solar particles to hit Mars's atmosphere and carry some of it off into space. In other words, the turnoff of the magnetic field, perhaps around 3,900 MY ago, might have been the very event that began the transition from the wet Noachian era of Mars to the dry Amazonian Mars of today.

In conjunction with this sort of reasoning, the archaeologists of alien landforms use stratigraphic superposition of craters and other features, clues from meteorites, and combinations of other data such as the magnetic patterns to understand the evolution of the planet and its landscape.

We've slowly cobbled together a picture of Martian geologic evolution, and we'll fill in more details as we visit other regions of the planet. Meanwhile, it's humbling to ponder planetary scars that survive from the days when asteroids the size of Massachusetts were drifting around the solar system and crashing into planets.

My Martian Chronicles

PART 3: DISCOVERING LUNAR BASINS, FENDING OFF ACADEMIC FEUDS

In the spring of 1961, just before I turned up for graduate work at the University of Arizona, President Kennedy had announced that we would try to land on the Moon before the end of the '60s.

Lunar formations that I had studied in my backyard telescope as a boy in New Kensington, Pennsylvania, suddenly became *places* we might visit.

My professor, Gerard Kuiper, had quickly established himself and his new Lunar and Planetary Laboratory as players in the new national lunar effort. My assignment in his program was to work on a new photographic lunar atlas. I lived and dreamed Moon.

In a Quonset hut–like structure named Temporary Building 6, Kuiper had set up a clever system for viewing the Moon from never-before-seen perspectives. The idea was to project the best lunar photos onto a three-foot white globe and then rephotograph them from directly above the globe's surface. In this way, we could "look down" on the Moon and see lunar formations as they would appear to an astronaut in orbit. You have to remember that the Moon keeps one side toward Earth, so we never see the back side, and we never see the portions around the edge, or "limb," except with extreme foreshortening. For this reason, observers on Earth had been able to map only the features in the central part of the lunar disk.

Our goal was to produce a rectified lunar atlas, showing all visible regions from directly above. Today, such a project would be carried out by computer, but in the '60s it was all done for the first time optically, using our globe and a large-format camera. The whole project was supported by NASA and the U.S. Air Force, the latter having an overly enthusiastic view that they would soon be claiming the "high ground" of the Moon. In my first summer in Tucson, Kuiper assigned me to work on the globe photography team, making photos and darkroom prints that would compose the atlas. It was a terrific job; I got to contribute to the Apollo program, and because the darkroom was held at 68° F, it was a welcome respite from the 105° F Sonoran Desert heat.

One day we were photographing the globe from the side to show the foreshortened regions on the extreme edge of the lunar disk, the regions never seen clearly from a terrestrial perspective. Suddenly, we noticed a huge bull's-eye-shaped feature on one edge of the Moon. An inner ring of mountains surrounded an obscure,

320-km-wide (200-mile-wide) mountain-rimmed lava patch known as Mare Orientale (Sea of the East). Surrounding that were two perfect rings of cliffs, and beyond that was a faint, outermost cliff-ring with a diameter of 1,300 km (810 miles). Curiously, through some physical or mechanical circumstance, each ring was about 1.4 or $\sqrt{2}$ times as big as the next inner ring. To be precise, we could see only the eastern half of the rings; the Orientale system was so big that the other half of the rings extended onto the never-seen back side. The Orientale lava patch had been mapped from Earth only as a small dark patch, seen nearly edge on, and the rings had been mapped only as ridges of mountains. The beautiful, circular, bull's-eye symmetry had never before been recognized.

This system rang a bell with me. As early as 1893, the founder of the U.S. Geological Survey, G. K. Gilbert, had written about the radial symmetry of grooves and ridges around the giant lava plain Mare Imbrium (Sea of Rains) on the northern front side of the Moon. In 1949, a lunar researcher named Ralph Baldwin had gone a step further, recognizing some circular rings of peaks around Mare Imbrium and several other circular lava plains. But none of the earlier workers had been able to see the near-pristine Orientale system, nor had they fully appreciated the repeatable concentric-rim symmetry of all basins as a class. Now, armed with knowledge of Orientale and our overhead views, we had a much clearer picture. I searched the images and began to recognize that almost all circular lava patches were

(continued on next page)

The most dramatic multi-ring basin in the solar system: the Orientale basin on the moon. The Hellas basin, roughly twice as big a feature, may have looked something like this before Martian erosion partially removed the rim and ring structures. Left: *One of the "rectified" discovery photos showing the eastern part of the Orientale ring system, from Kuiper's 1962 globe-photography program. (Lunar and Planetary Lab, University of Arizona.)* Right: *Overhead view of the Orientale system from lunar spacecraft Orbiter IV in 1967. (NASA.)*

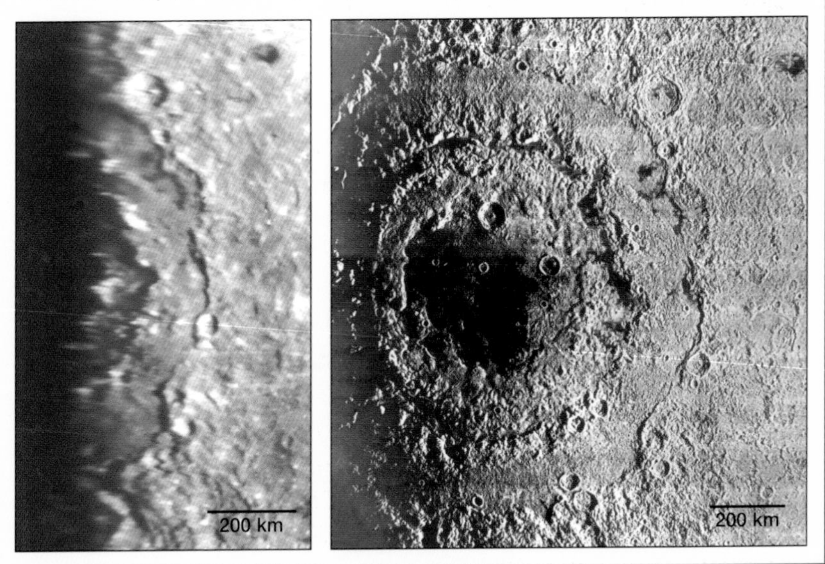

(continued from page 81)

surrounded by large, vestigial traces of concentric rings that showed the $\sqrt{2}$ diameter relationship. I also was able to find a handful of other multi-ring concentric systems that had never been recognized before, and to see that these formed a continuum of features with the largest impact craters.

I took the findings to Kuiper. He saw their significance, and we co-authored a paper, announcing the discovery in 1962. In a gesture that many senior professors would never consider, Kuiper graciously allowed his fledgling graduate student to be the first author.

There is a curious footnote. One of the few other scientists to write about the origin and history of planets in those days was the Nobel laureate chemist Harold Urey. Perhaps because they were the two dominant workers in the field, Kuiper and Urey had a notorious feud that lasted for decades. They could rarely agree on any planetary interpretation. Urey had a theory that the largest impact basin on the front side, the Imbrium basin, had been caused by a grazing impact that hit the Moon from the north and threw material along an axis of symmetry into the radial ridges and grooves to the south. On our "rectified" images, however, we were able to trace perfect circular symmetry around its central impact point, not Urey's axial symmetry around an off-center impact point.

One day not long after we published our 1962 paper, I got a letter from Harold Urey. I was very excited as I stood in the hall of the Lunar Lab, opening my letter from a Nobel prize-winner, but was soon deflated. The letter began: "As an older man to a younger man," and went on with stern criticism. This was not the way to start a scientific career, said Urey. He had studied the pictures in our paper and denied nearly all the circular symmetry we had mapped. I didn't have to read too deeply between the lines to see that Urey thought Kuiper was being a bad influence, filling my mind with mythical lunar features.

This was completely off the wall! Our discovery of multi-ring lunar basins was accepted immediately by U.S. Geological Survey lunar mappers in Flagstaff, including Gene Shoemaker, the pioneer of the lunar mapping system. Later, I had friendly contact with Urey during occasional meetings, and, although he never said so explicitly, I think he realized he had been hasty. I view his letter as a valuable lesson for daily life: Don't get so consumed by your own ideas that you can't accept new evidence.

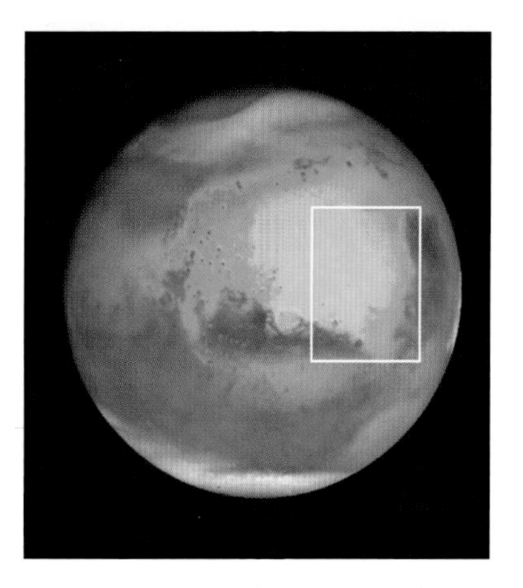

CHAPTER 6

ARABIA

Land of Ancient Violence

The sheer size of the Hellas basin testifies to the intensity of the primordial impact cratering during the Noachian era, but there is no shortage of other signs of that bombardment. Mars has ancient surfaces as heavily cratered as lunar regions that can be seen in a backyard telescope.

The Mariner 9 mapping mission in 1971 and 1972 revealed that the highest crater densities lie in and around a desert region named Arabia by Schiaparelli in 1877. Here, we see the scars of a thousand ancient explosions. Many of the craters are larger than 10 km across, and each marks a spot where an asteroidal impact unleashed an explosion with the energy of thousands of hydrogen bombs. Applying our rule—the more craters, the older the site—we can safely say that this is some of the oldest surface on Mars, in terms of the preservation of large-scale geographic features. The area is mapped as dating from the Noachian era, and most of the impacts occurred during the intense bombardment period.

According to lunar data, the intense bombardment of the Moon (a good marker for the cratering on all the nearby planets) lasted from the time of planet formation, 4,500 MY ago, until 3,800 MY ago. There are so many overlapping craters that it is hard to see the earliest one. We can guess that the oldest eroded craters in Arabia formed between 4,100 and 3,900 MY ago. The reason they

1000 km

are still visible is that Arabia is one of the rare Martian areas where nothing much has happened since then—no profound resurfacing processes to wipe out the large craters, no big volcanoes, no flooding by oceans or blanketing by lava flows or sediments.

DECIPHERING THE GEOLOGICAL CODE

Mother Nature kindly moves from planet to planet, stamping them with circular, cookie-cutter bowls of different sizes—impact craters. We get clues about Martian geological processes by

The most heavily cratered part of Mars (facing page), as mapped by the Mars Global Surveyor laser altimeter. (MGS MOLA team.)

Below: *Western Arabia, showing the highly cratered surface and the boundary between the bright desert and the dark marking known on historic maps as Sabaeus Sinus. Notice that materials are divided into three classes: the bright desert materials, intermediate materials, and the darkest deposits in crater floors and scattered dark patches. Precise differences among these three materials are unknown. (Mars Orbiter Camera Atlas, Malin Space Science Systems.)*

IN SEARCH OF MARS'S OLDEST REGIONS

When we first began mapping crater populations of Mars with Mariner 9 in 1971, the question in my mind was which kinds of areas had the most craters? The bright deserts? The dark markings? High areas? Low areas? The poles? Since I was "the crater guy" on the imaging team, one of my own projects was to examine crater densities. I took a topographic map, prepared from our photos by the U.S. Geological Survey astrogeology branch, and marked all the largest craters by hand so we could study their distribution. The map is shown here. The result surprised me. Since the Moon has high crater densities in the bright uplands and low densities in the dark plains, I thought Mars might show the same pattern. As it turned out, the number of craters in an area has almost no correlation with brightness, altitude, latitude, or other qualities of the area. The heaviest concentration of craters occupies the bright area of Arabia but spills over into the northwest and west parts of the darkest marking on the planet, Syrtis Major. All of which goes to show, once again, that the markings mapped by Schiaparelli, Lowell, and others are not the key to the actual underlying geology or topography. The markings aren't caused by the topography; they are simply draped across it.

It is true that, as on the Moon, the uplands (in the southern part of Mars) tend to be more heavily cratered than the lowland plains (in the north). That's because the lowland plains on both the Moon and Mars are covered by

comparing the appearance and numbers of Martian craters with those of the lunar crater population. The handy circular craters become erosion markers. Well-preserved lunar surfaces show that many small craters exist for every big crater. When we compare Mars with the Moon, we see that on Martian surfaces many of the expected smaller craters with diameters less than about 40 km are missing! It tells us that Mars, unlike the Moon, has had enough geological activity to wipe out the smaller craters but not the largest ones. This makes sense, of course, because if dust constantly fills in Martian craters, the small, shallow ones would be obliterated before the big deep ones.

The rate at which a world's craters are removed, and the relative losses of small ones compared to big ones, thus becomes a

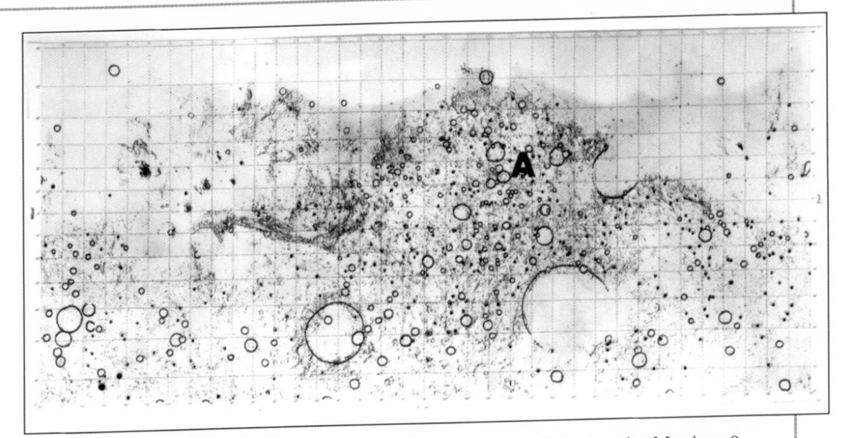

A map of Mars with the largest craters handmarked, prepared during the Mariner 9 mission. "A" marks the most heavily cratered region, in and around Arabia. Hellas basin can be seen below the "A," at the lower right center. The Valles Marineris canyon system is at left center, and the map shows that the northern plains generally have few craters.

enormous young lava flows that engulfed and covered the rims of all but the largest Noachian craters. But on Mars the highest regions don't have the most craters. They comprise something the Moon never heard of—enormous young volcanic mountains. The bottom line is that Mars has more complexity and many more varieties of landscapes than the Moon. Many Martian regions, light and dark, high and low, are so young that they are almost craterless. In these regards, Mars is much more like Earth than the Moon.

sophisticated index to the total net rate of geologic activity on that world. If you could fly around the universe in your starship, you could judge the level of geologic activity on every planet from a distance, just by looking at the number of craters. On Earth, you'd see far fewer impact craters than on Mars, because our restless planet has extraordinary geologic activity: shifting continents, mountain building, earthquakes, volcanoes, winds, rains, and floods. Lots of impact scars indicate a dead planet, where the craters remain behind like bullet wounds. Few craters indicate an active planet. Losses of small craters mean ongoing geological erosion or obliteration. Using this rule of thumb, we can see at once that the lunar surface is old and inactive, Earth's surface is young and active, and Mars is in between.

1 km

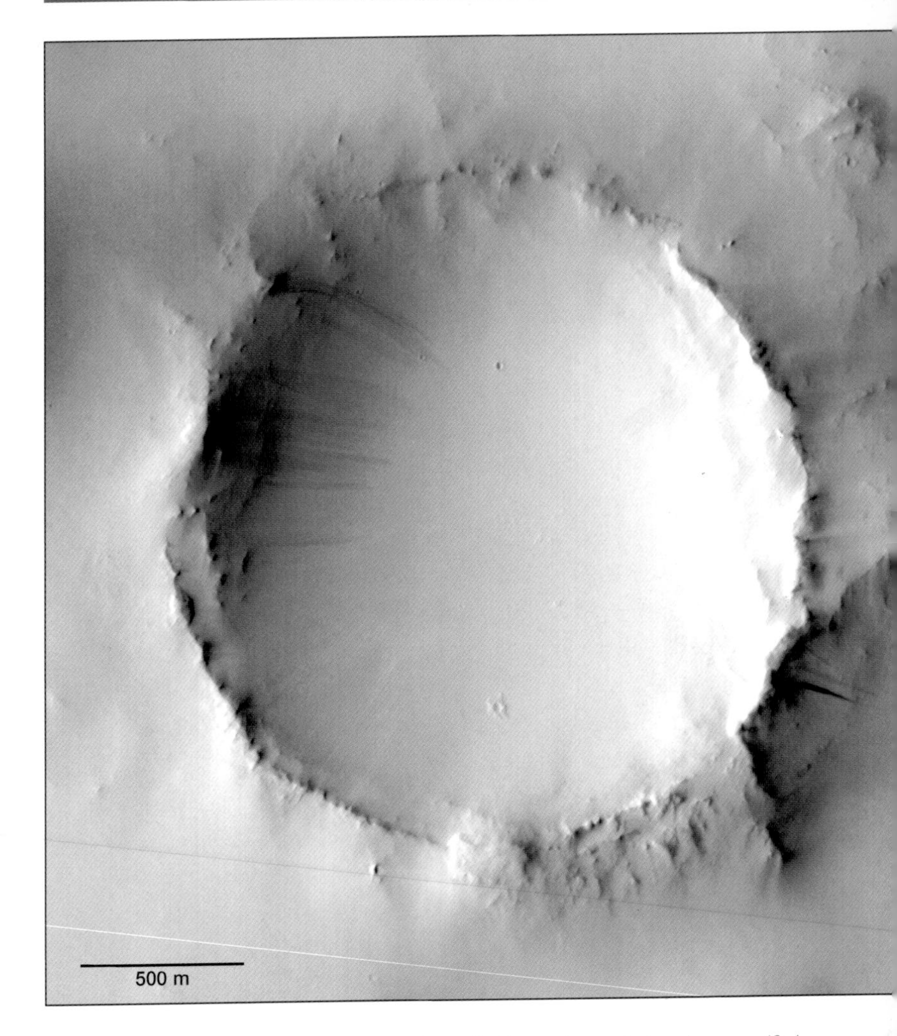

A typical Noachian landscape (facing page). This image shows several characteristic features of old terrain on Mars, when seen with the Mars Global Surveyor telephoto camera. Ancient strata are being exposed by erosion. The exposure of the old strata must be fairly recent, since few impact craters are superimposed on the landscape. Several old craters can be distinguished by their eroded rims, but they are partly buried under much more recent infill of dust. Near the bottom, erosion has created several stratified buttes. (336W 11N, M03-00758.)

Above: A nearly buried crater. Unaltered Noachian surfaces are hard to find because so much dust has engulfed the ancient landscapes. This old desert crater is nearly overwhelmed by deposits of smooth, loose dust. The dust fill in the area must be young because few impacts are superimposed on it. Dark streaks on the slopes are results of loose sand or dust slipping downhill. (331W 10N, M03-01115.)

WHERE TO LAND ON MARS

Arabia gives us a clue about how to pick a landing site to explore Mars. NASA committees, planning robotic probe missions for the next decade, have sought Noachian landing sites that would maximize what we learn about primordial conditions on Mars. The hope is that an ideal site might reveal sediments deposited in long-lost lakes or seas. If this strategy worked, a mission to an ancient seafloor might spot the first Martian fossils!

But it's not so easy. Landscapes that date back into Noachian times, when there may have been more water, have been damaged by the intense cratering that occurred in those days. Surfaces were being pulverized and churned by small impacts—a process picturesquely called "gardening." That's why the Apollo landing sites in the 3,600-MY-old lunar lava plains—to take a lunar example—do not show pristine lava flow textures but rather burial by about 10 meters (30 feet) of fine dust and rock chips. Impact gardening grinds up the surface. Yes, we can land in Noachian deposits, but the top 10 or even 50 meters may be ground into dust. Pristine samples might not be accessible except by drilling or finding recently exposed cliffside strata.

Worse yet, features that are used to classify a surface as Noachian, Hesperian, or Amazonian tend to be large-scale, such as numbers of 1-km craters and overlapping relationships among 100-meter-thick lava flows or sediments. If you are looking for old surfaces, you have to be careful, because you might land in an ancient crater filled with Noachian lake-floor deposits, only to discover that everything is covered by 10 meters of Amazonian sand dunes or a 5-meter-thick lava flow. You would be looking for a 4,000-MY-old surface, but you might be standing on 1-MY-old material.

The point is this: If you want to find an ancient surface on Mars, you need to seek special conditions that preserved that surface shortly after it formed and protected it from gardening and other processes. As we'll see later in this book, such protected sites do exist.

CHAPTER 7

CRATERS AND THE DEPTH TO THE ICE

As mentioned in the last chapter, the state of crater erosion in a particular region gives clues to that region's history. Craters formed before the Hesperian era seem to be smoothed, but craters formed after that time are well preserved, implying a change in erosive conditions. To read such clues, we need to know what a fresh impact crater looks like so we can search for signs of modification by other geological processes.

Fresh craters reveal awesome forces at work during planetary evolution. Remember that each Martian impact crater marks the place where a cosmic projectile crashed into Mars at literally astronomical speed, typically 12 to 20 km per second (7 to 12 miles per second), or around fifty times faster than a supersonic jet! Imagine the spectacular explosion that formed the kilometer-scale crater shown on the next page. An asteroid fragment the size of an office building plows into Mars at a speed of perhaps 10 km per second (6 miles per second). In the first few seconds, a massive fireball erupts from the surface, opening a cavity like a giant's soup bowl—a kilometer across and hundreds of meters deep.

The important thing to realize is that the projectile velocity is far greater than the speed of sound. This controls the explosion. The explosive impact energy must dissipate through the ground mostly as vibrations, which move outward through the ground at the speed of sound. Because the projectile's velocity is faster than the speed of sound, it penetrates into the ground faster than the energy can dissipate. Therefore, enormous amounts of energy concentrate in the volume of rock immediately in front of the projectile, like the shock wave in front of a plane breaking the sound barrier. A fraction of a second after the impact, the situation is the same as if a huge bomb had been set off at the front edge of the projectile. An enormous concentration of heat, pressure, and mechanical energy has been set loose and begins to propagate outward in all directions. Sheets of soil and rock are peeled back like the petals of an opening flower. In subsequent seconds, masses of debris are ejected from the crater floor, jetting high into the air. As the smoke clears, parts of the steep crater wall slump into the cavity until all is quiet.

This explains several things about craters. First, they are circular and symmetric because, regardless of the angle or direction of the

A very young crater the size of a football stadium has blasted rays of dark powder across the surface of Mars. Such ray systems (usually brighter than background surface) are common on the Moon but short-lived on Mars because wind blows the dust around. Here, the rays are dark material excavated by the crater. (10W 6N, M14-02051.)

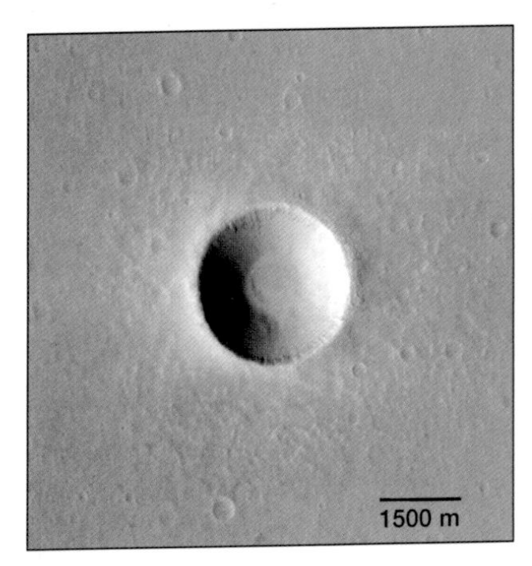

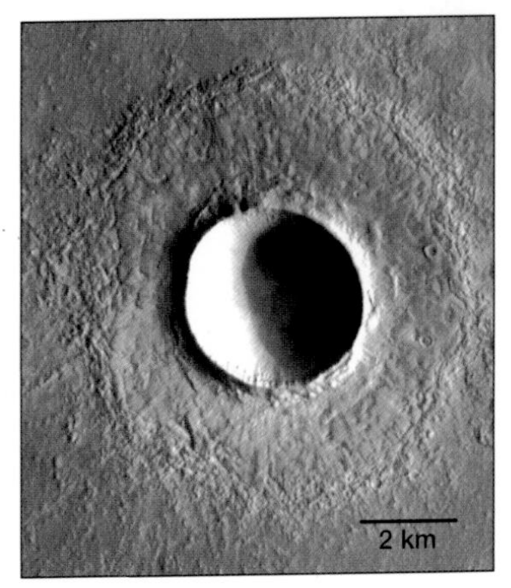

The effects of impacting into ice. Above left: A 2-km crater 14 degrees north of the equator formed where the upper layers of soil are ice-free; the rim was built of hummocky dry ejected debris, feathering out into the desert. (Mars Odyssey Themis camera image; 72W 14N, release 020510.) Above right: A 4-km rampart crater 33 degrees north of the equator formed in ice-rich soil and showing thick ejecta formed by muddy slurry. (239W 33N, SP2-43704.)

impact, they are formed mainly by the explosive energy concentrated at the point of impact, not by the projectile pushing material out of the way by its own motion. Second, while the initial cavity is almost hemispherical, material may slump down off the wall in the moments after the explosion and fill part of the interior, so that the final cavity may be about a third as deep as it is wide. (This may continue to fill in and become shallower later on if winds or water dump more material into the crater.)

Third, the explosion distorts underground layers of strata beneath the crater. Imagine the impact of a mountain-size asteroid into an area of flat-lying strata. The projectile penetrates before it explodes, and so the explosion occurs just below the surface of the ground, pushing the material upward and outward. At the resulting high pressures and temperatures, rock layers can actually bend upward and peel backward. The upward heaving of the ground creates part of the raised rim around the crater. We can see this effect in craters on Earth. At Arizona's Meteor Crater, which is similar in size to one of the craters shown here, the strata were initially flat;

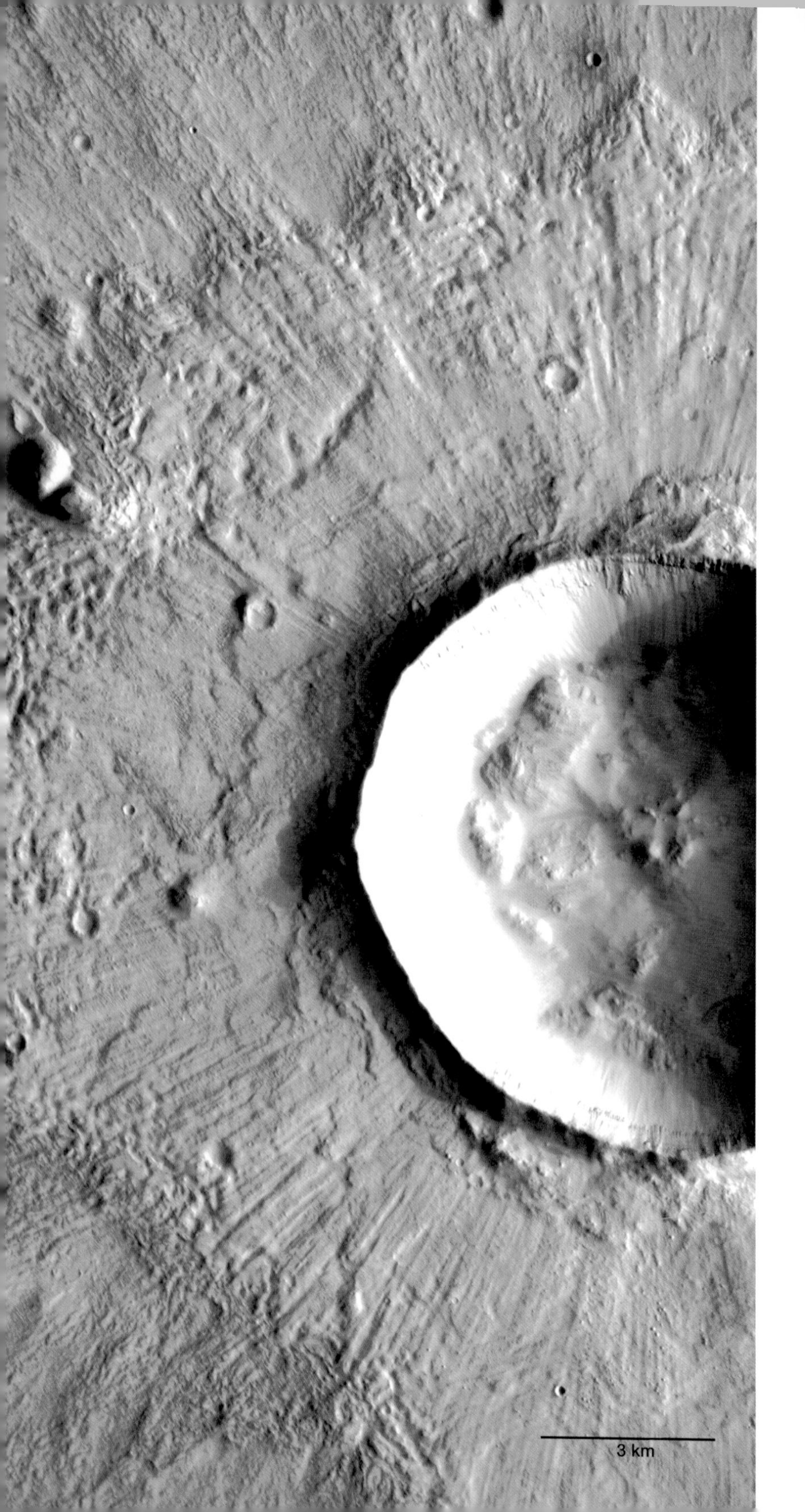
3 km

An 8-kilometer-wide rampart crater 29 degrees north of the equator (facing page) shows a spectacular ejecta blanket of muddy slurry with radial striations. (179W 29N, SP2-44305.) Right: *Detail from upper right of image hints at the process of ejecta blanket formation. The debris has flowed around a preexisting hill (at left), leaving unblanketed smooth terrain on the leeward side.*

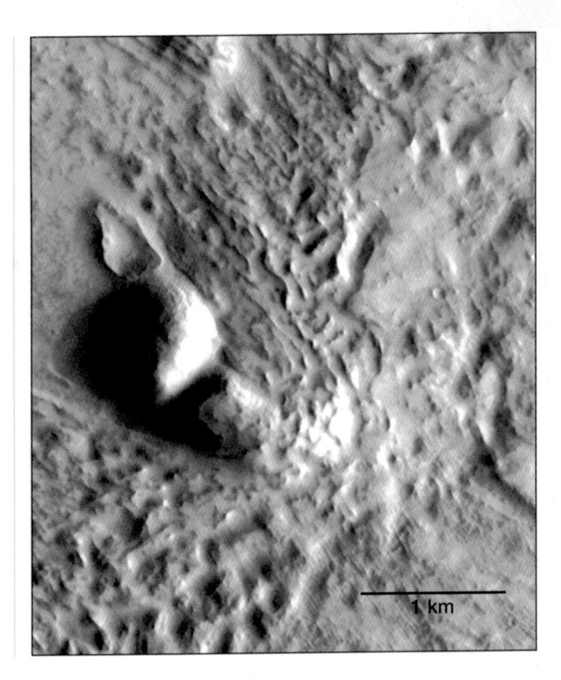

but if you visit the crater, you can see the rock layers jutting out of the crater wall at a steep angle, where they have been cracked and bent backward into position by the explosive force. Fractured and distorted rock extends into the ground far below the floor of the crater. Thus, even if a crater is obliterated by erosive removal of the top hundreds of meters of soil after thousands of years, it leaves a scar in the exposed underlying layers.

A fourth effect is that the explosion blasts material in all directions, just like a bomb burst, surrounding the crater with a layer of rubbly debris ejected from the pit, thrown upward and then dropped back onto the surface. Close to the crater, this debris piles up into a steeply crested rim, but it also extends out several crater-radii in a layer of rubble. This layer is called an **ejecta blanket**, and it forms a spectacular part of every fresh impact crater. The greater the distance, the thinner the ejecta blanket, till it feathers out to nothing. Ejecta blankets are easy to see among craters on the Moon, where fine bright **rays** of dust jetted outward from each explosion site. In addition, jets of rocky debris may fly out of the crater and fall back to the ground, making strings of small **secondary impact craters**.

My Martian Chronicles

PART 4: HIT RATES, CRATER COUNTS, AND INTERNATIONAL COLLABORATION

The more impact craters per unit area, the older the surface. But how old, exactly? During three decades, from the early Mariners to MGS, I've been trying to figure out the actual ages of areas with a given crater density.

How many million years does it take to get, say, ten 1-km-diameter craters in an area of a million square km? My general scheme for solving this problem has been to start with ages of surfaces measured on the Moon, based on rocks and soils brought back by Apollo astronauts and Russian robotic sample-return landers. If we count how many craters formed on a 3,500-MY-old lunar lava plain, for example, then we can figure out the cratering rate on the Moon, averaged over the last 3,500 MY. Next, we use orbital data measured for asteroids and comets to figure out how many of them hit Mars for each one that hits the Moon. Several scientists, such as Bill Bottke in Colorado and Boris Ivanov in Moscow, have estimated that Mars gets roughly two to three times the "hit rate" as the Moon. Using that, we can figure out how long it takes to accumulate the observed number of craters of any given Martian surface unit—that is, its age.

As early as the 1970s, after Mariner 9, I used this system to estimate that some of the giant volcanoes and young flows on Mars could have ages as low as a few hundred million years, which later turned out to agree with the ages of samples by the Martian meteorite lava rocks—an encouraging sign.

Another scientist, Gerhard Neukum, who headed a lab in Germany, has been the primary other researcher in the crater-count field. For years, we worked independently in a spirit of friendly but detached competition. Gerhard tended to get older ages for Martian features than I did. He concluded—correctly, I think—that the rate of volcanic and fluvial activity was highest before about 2,500 MY ago. I emphasized that some of the lava-flow surfaces must be younger than that. By the 1990s, Mars scientists were writing about the Neukum timescale versus the Hartmann timescale, and treated our work as if it represented a classic scientific feud.

Between 1999 and 2001, Gerhard and I joined forces for the first time, motivated by an important Mars-evolution workshop held at the International Space Science Institute in Bern, which facilitates international

Formation of a 100-m-scale crater on Mars by impact of an asteroid fragment. Large pieces arcing through the air will form secondary impact craters. Other debris will fall back outside the rim, making an ejecta blanket. (Painting by author.)

projects in space research. True to Switzerland's historic role as an international meeting ground, the Swiss organizers challenged Gerhard and me to try to reach some consensus on our two systems. We began drafting a report in offices and at café tables under medieval cathedral spires, just down the street from the apartment in which Einstein wrote his fundamental papers on relativity! With the brilliant Russian cratering expert Boris Ivanov acting as our diplomatic buffer, we combined some of our data, refined our approaches, and came up with our best estimates for the dates of the eras, as listed in chapter 1. As it turned out, we were really emphasizing different aspects of what now seems to be the truth. Many areas on Mars are older than 2,000 to 3,000 MY, but some areas are younger. Our new age estimates fall between estimates we had published earlier and confirm that early volcanism did not die out, but dwindled to a lower rate that continues in recent geologic time. The beauty of all this is that it agrees so well with the dates from Martian meteorites. So we think we have a rough handle on the outlines of Martian history.

During the writing of this book, a pleasant e-mail arrived, saying that Gerhard and I had been selected to share a prize called the Runcorn-Florensky Medal, given by the European Geophysical Society, for our work on cratering and planetary chronologies. What I liked best about the prize was its internationalism—the idea of a medal named for a Brit and a Russian, given to a German and an American by a European society. It came during a period when environmental and nuclear disarmament treaties were breaking down, and the world seemed to be splintering into warring factions that include medieval-minded fundamentalists. To me, it is important for scientists to stand up and show how international cooperation can pay off for everybody. The exploration of other planets makes us realize we share the tiny, fragile Earth as brothers and sisters, and a cosmic perspective makes the squabbles between ideologies and national factions on Earth seem tragically provincial and petty.

The Meteor Crater in Arizona is a kilometer-wide feature formed by a large meteorite impact roughly 20,000 to 40,000 years ago. Scale is shown by roads that lead to a museum and visitor center on the rim at right. (Photo by author.)

On the Moon, the fresh rays of pulverized dust are almost always brighter than the background surface; but on Mars, the craters tap into layers of different-colored materials, and the ejected dusty rays can be either brighter or darker than the background.

Specific qualities of the ejecta betray the nature of the underground material at the impact site. If the original meteorite hit in a dry, sandy area, then the ejecta blanket is mostly a fine dust sheet. If it hit a lava flow, the ejecta contains large, blocky masses of rock. An important distinction between the Moon and Mars is that some Martian impactors struck areas with underground ice, and the resulting ejecta blanket was like a muddy slurry of melted ice and soil. Such thick, muddy-looking ejecta blankets, like those shown in this chapter, are dramatic evidence of underground ice or water on Mars. A massive cloud of steam may result from such an impact, expanding and rushing across the ejecta blanket like a hurricane, leaving a pattern of radial grooves, as can also be seen in the photos. Martian craters of this type are unlike anything that had been seen before on the bone-dry Moon, where ejecta blankets are thin layers of bone-dry dust and rock fragments. Investigators called those on Mars **rampart craters**, because the muddy slurry formed rampart-like masses around the crater. Computer models suggest that the mud would splash out of the crater at a high angle and then fall back around the crater a minute or so later with a mighty *sploosh*, surging out in lobes and pushing up the bluff-like rampart at the outer edge of the ejecta blanket.

RAMPART CRATERS: A GUIDE TO UNDERGROUND ICE DEPOSITS

Rampart craters are more prominent in some areas of Mars than in others, but in any given region where they exist, they are generally larger than smaller craters that don't show rampart features. In 1980, Russian scientist Ruslan Kuzmin and NASA geologist Joseph Boyce independently discovered that although rampart craters can be found at all latitudes, a provocative progression exists: the farther from the equator, the smaller the critical size. This discovery suggests two important facts about Mars. First, much of Mars has an underground permanent ice layer, called **ground ice**, like that found in the arctic tundra on Earth. Second, the farther from the equator, the shallower the depth to the top of the ice.

This is a crucial discovery about Mars, and it makes perfect sense. Mars is a cold planet, and the normal subsurface soil temperatures are below freezing; barring other circumstances, water that sank into the ground would be frozen. The equator is the warmest region, on average, and the depth to a permanently frozen zone is greater here than elsewhere. As we travel from equator to pole, the depth to the top of the underground ice layer decreases, approaching zero at 60° to 80° latitude north and south. At the poles themselves, frozen water actually sits on the surface year-round in a permanent ice cap, just as on Earth.

According to the discovery of Kuzmin and Boyce, rampart craters are nature's way of giving us a tool to measure the depth to the top of the underground Martian ice layer! By 1988, Kuzmin and the French researcher F. M. Costard had completed a brilliant mapping of the depth of the smallest rampart craters in each part of Mars—the depth to the ice. This distribution is patchy, as with most geological properties, but the progression to shallower depths at higher latitudes shows up clearly. In most of the equatorial zone, the top of the massive ground ice deposits maps at a depth of 200 meters to 450 meters. The top 200 meters are relatively "dry," which in Martian parlance means ice-free. In the region from 30° to 55° latitude, either north or south, the depths to ice are more like 50 meters to 200 meters. Poleward of about 55°, the depth to the ice is less than 100 meters. As we'll see in the next chapter, this discovery may be crucial to understanding many phenomena of Mars, including the deformation of many craters from their pristine form into distorted or "softened" shapes.

Impact explosions with high energy create craters with more complex geometry than smaller explosions. At diameters above a few kilometers on Mars, the depth and weight of the high crater walls are great enough that the central floor of the crater rebounds, or is heaved upward, forming a central peak. This crater penetrated deep enough to hit ice, even though it is located near the equator; it formed a rampart ejecta blanket with flower-like petals created by surges of muddy slurry (bottom, top, and left). (277W 3N, Mars Odyssey Themis camera image V020614.)

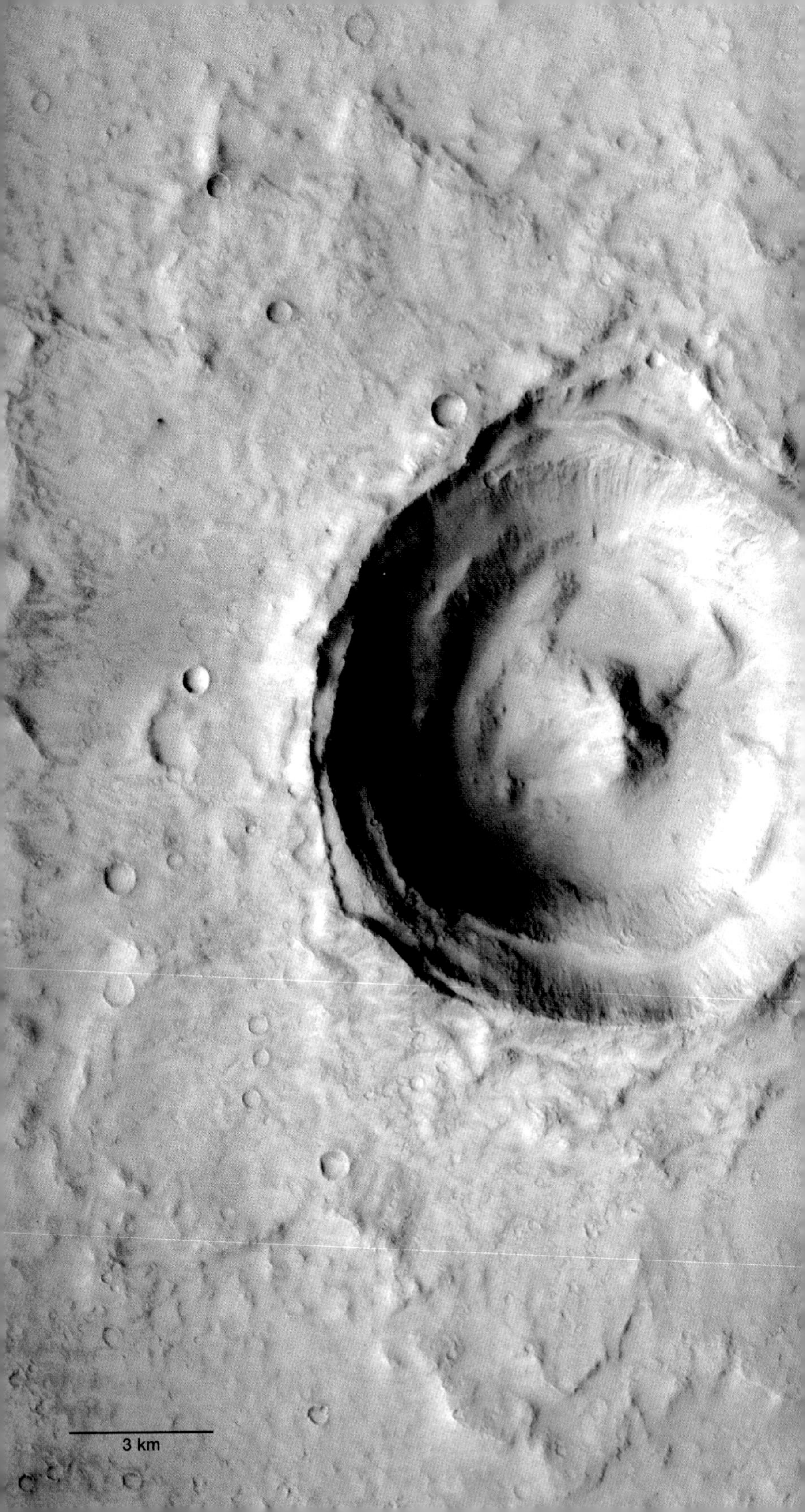
3 km

CHAPTER 8

NOACHIS AND THE SOUTHERN HIGHLANDS

The Mystery of Softened Terrain

The main concentration of ancient, heavily cratered Noachian terrain lies in the southern hemisphere, around a province called Noachis. This region gave its name to the Noachian era, and the craters there offer clues to the conditions on Mars during that ancient time.

To Earth-based observers, Noachis was a mottled and nondescript dark area. Schiaparelli assigned the name in 1877, picking a word that referred to Noah and legends of the floods. This was an inspired choice, because the dry and dusty craters and plains of Noachis hide ghostly hints of ancient floods, surging rivers, sediment transport, and perhaps long-gone lakes, now transformed into buried ice deposits. Today, the area is shown on topographic maps as Noachis Terra.

Topographic map shows heavily cratered terrain of Noachis, with part of the Hellas basin on the east. Red is high, yellow and green are mid-level, and blue is low. Most of the large craters appear relatively shallow and flat-floored, as if filled in with some material. (MGS laser altimeter data.)

It was in Noachis and adjacent terrain in 1969 that Mariners 6 and 7 took swaths of pictures that crossed cratered highlands and showed shallower and "softer" craters than on the Moon. Rims were flat or rounded instead of sharply crested, as if they had formed in wax that had then partially melted. From 1969 to the present, investigators have been puzzled by this unknown softening process, which came to be called **terrain softening**.

Detailed comparison between softened and unsoftened terrains turned up more facts. Large, Noachian-age craters in the southern highlands not only have flattened rims but also have shallow, flat crater floors that seem to have been filled with sediments, sometimes nearly to the rim. Instead of curving smoothly down to the floor, interior walls of old craters have shoulders or terraces at

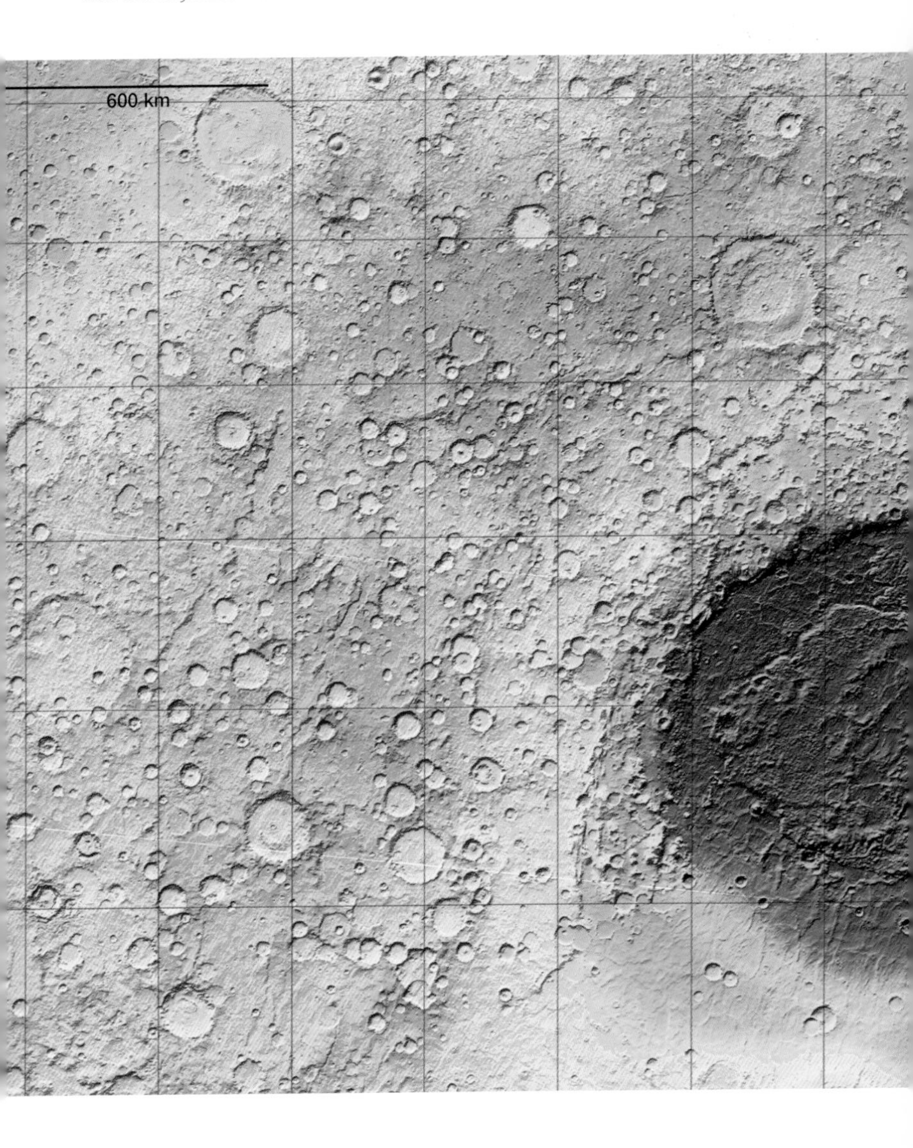

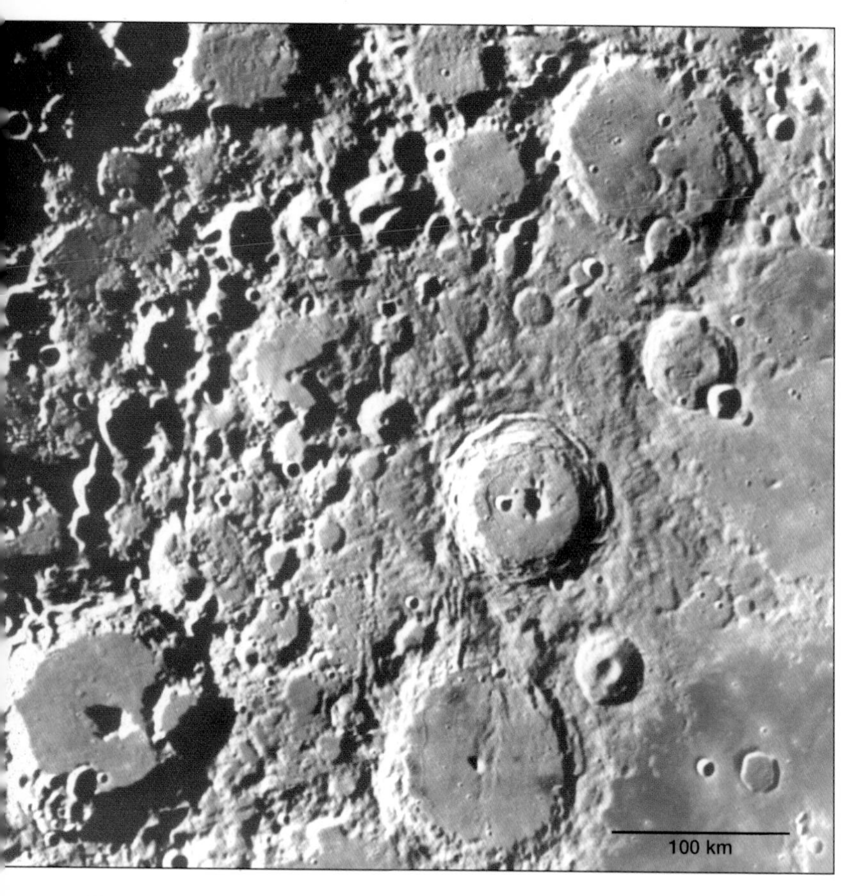

Comparison of lunar uplands as photographed from Earth (above) *and Martian-cratered uplands as photographed by Mariner 6 in 1969* (facing page). *The Mariner 6 images were the first to hint that Martian cratered uplands had seen different erosive processes than lunar uplands. Erosion has flattened the rims of the older Martian craters and filled in their floors with sediments. (Lunar and Planetary Lab, University of Arizona; Mariner 6N41.)*

the base, as if they had melted and slumped partway down into the crater. The landscape between the old, softened craters was not a surface crowded with smaller pits, as on the Moon, but often a flatter plain, sparsely cratered with remnants of old, softened craters and a younger population of small, sharply defined younger craters with sharp rims and bowl-shaped interiors. To summarize, these areas looked as if craters older than a certain age had partly

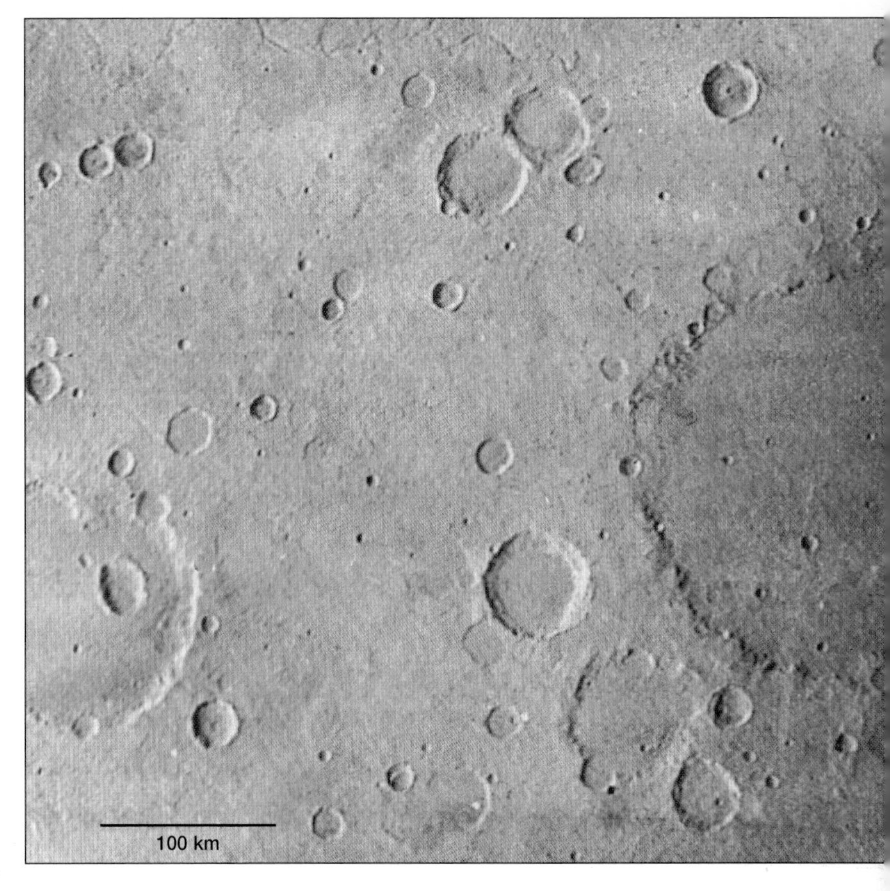

"melted down," and many of the smallest craters older than that age had disappeared altogether.

Terrain softening cries out for explanation. It's as if Texas-size regions have been subjected to some sort of melting process that removed all the rough edges. Complicating the puzzle, the process seems to have stopped at some time in the past, so that more recent craters retain fresh, sharp-rimmed profiles. How can we explain this? For the first researchers who first saw the Mariner and Viking pictures, the problem seemed overwhelming.

MARS'S BIG SECRET

Because the surface of Mars is so dusty and dry today, everyone thinks of Mars as a dry planet. But the secret of Mars is that it is not dry. Fundamentally, Mars is a very wet planet, except that the

Softened terrain on Mars at 36° S latitude, a common latitude zone for softening. This image shows an unusual crater in which the SE half preserves at least part of the original sharp-crested rim, but the NW half has undergone severe softening, so that the rim is reduced to a low, flattened plateau. Layers of strata can be seen in the SW corner of the crater. This image proves that terrain softening is not regionally uniform, but varies in degree from place to place. Bright "worm-like" dunes can be seen at the boundary between the flattened and unflattened terrain (lower left). (354W 36S, M03-01141.)

water is hidden in three places: the permanent polar ice caps, ground ice under the surface, and water molecules trapped in minerals within the soil. In fact, the more we learn, the wetter Mars seems to be. As a result of the MGS and Mars Odyssey missions, the evidence mounts that Mars has great reserves of water hidden in these three forms, especially ground ice. The new data support the idea that Noachian Mars may have been much wetter than present-day Mars, which would explain ancient river channels and other features.

One line of evidence starts with asteroids, which are surviving fragments of the planetary building blocks. The farther out in the solar system they are, the more water they have, because inner solar system materials were too warm to retain much water or ice. So it's likely that Mars formed from more water-rich material than Earth.

Another line of evidence is that terrestrial volcanoes emit a mixture of carbon dioxide, water vapor, and other gases in certain proportions, and based on known Martian lava properties, we expect Martian volcanoes to do the same. Thus, we can measure the carbon dioxide and other gases on Mars and use them to calculate how much water vapor should have been emitted by Martian volcanoes along with them. To do this, we can use the gas argon. Argon is a heavy gas that does not escape from Mars's gravity, and it is chemically inert, so it doesn't disappear into chemical compounds or minerals. This means that nearly all the argon ever emitted on Mars is still there, in the atmosphere. This makes argon an ideal marker of total atmospheric gases emitted during the history of Mars. For each amount of argon, there should be certain amounts of carbon dioxide and water. Based on this measuring technique, researchers have estimated that primordial Noachian Martian air may have been fifteen to seventy times more dense than it is today. In other words, it may have had 10 percent to 50 percent of the Earth's present atmospheric pressure, instead of the meager 0.7 percent that remains on Mars today. Much of this

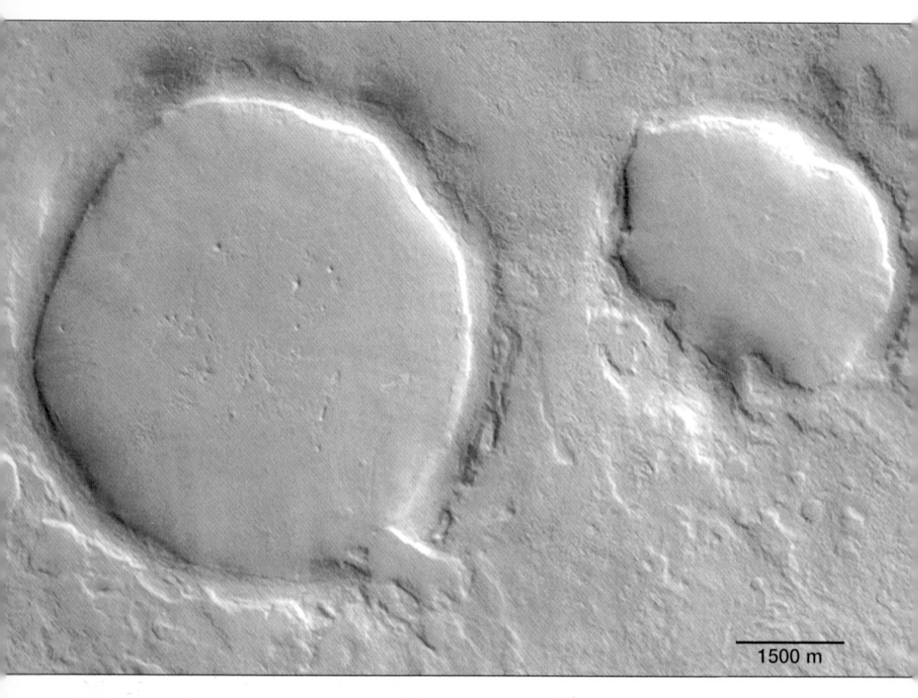

Terrain softening exists at moderate latitudes in the north as well as the south. This view from the Mars Odyssey spacecraft shows very similar softened rims on craters 4 and 6 km across, at 44° N latitude in the region of Arabia. (Themis camera visual image, 322W 44N, release 020425.)

would have been water vapor, which could have condensed into rain or snow. Hence, Mars may have had enough water to make substantial lakes and seas.

Another example of this same technique comes from work in 2001 by the Russian scientist Vladimir Krasnopolsky, working at the Catholic University of America in Washington, D.C., with University of Maryland colleague Paul Feldman. Krasnopolsky and Feldman detected hydrogen molecules on Mars and were able to use them as fossil tracers of broken-up water molecules that were there in the past. They concluded that the original water would have been enough to cover a level Mars with a global ocean at least 30 meters deep. (Scientists using the phrase *global ocean* are talking about an idealized ocean that would exist if the surface were level. More realistically, enough water was emitted to create many local lakes and seas, some deeper than 30 meters—and these may have had a crust of ice on the surface, like the Arctic Ocean on Earth.) This estimate does not count additional water that may have always been frozen. Thus, Krasnopolsky and Feldman were able to conclude that Mars may initially have had a higher fraction of water, relative to its total mass, than Earth!

Proportions of certain isotopes, the variant forms of individual elements, also act as similar indicators of the amount of water present at Mars's beginning. Each set of gases and isotopes acts as an independent tracer of early Martian water abundance. As summarized by Colorado researcher Bruce Jakosky and St. Louis researcher Roger Phillips in 2001, isotopes of Martian gases such as nitrogen and oxygen, as well as carbon from carbon dioxide, indicate that 33 percent to 90 percent of the original Martian atmosphere was lost directly into space. Additional losses may have occurred as various gases dissolved in Martian water and were eventually hidden among chemical weathering products within the Martian soils. According to these data, an initial atmosphere exceeding 7 percent of Earth's is not unlikely. Isotopic studies of various other gases indicate that during the Noachian era, Mars emitted enough water to cover the surface with a global ocean tens of meters deep.

Still another measurement comes from the oldest of the Martian rocks found on Earth—a rock called Alan Hills 84001 (or ALH 84001), which was collected in Antarctica. This rock, which we've already encountered in discussing the Hellas impact basin, is provocative in various ways. It formed 4,500 MY ago and contains Martian atmospheric gas that was trapped in the mineral and cavities of the rock since it was modified 3,900 MY ago. Excitingly, as shown by the University of California geochemist Kurt Marti and his colleagues in 2001, the gases in ALH 84001 do not show the kinds of gas losses discussed above, which means that the thick primordial atmosphere of Mars lasted until at least 3,900 MY ago. This supports the idea that a thicker, unmodified atmosphere lasted through much of Noachian time, which ended roughly 3,400 to 3,600 MY ago.

To summarize, Noachian Mars had lots of water (in the form of liquid, ice, and water vapor) and a thicker atmosphere that may have lasted until approximately 3,900 MY ago. Second, much of the water and other gases were lost into space as molecules broke up and the lighter ones drifted off into space under the weak influence of Mars's low gravity, while the remaining water ended up in underground ice deposits, in the polar ice caps, and as water of hydration in certain minerals.

Although the details are hazy, this transition probably had something to do with the history of terrain softening. Before the transition, water- or ice-rich soils may have flowed and deformed easily, like glaciers. During the transition, episodes of thawing and refreezing, especially during the Hesperian era, may have led to sporadic occurrences of terrain softening and water release. After the transition, soils near the surface may have been drier and more rigid.

Softened terrain on the oldest surfaces was one of the things that led to a division of Martian time into broad eras. Whereas the terrestrial system of geologic eras grew out of differences in fossils from one era to the next, the Martian system grew out of the striking differences between old terrain, such as Noachis, and young.

ANCIENT GREENHOUSE WARMING?

The ideas above lead us to questions about whether the Noachian climate was radically different from today's Martian climate. The debate over this issue has zigged and zagged.

The first, most obvious hypothesis is that a thicker Noachian atmosphere of carbon dioxide would have produced a greenhouse effect, which warmed the early Martian climate, helping to melt ice and producing rivers. It's the same effect that most knowledgeable researchers are concerned about today: global warming and climate alteration on Earth due to industrial release of carbon dioxide and other greenhouse gases. Under this hypothesis, Noachian Mars was definitely warmer than present Mars.

But would a thicker carbon dioxide atmosphere have really warmed early Mars enough to explain the kinds of river channels we saw flowing into the Hellas basin? Mars is farther from the sun, so there is a question of whether even a whopping greenhouse effect could warm it enough to melt underground ice or to make lakes and rivers.

Worse yet, astronomical theory based on observations of many stars indicates that the early sun was somewhat fainter than it is today, which tends to offset the theory of an early greenhouse effect. Thus, a second and less obvious hypothesis is that the *mean* Noachian temperatures were colder, not warmer, than the present-day temperatures.

Various scientists have attempted careful calculations of the greenhouse warming that might have resulted from thick Noachian carbon dioxide, taking into account the pale early sun. The first calculations suggested there was not enough heating to keep ice melted. Many scientists were disappointed that there seemed no explanation for the river channels and other signs of early fluvial activity. Noachian Mars wouldn't have been radically different than today's Mars. However, Penn State researcher Jim Kasting and others then pointed out that the proposed early thick atmosphere would have been cloudy, and the clouds would help hold in the heat (just as, on Earth, the clear desert air cools off rapidly at night, whereas a cloudy locale stays relatively warm). When this cloud effect is taken into account, the calculations suggest again that early Mars could have been warmer than today's Mars, and the idea of early Martian rivers begins to look much more feasible.

At a 2002 Houston conference, cosponsored by American and Russian institutes, Mars researchers debated the Noachian climate and divided into two main camps. A slight majority favored the "warm, wet" Noachian, the model described above, with a substantial greenhouse effect. The other camp cautioned that only the most optimistic interpretations of greenhouse theory could make Mars warm enough for abundant liquid water. In their view, Noachian

Mars was possibly colder than today and had only sporadic liquid water due to transient warming events. They proposed a "cold, wet" Noachian Mars, with water mostly frozen most of the time.

An additional effect comes from internal geothermal heat. Evidence from isotopes shows that primordial Mars, like other early planets, had short-lived radioactive isotopes. Most of them have decayed by now, but in the early years they produced higher heat flow from the interior than exists today. The underground layers 500 meters down would have been warmer, even if the surface and atmosphere were colder. This means that ground ice layers on Noachian Mars could have been thinner than at present. Underground intrusions of volcanic magma could have melted ice deposits from the bottom up, creating shallow underground rivers that occasionally broke through to the surface, producing short-lived rivers and lakes. This complex picture would explain why a cold, wet early Mars had more terrain softening (the ice was shallower and craters formed in ice-rich soil), and why there were more rivers and lakes (it was easier for water to reach the surface during transient conditions when the ice melted).

So the jury is still out on Noachian Mars's climate. The atmosphere was surely thicker, and the thicker atmosphere favored more fluvial activity on the surface; but it is unclear whether the normal climate averaged colder or warmer than at present. All we can safely say is that it was different.

GROUND ICE DEPOSITS: KEY TO TERRAIN SOFTENING

With each mission to Mars, the underground ice seems increasingly the key to many Martian phenomena. "Rigid" ice can flow if given enough time, as we know from glaciers on Earth. Topographic features formed in icy soil can "relax" into softened forms. A crater formed in ice-rich Mars might start out looking like the craters in the last chapter, but they can gradually soften and fill in, just as a crater formed in a terrestrial glacier would

HOW ICE BEHAVES ON MARS

A Martian explorer needs to understand the behavior of ice on Mars. First of all, remember that there are two kinds of frost and ice on Mars: frozen water (H_2O) and frozen carbon dioxide (CO_2). Both materials make white frost deposits that look about the same to the casual observer. Most days on Mars are cold enough to freeze H_2O, but the coldest days are cold enough to freeze CO_2, which is more commonly known to us as dry ice. For the moment, let's concentrate on frozen water, since water is crucial to Martian history and future Martian exploration.

An ice cube on Mars would appear to be stable, but gradually it would disappear if exposed to the sun and Martian air. That is, it goes directly from ice to gas, the same way that CO_2 dry ice on Earth turns to a misty vapor without melting into a liquid. This behavior is called sublimation, and we say that the ice will **sublime** directly into gaseous water vapor, instead of melting to form liquid water. Thus, we don't see much ice actually exposed on the surface of Mars except at extreme polar latitudes, where it is exceptionally cold. At moderately high latitudes, frosts occasionally form but burn off after days, weeks, or months, just as on Earth. On modern Mars, it's hard to form new water-ice because the air is so thin and dry. For all these reasons, it is unlikely you would find large masses of surface ice in most parts of Mars, despite the cold. Ice exists at the polar caps because that is the only place cold enough to stabilize it.

Underground conditions are better for ice, especially at high latitudes. Just as the temperature in caves on Earth tends to hover at the local year-round average temperature, such as 10° C (50° F), depending on latitude, so the temperature tens of meters below ground on Mars hovers at its annual average, which is about –60° C to –70° C (–76° F to –94° F)—sufficiently cold to keep underground ice solidly frozen. Also, if the ice is more than a few meters below the ground, the soil helps to protect it from subliming into the Martian air. For these reasons, underground ice deposits are believed to be stable over long geological periods, just as they are in arctic Canada. Even though Mars looks arid, frozen water may be not far below a visitor's feet in many areas.

deform and fill in by ice flow. A larger, deeper crater, penetrating through the ice-rich layer into bedrock, would tend to keep its shape. A crater rim of ice-rich material, suddenly piled onto the surface as impact ejecta, would tend to flatten. So the softening of

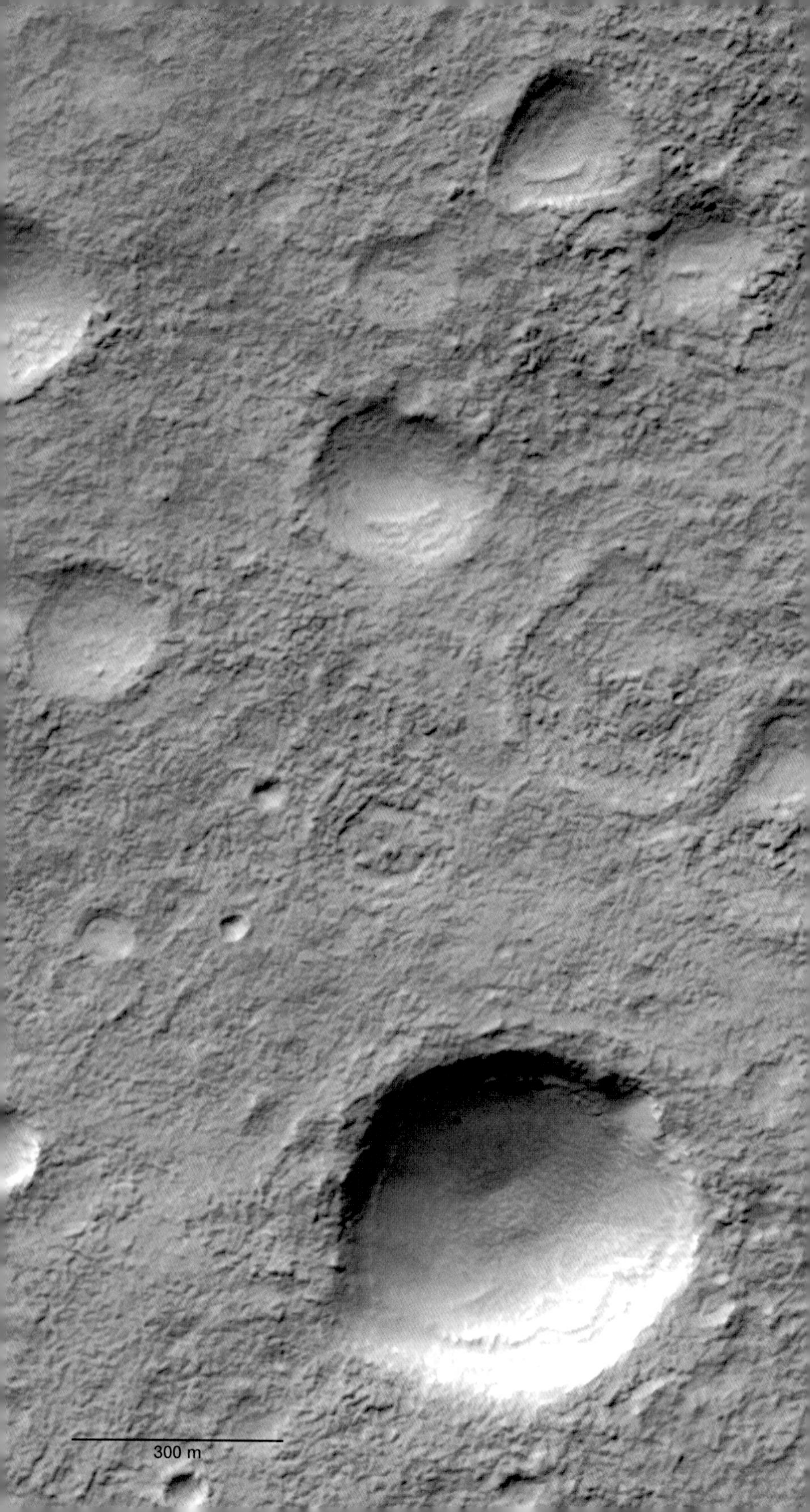
300 m

Craters with different stages of terrain softening are common in Noachis. In the middle are nearly flattened craters, in the upper half are moderately softened craters, and at the bottom is a somewhat better preserved, but not pristine, crater. The difference may indicate that softening in this area was severe up to some early date, acting on oldest craters, but then declined, leaving somewhat younger craters less flattened. This whole area lies on the floor of an old river channel, supporting the view that water or ice plays some role in this kind of softening. (337W 36S, M20-00860.)

Martian terrain seems to result not just from water's washing sediments into craters and eroding crater rims, but also from glacier-like flow of the abundant ice.

How long does it take such features to flow? Months? Years? Millions of years? Terrestrial glaciers sluggishly flow down mountain valleys on timescales of a few millennia. Calculations by Elizabeth Turtle and Asmin Pathare, working through the Planetary Science Institute (PSI), indicate that the timescales are similar on Mars, both for ordinary glaciers and rock glaciers. According to the calculations, craters that have been formed in such mixtures would deform substantially in periods of the order 10,000 years to 100,000 years. Large craters flatten more, in proportion to their initial depth, than small craters, because the stresses on their high walls are greater. The bottom line is that ice flow on Mars happens rapidly, in geological terms, so we would not expect ice-rich topography to remain pristine for millions or billions of years.

ENHANCED SOFTENING AT HIGH LATITUDES

In the last chapter, we saw that rampart craters gave one early tool for mapping the presence and depth of Martian underground ice. The results indicated that in most of the equatorial zone, the top of the ground ice is 200 meters to 450 meters down, but poleward of about 55°, the depth to the ice is less than 100 meters. According to this discovery, if terrain softening has something to do with ice in the soil, there should be more of it at high latitudes.

As early as 1978, Cornell planetologist Steven Squyres, who helped define the Noachian and other stratigraphic units, wrote about terrain softening as a possible example of glacier-like flow of ice-soil mixtures. By 1986, Squyres was collaborating with U.S. Geological Survey scientist Michael Carr, one of the most respected researchers of Martian water history. They mapped the distribution of softened terrain on Mars, proposing that these areas are rich in ground ice, and they confirmed that terrain softening, along with rampart craters, is more common at high latitudes than at the equator, especially in old, cratered uplands. Uplands at latitudes about 30° to 65°, either north or south, are rich in these features. On the maps prepared by Squyres and coworkers, the region of Noachis and western Hellas stands out as the main concentration of softened terrain. This strongly supports the idea that terrain softening on Noachian Mars had something to do with ground ice. Squyres's estimates placed the top of the ground ice at less than 100 meters below the surface in much of this region.

Direct support for the general idea of high-latitude, near-surface ice came in 2002 when NASA's Mars Odyssey mission went into orbit around Mars with three instruments that could detect frozen water, if present, in the upper few meters of soil. At near equatorial latitudes they found an ice-free layer in the top half-meter or so, with no more than a few percent ice in the first few meters below that. But in a sharply defined high-latitude zone, above about 60 degrees in both the north and the south, the instruments found a whopping 20 to 50 percent ice (measured by mass) in the top few meters of soil, below the more ice-poor top half-meter.. In addition, the instruments showed unexpected variation in the much smaller amounts of water bound in the soil minerals from one region to another, and the international investigators declared Mars to be a "hydrologically active planet."

The high concentrations of high-latitude ice have interesting implications. At a level of 20 to 30 percent, the ice would likely be only a condensate filling pore spaces between soil grains. However, amounts above about 30 percent imply ice not just filling pore spaces in soil, but existing in solid ice masses, with the soil grains separately embedded in the ice—a good recipe for glacial movement.

Important problems remain. For example, does the ice measured by Mars Odyssey really mark the top of a thick ground ice layer? Or is just formed by condensation of water vapor, in the upper few meters only? In either case, the ice detected by Odyssey in the top few meters lies in the latitude zones where rampart craters also indicate ice in the top 100 meters or so.

A question also exists as to why terrain softening concentrates at a slightly lower latitude zone, around 30° to 55° from the pole, where the depth to the ice may be greater. Perhaps the best expression of softened terrain arises not when the ice is at the surface (and everything flattens during periods of ice flow), but rather when the top of the ice is, say, 200 meters deep. In that case, the ice flow would occur at depth, and the overlying dry soil would shift and slide and settle, leaving the topography with a softened look. In any case, all the data indicate that underground ice plays an important role in Martian geological history at certain high latitudes.

DEPTH TO THE BOTTOM OF THE ICE

If the top of the ice layer lies at 1 meter or 100 meters or 200 meters, depending on latitude, how far down is the bottom of the ice? The answer depends on what controls the depth of the ice base. As on Earth, the interior of Mars is warmer than the surface, and the temperature increases as you go downward. At a depth of a few kilometers, ice would melt, so the ground ice layer probably ends there. In the past, as we saw above, this depth was shallower. This depth is important because it limits the total thickness and amount of ice hidden underground on Mars. Also, it controls the thickness of the layer that can flow like a glacier and cause terrain softening. Furthermore, if future astronauts could drill below this depth, they might encounter liquid water that could be pumped out.

My Martian Chronicles

PART 5: WHAT'S *THAT*?

One morning at the Planetary Science Institute in Tucson, one of our young research assistants, Gil Esquerdo, was looking at some of the first MGS images on one of our computers. Gil at that time was an undergraduate in physics, bright-eyed and bushy-tailed about anything having to do with planets.

In those first weeks and months after "our" spacecraft reached Mars, we were all aware that we were among the first humans to see details on Mars the size of a bus, and Gil spent many days poring over the pictures.

Suddenly, Gil exclaimed, "What's *that*?"

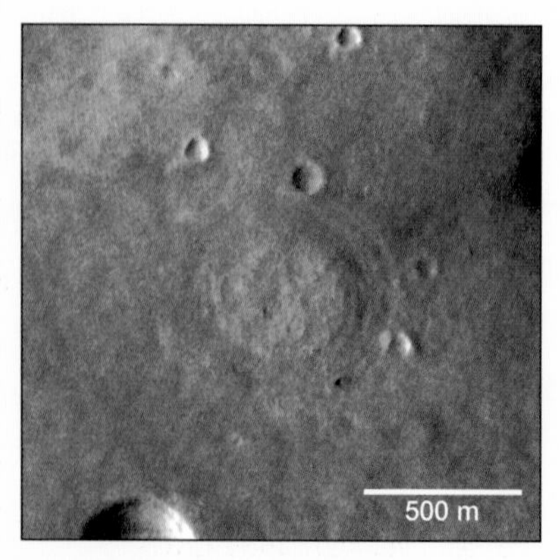

The first "melted craters" found by Gil Esquerdo. (19W, 35S ABI-03005.)

When someone in planetary science says, "What's *that*?" it's usually a good sign. I rushed to look over Gil's shoulder. There, in the middle of Noachis, near the bend of an old, undistinguished riverbed, were two mysterious features: circular rings of low hills and concentric ridges. To maintain my credibility as a senior scientist at the Institute, I managed to find a third, fainter example in the same frame. The rings were delicate as Mars features go, and only a few kilometers (a couple of miles) across.

An advantage of senior scientist status (i.e., middle age) is that you may remember obscure old papers that no one else knows about. In this case, Gil's discovery reminded me of an image I had seen two decades earlier in a paper about crater erosion processes. A researcher named R. F. Scott had made tabletop-scale artificial craters of different sizes in asphalt, and then he'd heated the asphalt so that it gradually flowed and filled in the craters. This process left a ghostly, flattened system of rings that looked astonishingly like Gil's craters.

We wrote a paper about it, suggesting that the features Gil had found were fossil remnants of craters that had formed in ice-rich mixtures, then flowed, deformed, and filled in, like the asphalt lab craters. These features might be remnants of the

Another of Gil Esquerdo's "melted craters." In the center, amid a field of scattered impact craters with odd-shaped dunes on their floors, is an almost completely flattened circular feature, which appears to be a remnant of an ancient crater that has undergone severe terrain softening. (18W 35S, M18-00848.)

ancient terrain-softening era, when many of the smaller, kilometer-scale craters were wiped out entirely.

Still, many questions were left. Although there were three of these ghost rings in one image, we found very few duplicates when we searched in other areas. The best additional example, which we found later, was also in Noachis. Why weren't these things seen in many more parts of Mars? Was there something special about this area, which made the softening more active here? Or was the process elsewhere so complete that all ancient kilometer-scale craters were obliterated? Or are these not degraded impact craters at all, but some other type of geological feature? We still don't know the answers.

A MARTIAN ODYSSEY

The head of the Mars Odyssey instrument team was Bill Boynton, an affable Mars scientist (and excellent poker player, as I happen to know, much to my regret). Boynton had his own personal odyssey getting to Mars. His first spacecraft instrument was on NASA's early '90s Mars Observer, a spacecraft that blew out a fuel line and failed just before reaching the planet. Next, Boynton built an instrument for NASA's Mars Polar Lander . . . which crashed on the polar cap of Mars due to a spacecraft engineering failure. In the meantime, he had developed instruments for an ill-fated comet mission for which funding was canceled in midstream. Next, he was selected as team leader for a Mars Odyssey package of three ice-seeking instruments, one built in Russia, one at Los Alamos, and one by Boynton's own group in Arizona. This time, after nearly two decades of work, Boynton was dealt a winning hand. The instrument flew to Mars and worked perfectly.

Boynton had expected that after the instruments reached orbit around Mars, his team would have to add up weak ice detection signals for many months before having a strong enough signal to confirm or deny ice in the soil layers. Instead, within days of being turned on, all three instruments detected whopping amounts of underground frozen water in the top 2 or 3 meters of soil, poleward of about 60° south latitude, at all longitudes.

TRACES OF ANCIENT MARTIAN GLACIERS?

If Noachian Mars's surface topography was deformed by the subtle flow or creep of ice-rich soils, why not full-fledged glaciers, filling ancient highland valleys with ice? Many researchers, such as Robert Strom, Baerbel Lucchitta, and Jeff Kargel, all in Arizona, began as early as the 1970s and '80s to point out various features that they interpreted as remnants of Noachian Martian glaciers. These include soil lineations and possible **eskers** and **moraines**, or winding, ridgelike deposits carved by streams running under glaciers or left as glaciers melt and drop loads of gravel that have been picked up and accumulated in the moving ice. In 2001, Lucchitta pointed out striking similarities between Martian features and glacial features in Antarctica, noting that the Antarctic ice flows over loose gravels that are saturated with water, with the ice flow eroding the gravel into long grooves parallel to the direction of ice flow. While some ancient Antarctic glaciers have disappeared, the striated gravel surfaces remain—and match Martian features.

Such Martian glacial activity may have declined. As the Noachian atmosphere thinned, Martian glaciers may have sublimed and disappeared, leaving their ghostly imprints on the landscape. Glaciers on modern Mars would thus be short-lived, although we will later visit a site where a glacier may still be active.

INTENSE CRATERING IN THE EARLY NOACHIAN ERA—A KEY TO THE ICE DEPOSITS?

The heavy cratering of Mars during its first few hundred MY—due to the interplanetary debris and asteroid-like bodies left over after planet formation—may be a key to understanding how the massive Noachian ground ice deposits formed in the first place. For the first 600 MY of Mars history, Mars and the other planets swept up these bodies. The earliest Noachian surfaces were repeatedly blasted by impacts—essentially saturated with craters. This means that the crustal layers were pulverized down to depths as deep as the deeper craters, probably at least 1 or 2 km (a mile or so). Here, then, was a surface of gravelly material—the perfect porous medium to absorb ancient water, which may explain how so much of the ancient water ended up in underground ice deposits.

CHAPTER 9

CRATER BAKHUYSEN

Visiting Valley Networks

Closely linked to the debate about the causes of terrain softening and the climate in Mars's thicker Noachian atmosphere is the debate about the origin of Mars's dry river channels. Researchers divided them into two broadly different categories. The classic riverbeds, or **outflow channels**, are several kilometers across and relatively isolated; tributaries are either absent or short and stubby. Many of these channels emanate from a restricted source area, sometimes a box canyon that seems to provide water sources, or a localized region of collapsed terrain, which is the reason for the adjective *outflow.*

A second type of channel system is smaller, subtler, and common among the craters and plains of the old Noachian uplands. These are delicate filigrees of small channels with widths of perhaps a couple hundred meters, branching upslope into many fine tributaries. They don't originate in a single specific source area but rather fan out into the Martian desert. The general pattern is tree-shaped, or **dendritic**, to use geologists' terminology. Because of this filigree pattern, these channels are called **valley networks**. An intriguing example, shown here, lies west of Terra Tyrrhena, along the north border of Noachis, on the inner slopes of the giant crater Bakhuysen, named after a Dutch astronomer who investigated Martian dark markings in the 1880s. The northern inner wall is laced with a network of fine channels, each filled at its bottom with a narrow deposit of brighter dust, as if a white line had been painted down each one. Tracing the channels upslope to find their origin is frustrating. They disappear into the rolling hills on the crater rim.

Valley networks can be found in many of the most ancient parts of Mars. The region known as Terra Tyrrhena, which we visited in chapter 2, lies northeast of Noachis Terra and is especially richly laced with valley networks. As seen in the Viking image below, slopes between craters in Terra Tyrrhena are often laced with tributary patterns of streambeds converging into larger river channels. Martian valley networks are densely scattered across old, cratered surfaces and are virtually absent on younger surfaces. They are essentially features of the Noachian era and did not form much later.

Because valley networks look like terrestrial stream drainage systems in hilly or rolling country with moderate rainfall, some researchers took the next logical step and proposed that they really do indicate Martian rainfall during the Noachian era.

Mosaic of Bakhuysen crater shows the location of drainage channels on the inner wall (white box). (345W 22S, mosaic of MGS context images.)

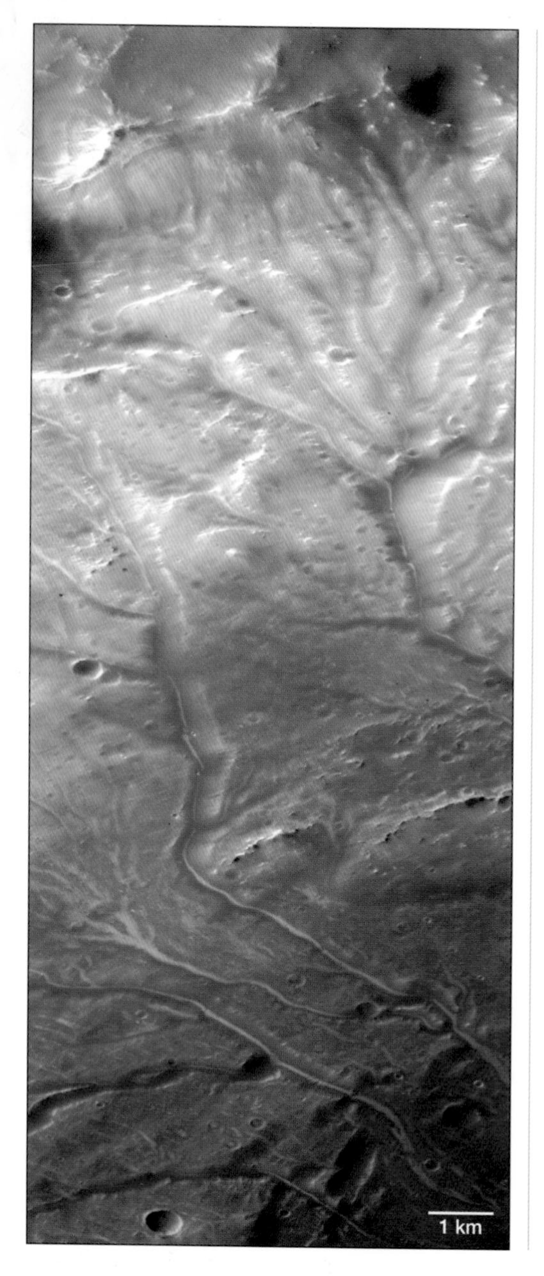

Close-up view of the north inner wall of Bakhuysen crater shows dendritic patterns of valley network channels draining off the rim. Crater rim crest is at the top. The source of the water remains a puzzle. Does it originate from inside the walls, in springs, or from precipitation upslope on the crater rim? (345W 22S, AB1-10605.)

Enlarged view of the channels on the wall of Bakhuysen crater (facing page) *reveals 400-meter-wide streambeds with a narrow deposit of bright dust running down the axis of each one. (345W 22S, AB1-10605.)*

This idea continues to generate controversy. Others say valley networks are merely drainages created by underground sources of water, creating runoff from springs or seeps. The answer would make a huge difference to our understanding of early Mars—if only we knew what it was.

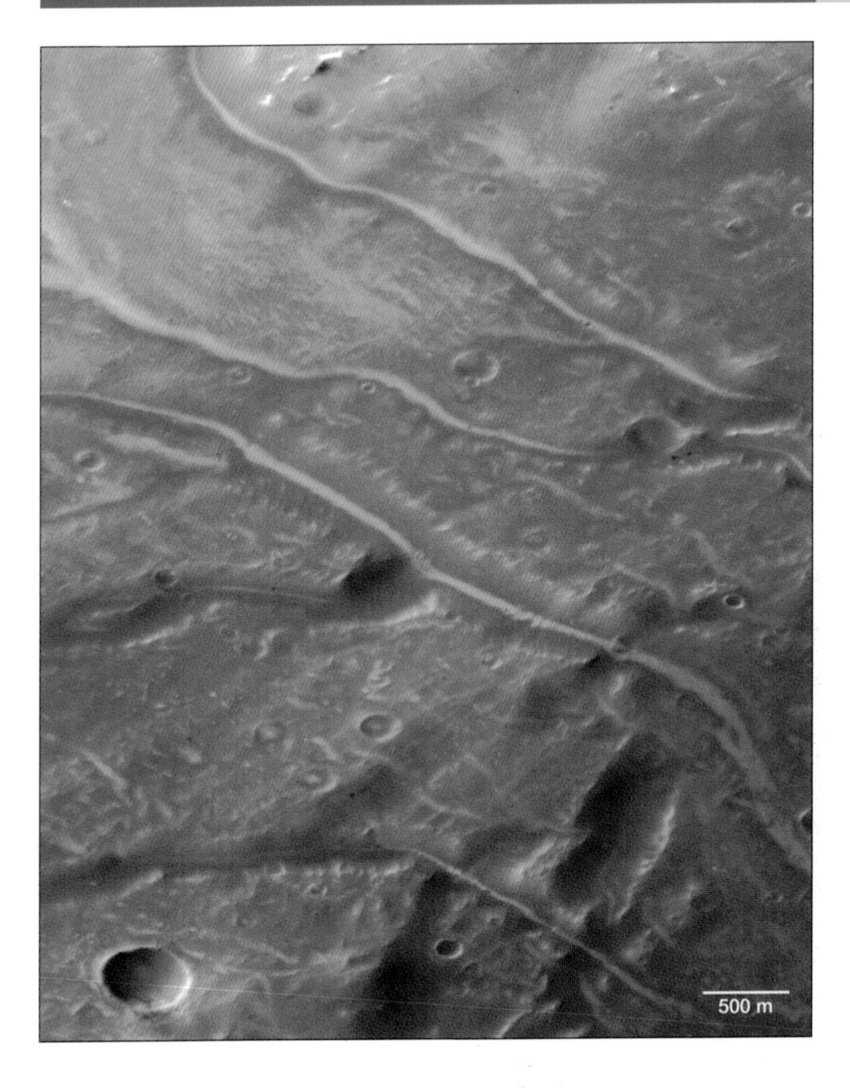

Michael Carr, a U.S. Geological Survey researcher who published a book about the history of water on Mars, was a member of the MGS imaging team and used MGS images to test whether the valley networks were caused by Martian precipitation. If so, he reasoned, the channels should branch out into finer and finer gullies, in the uphill direction, and therefore even the highest resolution MGS images should show tiny rivulet channels in the soil. If they were absent, it should mean that the valley networks did not form by runoff of rainfall or snowmelt. In general, Carr's survey of MGS images did not show tiny rivulets. In many cases, the images showed valley network channels ending abruptly in stubby gullies, not fine branching systems. Carr concluded that rainfall was not the culprit, and that the valley networks were formed mostly by water coming not from above, but from below, out of the ground, perhaps by melting of underground ice.

Still, it is hard to be sure. Some of the filigrees of valley networks are so spread out that they would seem to have required a dispersed source of water. It would not have to have been rain. It might have been melting snow or ice deposits. Noachian Mars may have had a thick enough atmosphere and cold enough conditions to have produced massive snowfall or ice-age-style ice packs. These could have produced networks of runoff channels as they melted. As the accompanying photos show, many valley networks are overlaid by smooth deposits of windblown dust or other fine material—so it is possible that the finest-scale runoff channels are missing only because they are buried and hidden by dust. Mars researchers seem to agree that the valley networks were formed by water, but the source of the water is controversial, with some still holding out for rain, snow, or melting surface ice. In any case, the dendritic-patterned valley networks remain the best support for the idea that early Mars may have experienced a period of rainfall or snowfall that could produce uniform water runoff dispersed over very wide areas.

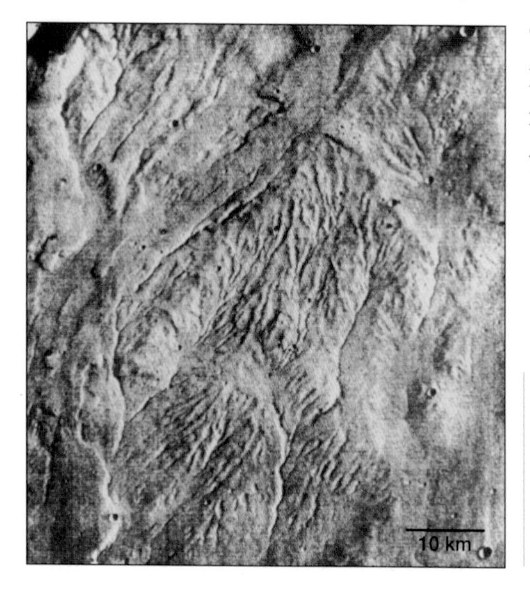

Valley networks were discovered by Mariner 9 in 1972, as seen in this image of the slopes of the volcano Alba. (116W 45N, Mariner 9.)

CHAPTER 10

A GREAT WATER-WAY?

If a planet has liquid water, that water forms a dynamic system with a climatic cycle of its own. On Earth, this cycle is familiar. Water evaporates from the oceans primarily at the warm equator and goes into the atmosphere, where air currents carry it aloft. Some of the water vapor rains out of the air at mid-latitudes, drains along rivers, and empties into the sea to start the cycle again. Other water vapor reaches high, polar latitudes, where it freezes out as snow and is dumped onto the polar ice pack. As more snow and ice are added at the poles, the ice spreads laterally. Icebergs break off into the sea. The icebergs melt and add their water to the oceans, restarting the cycle.

Water on Mars today is fairly static and is hidden in the three forms we've mentioned: ice caps, underground ice, and hydrated minerals. Once scientists realized that "dry Mars" has abundant water and that the water may have been liquid in the past, they began to think about Mars in terms of its climatic cycle. In particular, Stephen Clifford, a Mars geologist based at the Lunar and Planetary Institute near Houston, envisioned a Martian global water cycle. Clifford once ran for the House of Representatives as a Republican, attempting to change the notorious dearth of scientists and technically trained people in the U.S. Congress. He lost, being a little too centrist for Texas politics, but his loss was Mars's gain. Contemplating the big picture of Martian water, Clifford

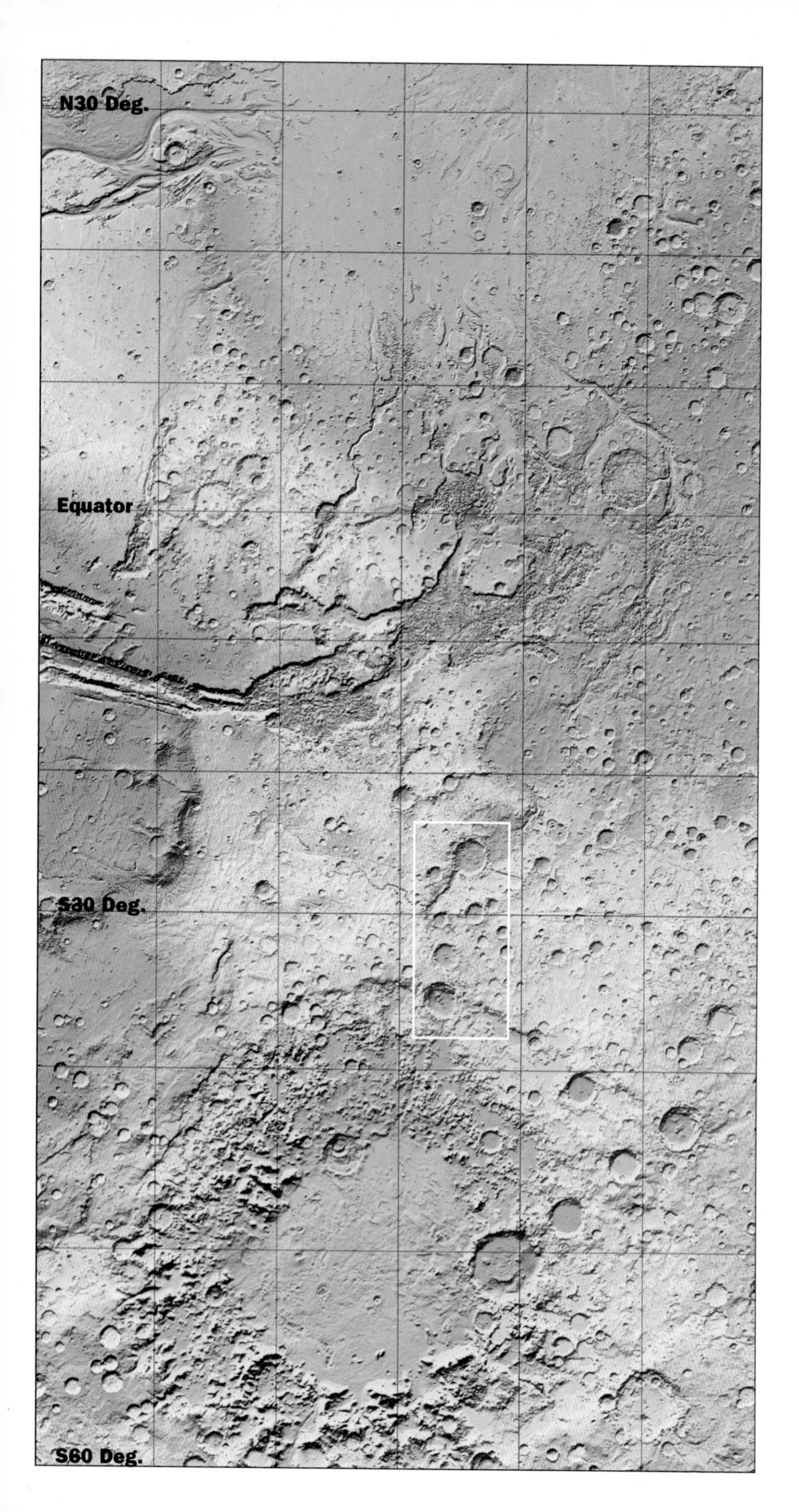
N30 Deg.
Equator
S30 Deg.
S60 Deg.

Topographic map shows the proposed ancient waterway leading from the highlands near Argyre impact basin (large green circular feature, bottom) through various craters and channels (green, middle) into major river outflow channels (blue, upper right). Red and orange are highest terrains, green is lower, and blue is lowest. The east end of the huge Valles Marineris canyon system is at the upper left, emptying into additional outflow channels. All the river outflow channels at the top empty into the low (blue) northern plains (top edge), which may have once been a huge Martian sea. (MGS laser altimeter team.)

started with the fact that the south polar ice cap and the southern highlands, such as Noachis, are several kilometers higher than the northern plains. The southern highlands are also heavily cratered, ensuring that they have been fractured and broken into loose material. Thus, whatever early liquid water sank into the porous soils of the southern cratered uplands, there was a tendency for it to migrate downhill through the porous material, one way or another, from the southern highlands to the northern lowlands.

In Clifford's grand vision, ice and snow in a once-thicker Noachian atmosphere tended to precipitate and collect on both ice caps, where permanent water-ice deposits even today. Ice and snow can melt more easily under high pressure at the base of the ice pile than at the top, and so, as snow and frost were added at the upper surface of the cap, "basal melting" produced liquid water under the cap. Meanwhile, as ice accumulated, it also spread laterally by glacial flow, aided by the lubrication of the water layer at the bottom. We see this effect on Earth in Antarctica, where ice is added at the pole, and where many glaciers flow outward from the pole until they break off at the sea.

According to this view, as fast as Noachian snow and frost collected on top of the south polar ice cap, water seeped out from under the cap and crossed the uplands either through underground **aquifers** or as rivers, ponding in craters, overflowing occasionally, and eventually flowing out onto the northern lowlands. This hydrologic cycle continued throughout the Noachian era, until Mars lost atmosphere and the pressure dropped, the climate changed, evaporation increased, and river activity declined. Water ended up frozen in underground deposits, locked in hydrated minerals, and trapped in the permanent polar ice cap.

In accord with this theory, we can see that many of the now-dry river channels run from southern uplands northward, emptying onto low northern plains. As new altitude and topography measurements accumulate from the latest Mars missions, the gross outlines of this theory look better and better.

A GLOBAL-SCALE WATERWAY?

Armed with new photos and new topographic data, geologists such as Clifford and his Caltech colleague Tim Parker emphasized the possibility of a particularly impressive drainage system, a great waterway that stretches all the way from the south polar cap to the northern lowlands some 10° N of the equator. According to the waterway's mappers, water that escaped from under the south polar ice cap drained into the great impact basin named Argyre, which lies west of Hellas and is not quite as big. It can be seen dominating the bottom third of the accompanying color topographic map. The high rim of this basin, piled up by the basin-forming explosion, would originally have been a barrier to water entering from outside, but it was apparently eroded away by winds or very early Noachian rainfall (we saw this also with the Hellas basin). In any case, channels can be seen at the southeast edge of the basin, entering it from the direction of the polar ice fields.

According to some reconstructions, the water filled the Argyre basin until it cut through the north rim and began to drain down the north-sloping outer rim, carving a winding channel system across

Oblique view showing the southern cratered uplands and southern part of the Argyre impact basin (lower left quadrant), as seen from the west by the Viking 1 Orbiter in the 1970s. According to some interpretations, water may have flowed from the south pole into the basin through the hills at right. Notice the thick dust layers seen in profile high in the Martian atmosphere at altitudes of tens of kilometers above the surface. (NASA.)

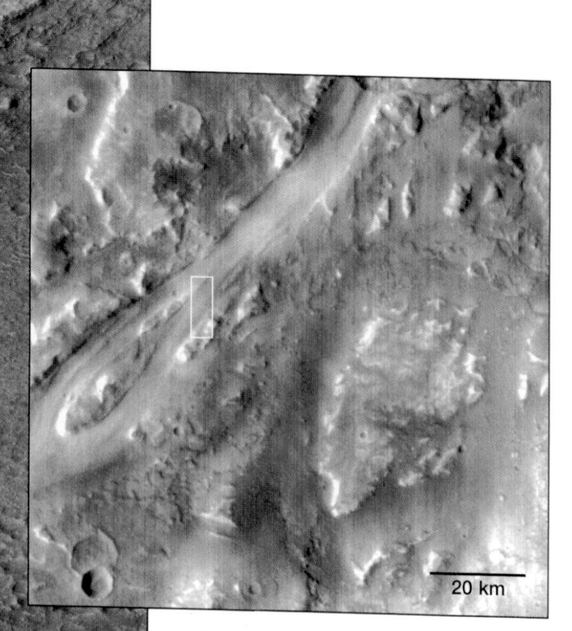

In some areas, the waterway is defined as an individual channel (above), containing a beautiful teardrop-shaped island streamlined by water flow. White box shows area in adjacent view. (23W 12S, M07-03773.)

500 m

Telephoto view shows the region from the streamlined island (top) to the SE bank of the channel. The floor of the waterway channel is highly cratered, indicating that the last water flow down the channel was not very recent. (23W 12S, M07-03772.)

the cratered uplands, which are orange and yellow on the topographic map. This main channel is prominent on the altimeter map, creating a valley (green) that winds north through Noachian- and Hesperian-age cratered regions, across crater floors (also green), and into the equatorial lowlands (green cratered and broken zone in the upper half of the map). The last flows in the channel floor itself are mapped as dating from the Hesperian era, but the original drainage system is probably Noachian in origin. The equatorial lowland region, at the terminus of the waterway, is cut by many larger, younger, and more localized channels that we

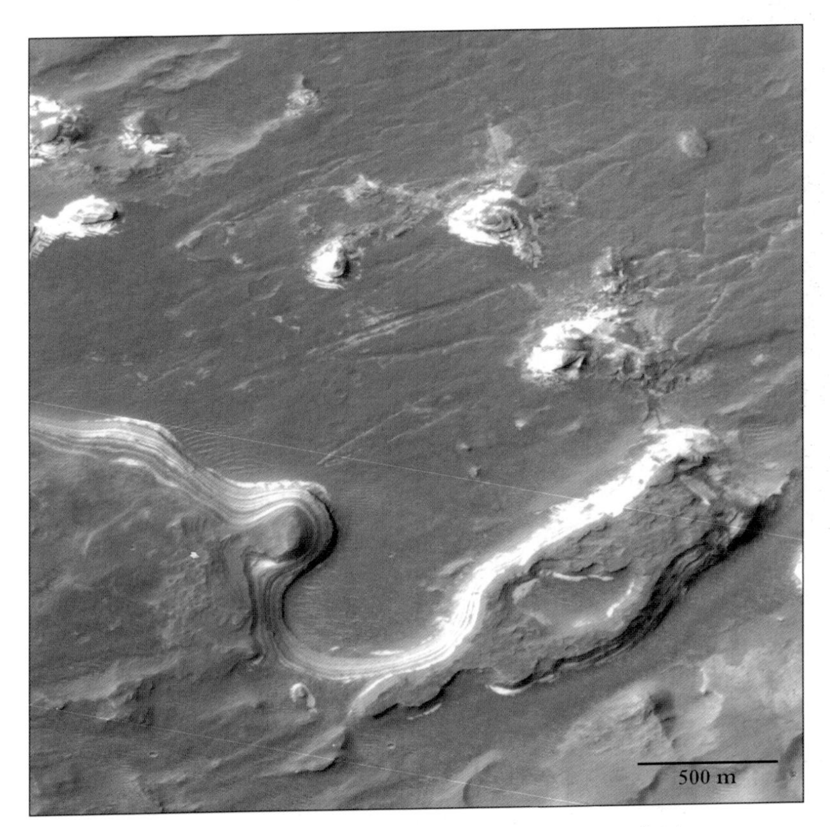

The floor of Holden crater—where the water ponded? Just inside the point where the water entered Holden crater are buttes composed of stratified layers. These may be remnants of loosely consolidated layers of sediments deposited on the floor of the crater by the floodwaters. Bright, overexposed surfaces are probably covered by frost deposits. (35W 27S, M03-02733.)

will visit later. All of them empty at the top of the map into low northern plains (blue) that some researchers, such as Parker, believe were ocean-covered in Noachian times.

Writing in 2001 in the planetary journal *Icarus,* Clifford and Parker noted that this drainage from near-polar regions to the equator exceeds 8,000 km (5,000 miles) in length, making it "the longest known fluvial system on either Earth or Mars." Mars-ocean proponents think big.

Telephoto images along the channel system confirm parts of it, although Clifford and Parker admit that the drainage pattern is obscured along other parts by later impacts and other processes. In the main channel, beautifully streamlined, teardrop-shaped "islands" were shaped by the water flowing around them. By mapping the channels on topographic maps, geologists have traced the path of the water in detail, as it flowed out of the uplands, broke into a crater here or there, ponded, then broke out of the northern crater rim and continued downhill to the north.

One of the larger craters along the waterway is called Holden crater. It was named after an American astronomer, Edward Singleton Holden, who directed Lick Observatory in the late 1800s. He was known for early telescopic photography of the Moon. The channel system breaks into Holden crater from the southwest side. Some researchers believe Holden was once occupied by a lake. MGS images show that the floor has some mesas composed of flat-layered sediments, which may be eroded remnants of silt and other waterborne material that settled to the bottom of a lake that once occupied Holden crater.

Ancient water-flow features, such as this waterway and the valley networks in the previous chapter, serve as our introduction to fluvial features of Mars. Soon, we will encounter much younger and smaller examples of such features.

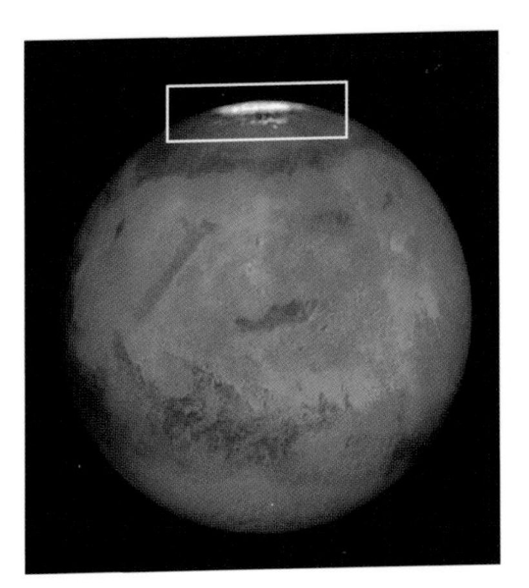

CHAPTER 11

VASTITAS BOREALIS

The Secret of an Ancient Ocean?

Perhaps the most controversial question about Noachian Mars is whether there was ever enough water to create ancient oceans in low-elevation areas such as the northern lowlands. A whole progression of questions leads to this subject. How much water emptied out of the southern uplands onto the northern plains? Was the atmosphere thick enough to make rain? If there was rain, could there have been enough runoff to make lakes? If lakes, why not oceans? Was the red planet once blue, like Earth?

The best evidence for an ocean comes from a depressed area in the far north, with the impressive name Vastitas Borealis. This name, meaning broad plain of the north, was bestowed in 1973 after the low northern flatland was recognized from Mariner 9 mapping. The plains ring the north polar ice cap. Some researchers claim Vastitas Borealis is the floor of a once-great Martian ocean that engulfed the whole north polar region—an ocean they call Oceanus Borealis.

One of the best-known proponents of this controversial idea is Tim Parker, a flamboyant researcher from Caltech's Jet Propulsion Laboratory in Pasadena, who is a master of amusing quips and is given to wearing audacious Hawaiian shirts on (as far as I can tell) all occasions. Parker began advancing the idea of Martian oceans in the late 1980s and developed the idea with several colleagues in a 1993 paper, and still further in his 1994 Ph.D. dissertation. He thought he could see two sets of ancient shorelines winding across arid hillsides, in the same way that the prehistoric Lake Bonneville left prominent shorelines on desert hillsides in Utah. Parker's whole enterprise was unpopular during that period. Everyone else "knew" that Mars was a dry planet, which had only modest traces of water that cut valley networks and outflow channels here or there, but that quickly evaporated or sank into the desiccated soil.

North polar region, charted by the MGS laser altimeter, dramatizes the low, featureless plains of the region, possibly the floor of an ancient ocean. Consistent with the other laser altimeter map views, the dark blue has elevation of about –4 km below the global mean reference level, and the green "island" represents polar cap deposits of sediment layers and ice rising to about –2 km below the reference level.

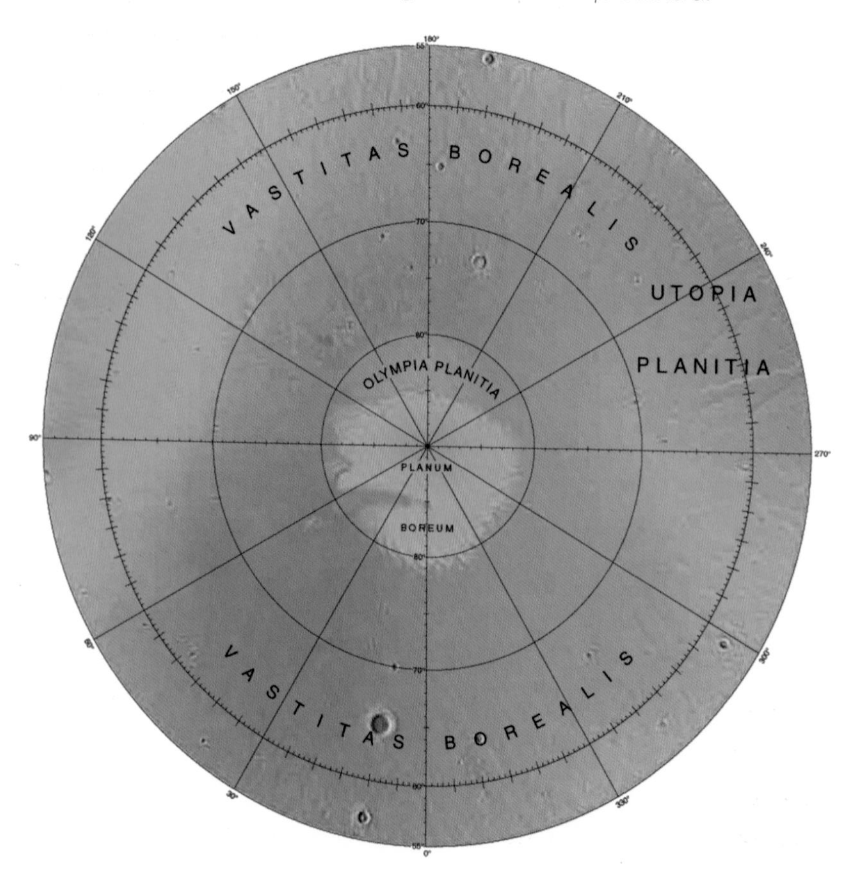

But Parker stuck to his guns, even though the possible shorelines and the source of water in the supposed ocean were serious question marks. One obvious potential source was that the water simply poured out of the many riverbeds onto the plains. Writing in 2001, Parker and his colleague Stephen Clifford took a different tack, arguing that the northern ocean was primordial, having condensed out of the initial atmosphere soon after the planet formed. Remember that isotopic studies of the atmospheric gases support the idea that the early atmosphere was thicker, and the total water vapor in it may have indeed been enough to condense into broad regional seas tens of meters deep or more.

Parker and Clifford made a different calculation of the total water amount—they assumed that to create the outflow channels, all the underground available pore space in the soil was filled with ice or water and that this water was out on the surface in Noachian times, making an ocean. Assuming 20 percent to 50 percent pore space in the soil, they calculated that the total water was enough to make a "global ocean" with a depth of 550 to 1,400 meters. Again, global ocean is an idealized term that scientists use to discuss the total amount of liquid water on Mars—how deep the ocean would be if Mars were flat. In reality, this ocean would not have been global, but would have covered the vast low area north of about 20° N to 40° N latitude—Oceanus Borealis—and would also have filled the Hellas and Argyre basins. Parker and Clifford picture the ocean being fed and maintained, at least in part, by Clifford's global-scale flow of water from the southern highlands, through underground circulation of water as well as in river channels.

Victor Baker, a scientifically adventurous Arizona geologist who likes to challenge existing views, independently propounded similar ideas after writing numerous papers and a book on Martian channels and water systems. As early as 1991, Baker hypothesized the existence of oceans on early Mars, and he argues that watery conditions on Mars lasted longer than other scientists have accepted. Baker's scenario is that during cycles of warmer, wetter

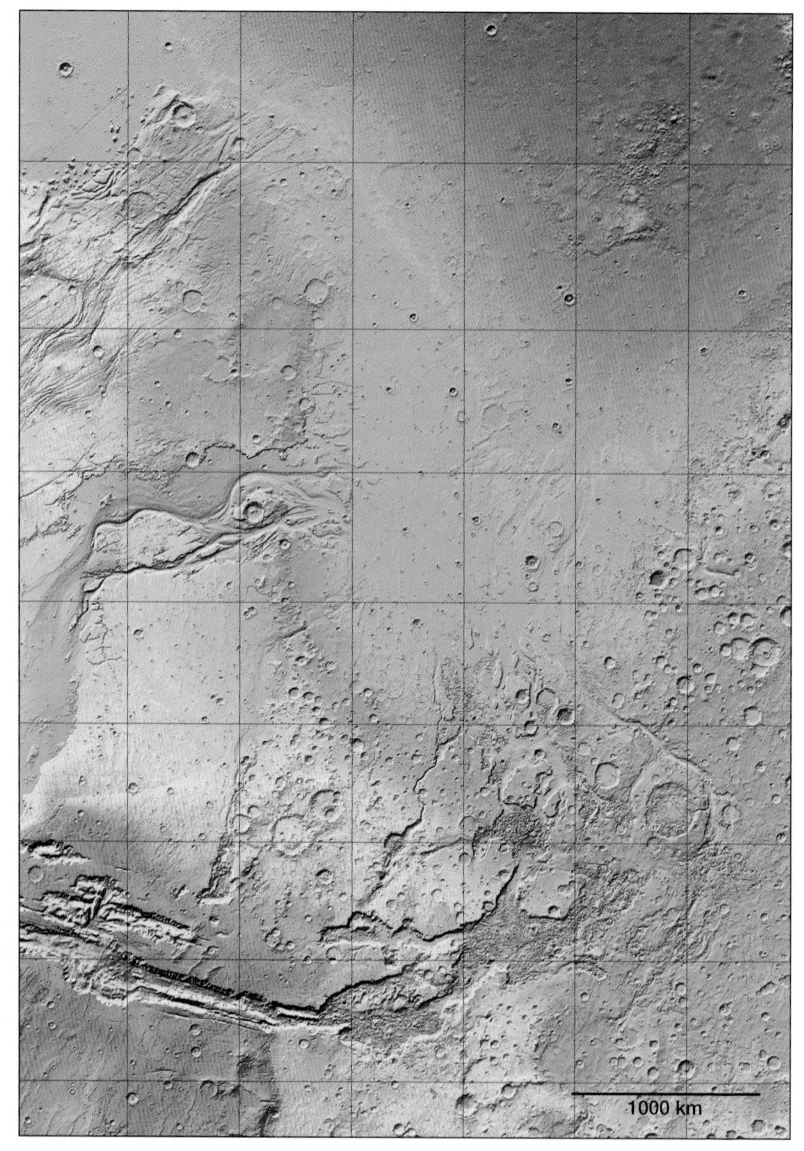

Where the rivers emptied into the sea? MGS laser altimeter map of the south margins of the northern plains shows a number of places where huge outflow channels dumped water into the plains (blue) from the cratered uplands (green). Whether the plains were ever covered with water remains controversial.

climates, an episodic northern ocean repeatedly formed and then dissipated in Vastitas Borealis, not only during the Noachian era, but during the Hesperian and into the Amazonian eras as well.

If Oceanus Borealis ever existed, it probably looked more like our ice-choked Arctic Ocean than our Pacific. To ocean theorists, ice is a boon, not a hindrance, because it would protect the ocean from the atmosphere, hence inhibiting evaporation and stabilizing the water long enough to allow the water to carve telltale geological features that are still visible today, such as Parker's shorelines.

RAMPART CRATERS AND PATTERNED GROUND

In the northern plains of Vastitas Borealis, orbital photos show other lines of evidence that might support an early ocean. We've already seen that these regions are full of rampart craters, which indicate underground ice deposits; but in addition, there are systems of cracks and grooves arranged in crudely polygonal, fingerprint-like or other patterns, which also imply underground ice. This phenomenon is called patterned ground. It is familiar in the Canadian arctic tundra, where freeze-thaw cycles cause ground ice to melt, then expand as it refreezes. Remember that water is one of the few substances that expands when it freezes, which is what causes ice cubes to float instead of sink. Each expan-

Patterned ground, common in the northern plains of Mars, is thought to be produced by freeze-thaw cycles in underground ice deposits. They suggest that massive amounts of ice, possibly remnants of ancient oceans, are buried under these plains. Note the rampart ejecta blankets around the craters, also suggestive of underground ice. (Acidalia Planitia, 6W, 48N, Viking I Orbiter photo 37A14.)

FOSSIL SHORELINES: THE PLOT THICKENS

The fossil shorelines proposed by Tim Parker remain controversial. During the '90s, Parker mapped the best examples with Viking orbiter photos, then waited to see whether the MGS telephoto camera would confirm them. After MGS pictures started to come back, Parker worked with geologist Ken Edgett, one of the managers of the MGS camera, to examine the first images and to see if the features were clearer on the MGS images. They weren't. In the best scientific fashion, Parker swallowed hard and concluded that the initial evidence did not support his theory, and bravely wrote a 1997 article with Edgett noting that the detailed pictures did not readily confirm his ideas. Writing in the *Journal of Geophysical Research* in 2001, MGS camera designer Mike Malin, along with Ken Edgett, summarized a further examination of the proposed shorelines. Noting that the difference between "unverifiable speculation and testable hypothesis is often very small," they concluded that there is an "absence of coastal landforms" and "no evidence supporting the hypothesis that there was a northern ocean." Seeing themselves perhaps as the bearer of bad news in the face of enthusiasm for an ocean, they added, somewhat ruefully, that it is often harder to disprove popular ideas than unpopular ideas.

Nonetheless, to assume that the MGS camera ends the story presupposes that the guy with the biggest camera has the last word—which isn't necessarily true. The problem is that if you get too close to a thing, you can't see it. You can see the Lake Bonneville shorelines from an airplane, but if you walk over them with a 35-millimeter camera and shoot vertical close-ups from waist level, the pictures won't reveal them. Concerned about this effect, Parker continued to examine some of his proposed Martian shorelines in additional new pictures, and writing in 2002, he presented new MGS images in which some of the proposed shorelines look convincing after all, resembling aerial photos of the Bonneville features at similar scales. The whole subject seemed more confusing than ever.

RADAR MAPS A COASTLINE?

Parker had another ace up his sleeve. MGS sent back a completely different set of data that seemed to support the ocean theory. Unlike photographic images, the laser altimeter maps allowed topographic altitudes in one region to be compared directly with altitudes in other areas far away. This gave a new test of proposed shorelines. If a coastline truly existed around a long-lived ancient sea, it would have had to lie at a constant elevation all the way around the sea. After all, water creates a level surface. As Parker quipped, "That's why you need a motorboat to go water-skiing—there are no slopes." Sediments washing into the sea from the shoreline regions would form a shelf around the edge, at constant elevation. Assuming ground elevations have not been radically altered since the Noachian by earthquakes, faults, and mountain building, such a shelf should be visible. The altimeter data dramatically revealed that one of Parker's shoreline levels around the Vastitás Borealis basin lies at exactly constant elevation, about 3,700 meters below the mean Martian surface, all the way around the basin. Below this level, the radar shows a broad shelf and a strikingly smooth basin floor. This evidence for a sea seems strong, because there is no other good explanation for coastline-like features to be at constant altitude over distances of hundreds of kilometers.

The hypothetical ocean below this possible coastline would have averaged 570 meters (1,850 feet) deep, but reached 3 kilometers (nearly 10,000 feet) depth in the basin now known as Utopia Planitia. Expressed as an average "global ocean," it would have had a depth around 100 meters (330 feet)—a plausible value.

WHAT LIES BENEATH?

Today, Vastitas Borealis is not a wave-tossed sea, but a silent plain pocked by scattered craters. The existing kilometer-scale surface structures are not Noachian, but Hesperian, based on the number of impact craters formed on the surface layers. So the

Possible shorelines of an ancient ocean, as proposed by Mars researchers Tim Parker and Steve Clifford, can be glimpsed in this MGS image east of the Cydonia hills of Mars. At the top are hills rising about 600 meters (1,900 feet) above the plain. On the lower flanks of the hills (center) and in parallel lines across the plain (bottom, left to right) can be seen faint lines that have been interpreted as shorelines left by a retreating ancient ocean. (9W 36N, south at the top to emphasize uphill regions in top part of image, M07-04326.)

present surface layers are probably not sediments laid down in the Oceanus Borealis, but postdate the end of the hypothetical ocean. We can estimate that these layers, the top couple of hundred meters, formed about 2,000 to 3,000 MY ago.

Two studies of MGS data support the idea that any original ocean floor is well buried. First, Brown University researcher James W. Head III and his colleagues pointed out that the Vastitas Borealis plains are laced by faint "wrinkle ridge" formations typical of lava flows. This finding implies that much of the visible surface structure involves not sediments of an ocean floor, but lava flows.

Second, an additional study of the laser altimetry revealed faint circular depressions scattered across the plains. They are so shallow that most of them could not even be seen in the conventional photo imagery. Baltimore researcher Herbert Frey and his colleagues (including his daughter) counted them and found that they have the size distribution typical of surfaces very heavily covered by impact craters. So they seem to be ancient craters on a buried terrain surface, partially covered over by later lava flows. This creates a question for the ocean theory: Why didn't the ocean erode away the craters? Maybe it almost did. Maybe that explains why they are such faint depressions in the existing surface, instead of appearing as craggy rim crests, sticking up through the lava- and sediment-covered plains, as seen in some other areas. The large numbers of craters imply that they date back to the first few hundred MY of Martian history—Noachian times during which the intense bombardment quickly battered the surface—and this, in turn, suggests that the ocean did not exist long after that; otherwise, early, densely crowded craters would have been washed away.

These findings support the idea of an ancient surface under Vastitas Borealis, but at the same time indicate that the original ocean floor—if it exists—is buried under lava in many areas. This complicates the interpretation of the ice revealed by the rampart craters and patterned ground. It may not be a direct remnant of an early ocean, but rather a remnant of later waters that got into the area after the lavas flowed.

If proponents of early oceans are right, Noachian Mars was extraordinarily different from the planet we know today, and the excitement of Mars exploration is ratcheted up a notch. If early Mars of 4,000 MY ago had bodies of water similar to Earth's oceans, in which the currents of oceanic organic chemistry moved in mysterious ways to create life, then we can reasonably ask: Did life form on Mars, too? In Vastitas Borealis, future explorers might be able to search cliffside exposures of ancient Martian seabed sediments for the Holy Grail of the Martian quest—fossils of ancient Martian organisms.

CHAPTER 12

TERRA MERIDIANI

Hematite Deposit Number One

Not long before MGS went into orbit in 1997, I foolishly wrote an article trying to predict what its findings might be. The most important instrument on board, I claimed, was not the camera (an act of disloyalty, as I was on the camera team), but rather the spectrometer, because it could look for unusual soil compositions and might discover the mineral deposits left by the putative ancient lakes and oceans.

The instrument is technically called a thermal emission spectrometer (TES). The name denotes that this spectrometer doesn't analyze visible sunlight reflected by the Martian surface, as would a more traditional spectrometer; instead, it analyzes the extreme infrared "thermal radiation" emitted by the rocks and soils themselves as a result of the scanty heat they absorb from the weak Martian sunlight. The qualities of this radiation depend on the composition of the material, and TES was thus supposed to measure the composition of rocks and minerals. Because the rock radiation comes from fractions of a millimeter below the rock surface, TES could even identify rocks that had a stain of different composition on their surface.

TES is a beautiful instrument. Its builder, Phil Christensen at Arizona State University, took us on a tour of an Arizona desert over which TES had flown in an airplane. Phil was able to show us

10-meter-scale outcrops of various rock types that showed up on the TES-generated maps but had been entirely missed by earlier ground mappers! Hopes were high that TES would yield similar revelations about Mars.

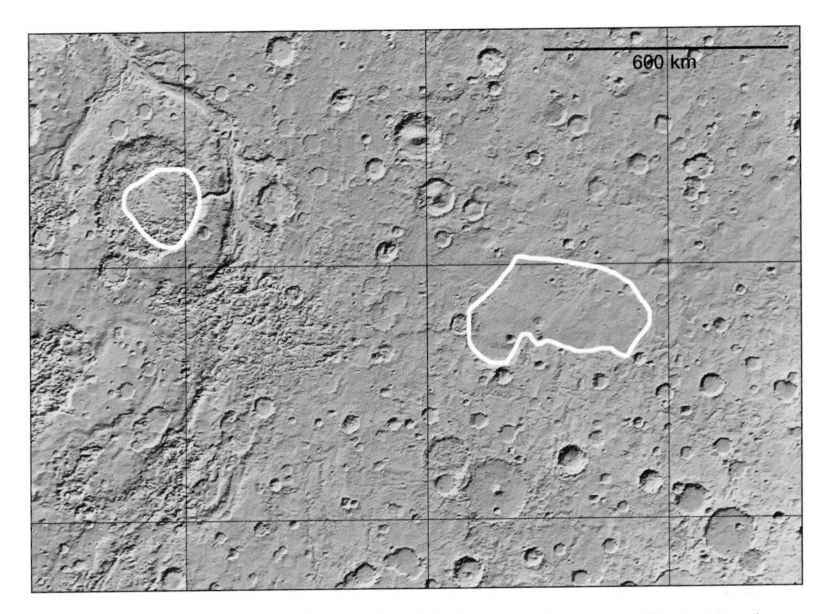

MGS laser altimeter map of the two regions identified as containing strong hematite deposits. The first one, discussed in this chapter, is a depressed plain (right center). It may be an area where water ponded in the past. The second is a crater filled with collapsed ground known as Aram Chaos (left) and is discussed in Chapter 22.

TES AND THE MISSING CARBONATES

The most dramatic discovery of TES, we all reasoned, would be deposits of minerals such as salts and carbonates—the minerals that should be left behind by any hypothetical ancient Martian lakes or seas as they evaporated. When water runs across the ground, it dissolves these salts, carbonates, and other minerals from soils. On Earth, rivers normally carry these materials to the sea, which is why the oceans are salty. But in desert areas, when the water flows into shallow basins and evaporates, it leaves these minerals behind as whitish deposits. Soils throughout desert regions tend to get rich in salts and carbonates due to these evaporation processes. (You've probably seen similar deposits in your house; if you live in a region of mineral-rich or "hard" water, carbonates from evaporating water leave whitish rings in your tubs and sinks.) Our prediction was that TES would reveal deposits of salts and carbonates marking Martian dry lake beds, formed as Noachian water evaporated on arid Mars, providing us with grand proof that ancient lakes or oceans had existed.

Alas, the prediction was wrong. Many of us were disappointed with TES's first discovery in 1998: fine Martian dust is so pervasive, both in the air and on the ground, that the spectrum of every region is dominated by a bland dust mixture of various rocky powders rather than specific rocks or exotic mineral deposits. Martian dust from different regions has been so thoroughly mixed by the wind that, to the eye of TES, the surface appears fairly uniform—a sort of smeared-out average composition for all of Mars.

Finding carbonate deposits may depend on being able to analyze localized areas. The area seen on Mars by any camera or spectrometer is called the footprint of that instrument. A typical footprint for one of our MGS camera images was a couple of km wide and maybe 20 km long (depending on the camera settings). For TES the footprint was a patch of ground typically several kilometers on a side—a village-size piece of real estate—and TES did rule out 20 percent concentrations of salts that might mark preserved seafloor or lake-floor deposits over such large regions. However, it couldn't rule out hundred-acre-scale concentrations that might have been left by evaporating ponds. Instruments on the Mars Odyssey orbiter should be able to find such deposits if they exist.

An example of the surface in the general region of the Terra Meridiani hematite deposits shows ghostly "fossil crater" rings. Two of these are outlined by white rings (lower right corner, upper right), and a third is a faint depression with light-toned sand dune deposits on the floor (upper left center). They seem to be flattened traces of early impact craters. A few sharp, small, recent craters can be seen with bright rims and dark interiors. Their sparse numbers indicate that the old "fossil crater" surface has only recently been exposed. (5W 2S, E03-01203.)

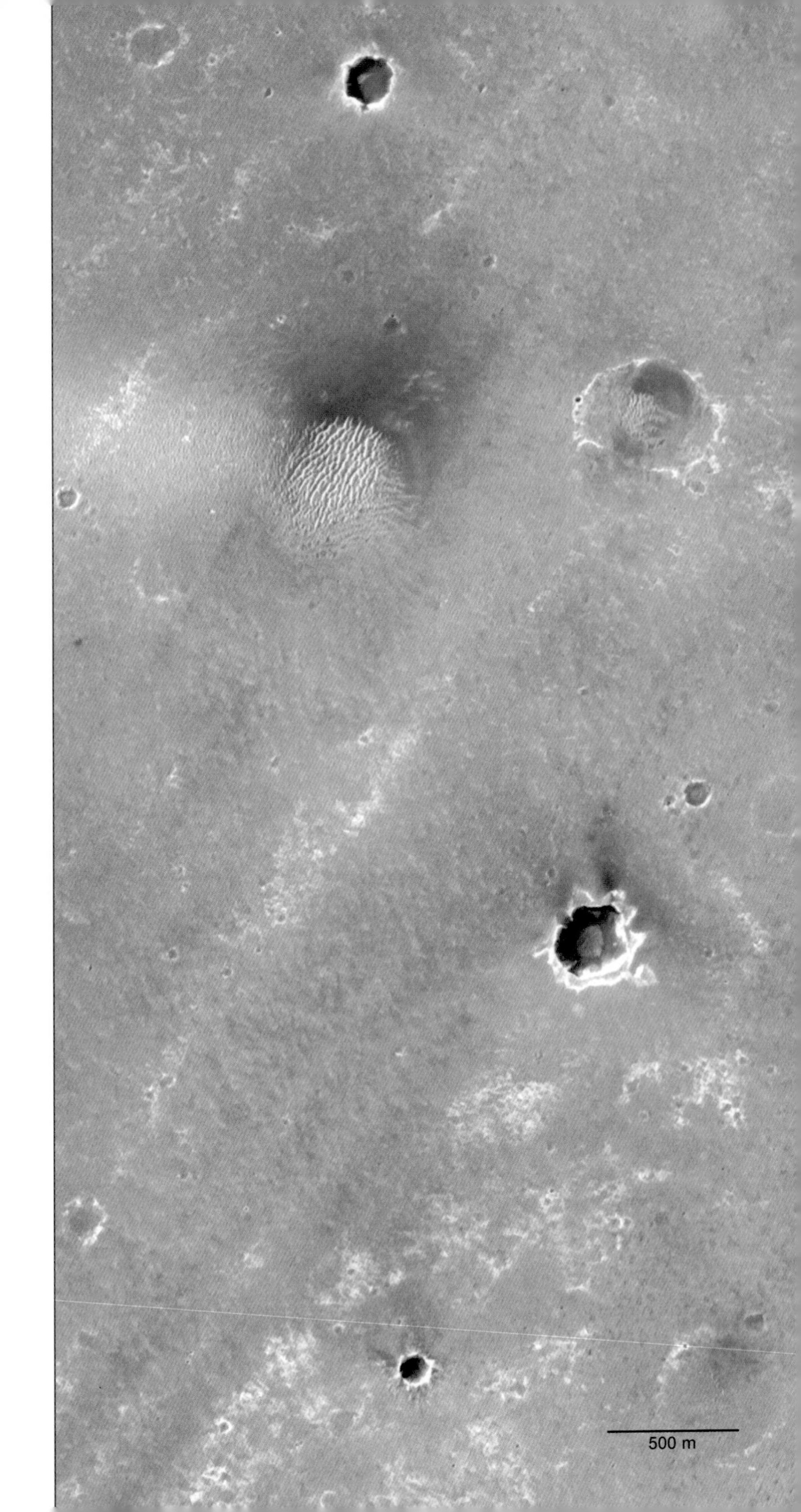
500 m

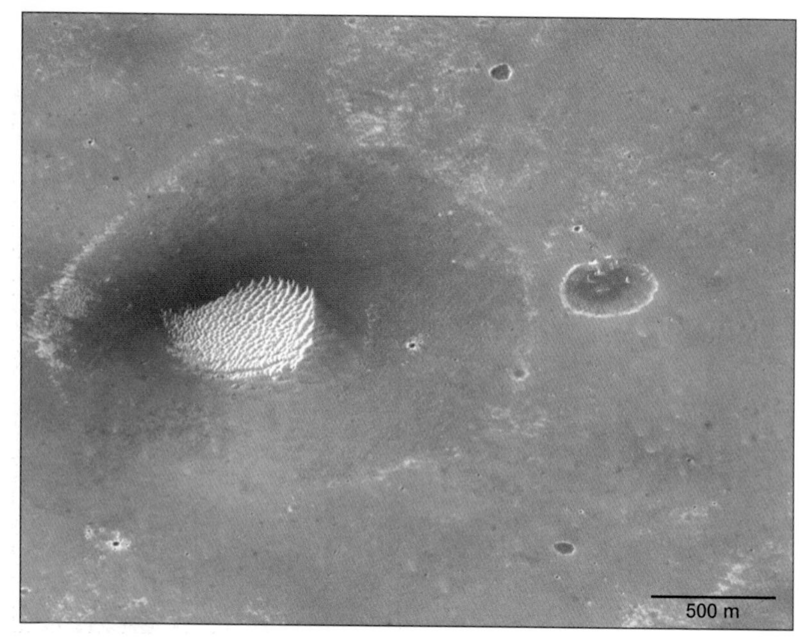

A large, barely traceable fossil crater in the hematite-rich area is defined by light rim deposits, with a bright patch of dunes on its floor. A smaller example is at right. (6W 2S, E11-01328.)

A MINERALOGICAL DISCOVERY

After its underwhelming start, TES made a comeback in 2000. As the data accumulated, TES was able to make finer-scale maps and to confirm that a few anomalous areas of mineral concentration really did, after all, exist. The first one found was not the expected salt or carbonate deposit, but something else. Oddly enough, it was almost exactly at 0° latitude and near the prime meridian at 0° longitude, in an area called Terra Meridiani—the Land of the Prime Meridian. TES showed a Kentucky-size patch of ground about 150 km by 500 km, with patchy concentrations of the iron-bearing mineral hematite. Hematite isn't a particularly valuable or unusual mineral, though polished samples of a variety called specular hematite are used to make attractive shiny black jewelry.

The region itself was nondescript, a slightly depressed area in the midst of heavily cratered Noachian plains. Years before the MGS mission, the depressed area had been mapped as a bland,

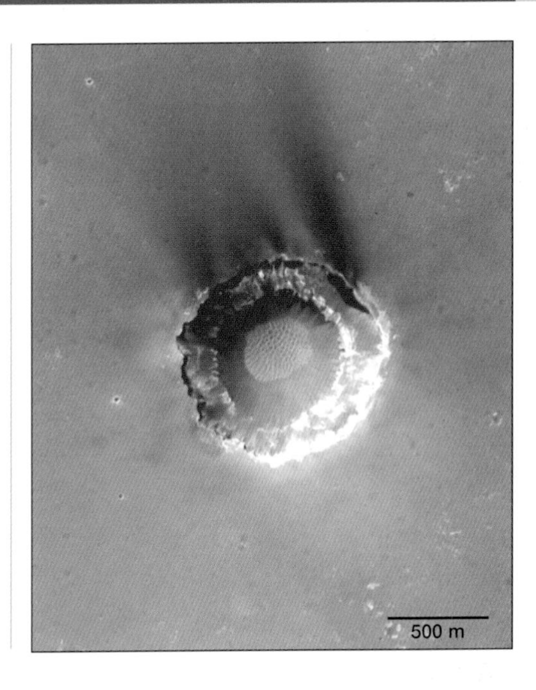

A 1-km crater in the Terra Meridiani hematite area shows dramatic inner wall structure composed of two distinct layers—an upper stratum marking the rim and a lower, inner stratum halfway down the wall. A deposit of light dunes is on the floor, and dark wind streaks radiate to the north. The layering may be an indication of massive layers in the region—possible lake bed sedimentary deposits? (5W 2S, E03-01203.)

slightly younger than average, "subdued" Noachian plains unit, containing patches of "etched" Noachian plains, interpreted as eroded by wind action.

If hematite was concentrated in such an area, how did it get there? On Earth, hematite deposits are often associated with hot spring activity, where geothermally heated water dissolves iron-bearing minerals, then evaporates and leaves the iron-rich material behind. An example of this process occurs in Yellowstone Park, where hot springs and geyser vents are encrusted with exotic, colorful mineral deposits. The same thing may have happened in Terra Meridiani. The process need not imply a large lake. Hematite deposition could happen either in transient lakes or in soils only briefly moistened by episodes of geothermal water release. But the TES discovery is usually considered to be evidence of active water transport and creation of iron-rich solutions.

As soon as the TES discovery was announced, those of us on the MGS camera team raced to see what the area looked like in our images. It turned out to be a very peculiar area. Part of it is covered by dune-like materials. Other parts are a smooth-looking plain dotted with strange rings—faint, doughnut-like annuli that give an impression of their being crusty material. They are sometimes darker than the background, and sometimes lighter. Each ring is the raised, gently rounded rim of a shallow, bowl-shaped depression. The rings seem to be very old, softened impact craters. They are not ordinary craters, filled in with dust or sediments, in which case the sharp rims, perhaps somewhat eroded, would stick up from the filled-in floor like the roof of a house rising above a Mississippi flood. Here, it is as if the rims had mysteriously flattened, or as if a pressure from above had squashed them nearly flat. I've called them fossil craters, and I think that's what they are.

We are not without clues about the origin and history of this area. First, when our group at the Planetary Science Institute counted up the numbers of the fossil craters per square kilometer, the numbers fell very close to the maximum number per unit area observed on any planet or moon—the number believed to correspond to surfaces that are approximately 4,000 to 4,400 MY old—the early Noachian period, when Mars may have had lots of surface water. Second, the MGS laser altimeter confirmed that the area is a slight depression. Third, there is a distinctive second population of craters in the area, very small, sharply defined, and fresh-looking. They must reflect recent cratering, but there are so few of them that the surface could not have been exposed to cratering for more than perhaps 10 MY. These facts suggest that the area is a very old depression, possibly a lake, that was covered by sediments long ago and then uncovered again in the last 10 MY.

There is still another line of evidence that supports this interpretation. Melissa Lane, a member of the group that devel-

A region where sedimentary layers are being stripped away, a few hundred km northeast of the main hematite area. If you cover the right two-thirds of this image you see half of a fairly normal-looking, somewhat eroded crater. But the image as a whole shows that the right side of the crater rim has been completely removed, exposing deeper layers. This picture and the following ones suggest that thick layers of material are being stripped off this area. (359W 3N, M03-06290.)

oped the TES spectrometer and that is currently working with our Planetary Science Institute, analyzed the TES spectrum of the hematite and demonstrated that it matches a platy form of that mineral, which (according to the spectral data) must be oriented in a horizontal plane; this suggests that it had been flattened and compressed from above.

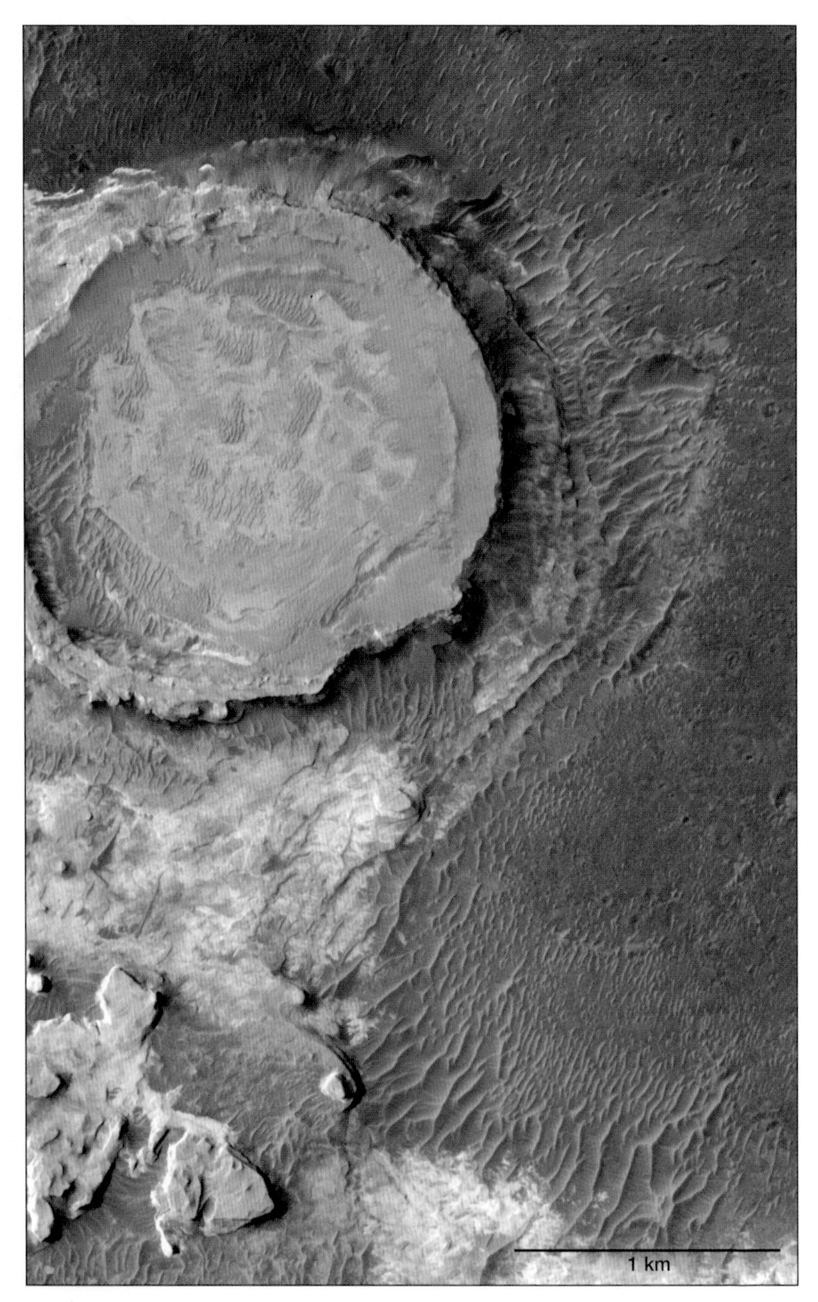

A crater turning into a mesa? The circular structure, 2 km across, resembles partly exhumed craters and has a rim structure visible on the left (partly cut by the camera frame). It may be the remnant of a 2-km impact crater that formed in loose gravel, was filled by cemented sediments, and was left standing as an elevated feature when surrounding weakly consolidated soils eroded away. (3W 5N, M02-01847.)

Images and spectra taken in 2002 by the Mars Odyssey orbiter mapped the hematite-bearing unit as a flat surface. In some areas, it is partly covered with dunes or other deposits; but in other areas, erosion has eaten down through it to a still lower area that is hematite-poor. So, it seems that the hematite is concentrated in one stratum, which fits the idea that it might be a deposit from an ancient lake.

EXHUMATION

Starting with his doctoral thesis in 1976, the year of the Viking landings, MGS camera team leader Mike Malin emphasized another subtle but extremely important Martian process: Some sediment layers are being removed. This process is known by the morbid geological term **exhumation**, referring to the fact that long-buried surfaces and even topographic features such as hills and craters are being exposed in some areas, like a long-buried Egyptian temple floor being exposed by shifting sands.

This process completes the story of Terra Meridiani. In the hematite-rich area, we may be looking at an ancient lake bed or geothermal hot spring area where iron-rich waters ponded and left a concentration of hematite grains roughly 4,000 MY ago, only to be covered by sediments. After the water evaporated, the whole area sat under these sediments for, say, 3,950 MY, which acted to align and orient the platy hematite minerals. Then, under the most recent climate conditions, the capricious winds have begun to strip away the overlying layers, exhuming the ancient deposit.

Perhaps the very reason that we have not found more obvious mineral deposits marking ancient lake beds and ocean floors is that most of them were buried long ago. Like a paleontologist lucky enough to find an exposure of dinosaur bones, we have to be very lucky to catch an area that is being exhumed within the last few MY, freshly exposing the fragile but telltale deposits, but not allowing them to be exposed long enough to be destroyed by fine-scale cratering that would churn up the surface layers.

CLOSE ENCOUNTERS OF THE WRONG KIND: WHEN ASTEROIDS HIT EARTH

If asteroids are crashing into planets all the time, when will Earth get its next big hit? Many small asteroids (half a kilometer to 30 kilometers across) are now known to pass near Earth's orbit, but none has an imminent chance of hitting Earth because their orbit planes lie at slightly different angles. Nonetheless, long-term shifts in these orbits can, and do, bring some of them onto a collision course, and so astronomers in the past two decades have become more and more interested in the timescale between major impacts.

Our last dramatic—and traumatic—hit on land occurred in 1908, when a 50-meter-scale asteroid fragment crashed into the atmosphere over Siberia and exploded high above the ground so violently that trees were toppled out to a distance of roughly 30 km (almost 20 miles). The explosion was comparable to that of a small nuclear weapon. Fortunately, the ground-zero area in 1908 was unpopulated, although one or two reindeer herders may have died due to injuries. At a trading station 65 km (40 miles) away, a witness was blown off a porch and momentarily knocked unconscious; and witnesses, not knowing what they were seeing, said, "The sky split in two and . . . the whole northern part of the sky appeared to be covered by fire." Interestingly, the explosion blasted the asteroidal visitor to dust; no solid rocks reached the ground, and no crater was formed. Fir trees at ground zero had their lateral branches stripped off from above and were left standing like zombie telephone poles.

How often do such explosions occur on Earth? Estimates of the frequency range from about once per century to once per 800 years. Remember that about six out of seven occur over oceans. These numbers explain why we have few reports of celestial explosions during the recorded history of civilization. In most of the world, the occurrence of such events more than a few hundred years ago probably would have been attributed to mythological forces.

A record of another moderate asteroid "hit" may have turned up. A few years ago, when I was serving as an associate editor of the journal *Meteoritics and Planetary Science*, a manuscript turned up from Siim Veski and four colleagues in Estonia and Sweden. They had studied an impact crater cluster named Kaali on the 50-km-wide Estonian island of Saaremaa in the Baltic Sea. The craters were formed when an iron meteorite exploded overhead with, again, approximately the energy of the Hiroshima A-bomb. The largest crater, at 108 meters across (117 yards), could hold a football field. Eight smaller craters lie nearby. Using radiocarbon dating of peat bogs that contained crater ejecta, the research team dated the craters as having formed between 800 and

400 B.C. Wildfires spread at least 6 km (4 miles) from the craters. Pollen samples indicated that fields cultivated before the impact were abandoned for several generations thereafter.

As described in their 2001 article, this is almost surely a rare record of crater-forming impact in a populated area, which probably killed some inhabitants and drove the rest away for many years. Such an event would have been recorded in local myths, perhaps as a judgment rendered by the gods. Northern European mythology from that period is hazy, but, sure enough, around 330 B.C., the ancient travel writer Pytheas of Massalia (Marseille) wrote that in this area a "barbarian showed me the grave where the sun fell dead."

It's important to remember that small asteroids are much more common than large ones, and therefore there are many more small impacts than big ones. Conversely, the larger the impact, the rarer. How frequent are the even bigger impact-explosion events that might disrupt agriculture, the climate, and hence national economies or global civilization? Researchers such as NASA's David Morrison and Colorado researcher Clark Chapman have tried to assess the size of impacts with resulting damage and with frequency. The best estimates suggest that Earth gets a hit sizable enough to interrupt continent-scale agribusiness and disrupt modern civilization perhaps on the order of once every 10,000 to 100,000 years.

There have been talks about expensive plans to divert asteroids (H-bombs have been discussed) in case one is found approaching Earth. Most discussions have centered on how to build defensive systems to deflect or destroy any dangerous asteroid before it can hit Earth. This is a very expensive initiative for a threat that might not materialize for 10,000 years.

My own view is different. Within a hundred years, if we focus our creative and engineering efforts, we will expand our capability to operate in space, pursuing not only science but also resources such as pure nickel–iron asteroids and the 24 hours a day of free solar energy flowing through space. A proactive program to pursue such resources has its own direct value—it would reduce our need to dig ever deeper for lower grade ores and fossil fuels on Earth and would let our planet begin to relax back toward its more natural state. At the same time, as a side benefit, it solves the asteroid threat. We would have the capability to fly to any threatening asteroids and deal with them directly. A small nudge, months in advance, could deflect an asteroid from a collision course. By going on the offense with a proactive program, we would at the same time create a defense to shield future civilization, without wasting resources on an overly hasty anti-asteroid defense system.

CHAPTER 13

THE SEDIMENTS OF CRATER CROMMELIN

Our complex Earth has three basic rock types: igneous (crystallized from molten magma), sedimentary (cemented layers of debris deposited by water or wind), and metamorphic (either of the other two types, altered underground by high pressure or temperature). Mars, by contrast, was supposed to be simple—basically an igneous/volcanic planet with lots of loose gravel and dust.

To the surprise of many, MGS photos revealed many places on Mars where massive layers of sediments fill low areas such as craters. Some of this material may merely be layered lava flows, but much of it appears to form weakly resistant layers, like loosely consolidated sedimentary rock. Some of the layers may be deposited by wind, but others by water. This makes Mars more Earthlike than a merely igneous planet, like the Moon. The ancient sedimentation covered and hid many of the Noachian features, just as sediments on Earth conceal most of the oldest rocks. This helps explain why it is hard to find evidence of the putative ancient Martian oceans. Not only have sediments filled in low areas, but in many cases they are now being removed, as in the hematite area of the last chapter.

Wonderful examples of these sediment layers, brought in by either wind or water, can be found on the floors of many large, low-latitude craters. In this chapter, we'll visit a striking example of sedimentary layers in and around the crater Crommelin, named

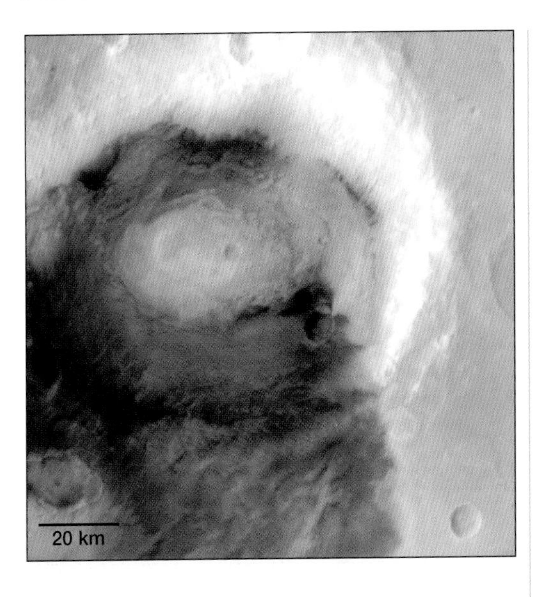

Dark deposits (extending in a tail to the southeast) and layered masses can be seen on the floor of crater Crommelin in this overall context view. (11W 5N, M03-02717.)

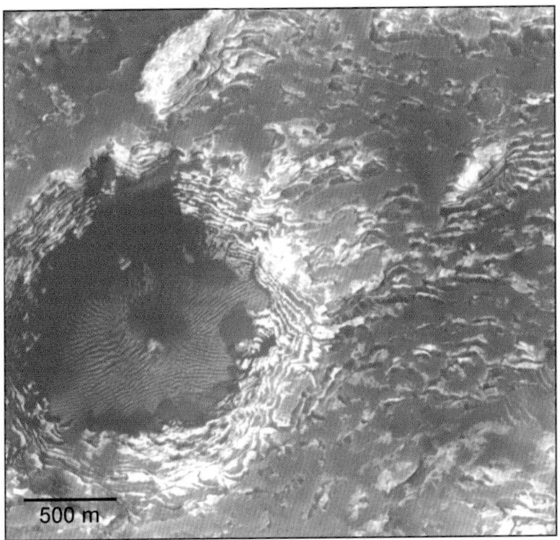

An old impact crater on the floor of Crommelin has punched through flat-bedded layers of sediment, exposing the bright layers in the crater wall, like contour lines on a giant topographic map. This crater has probably been recently exhumed, since no rim structure is visible. Still more recent dunes cover part of the crater floor. (11W 5N, M03-02716.)

after a British astronomer. Crommelin is about 64 km (40 miles) across and lies near the equator at about 5° N, in an old, cratered area mapped as early Noachian plains. Its interior is plastered with layered deposits. In one striking place, a small crater appears to have punched through as many as a dozen neatly stacked layers that are exposed in stair-step order in the crater's walls.

The MGS discovery of these sediments was announced in 2000 by the camera builder Mike Malin and his colleague Ken Edgett. Their paper in the premier American scientific journal

Science caused a stir, because they argued that at least some of the layered deposits were lake-floor sediments. Fitting this theory, many of the layers seem to be flat-lying rather than draped across hills and valleys. (Draped deposits would more likely be composed of dust layers dropped by the Martian winds.) Malin and Edgett identified two interesting types of deposits: "Layered" units, which are typically light-toned material of individual layers that can be as thin as a few meters, were the first. These layers pile up to form thicker composite units, creating so-called "massive" units, in which individual layers are not easily visible. These deposits can be hundreds of meters to as much as a few kilometers thick.

EVIDENCE FOR CLIMATE VARIATIONS

Why are sediments laid down in distinct layers? Malin and Edgett emphasized that, in either waterborne or wind-

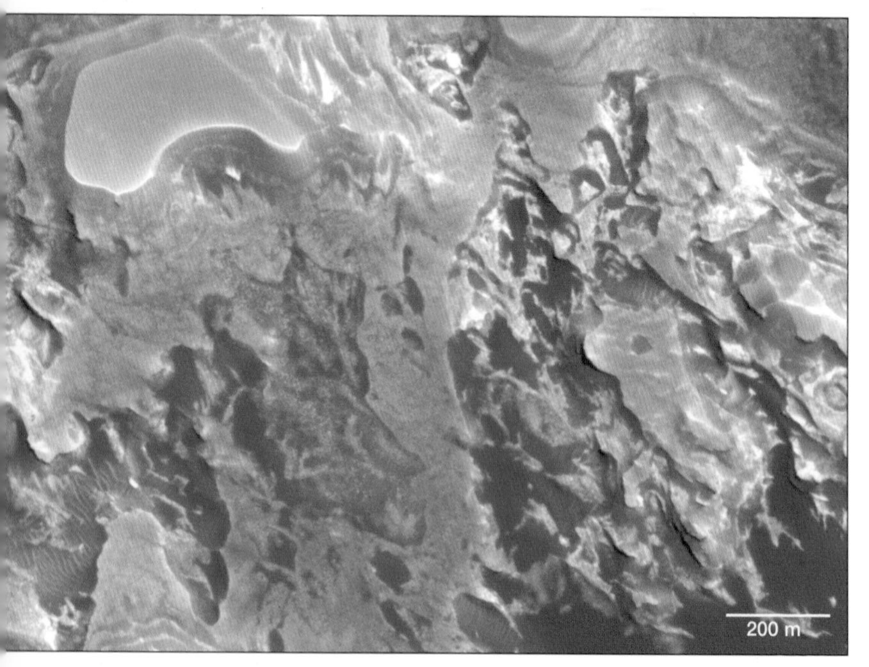

A mesa (upper left) has been left on the floor of Crommelin as surrounding layers are being stripped away. The flat top of the mesa is probably the original surface level of sedimentary deposits in the crater. West is at the top. (11W 5N, M20-01287.)

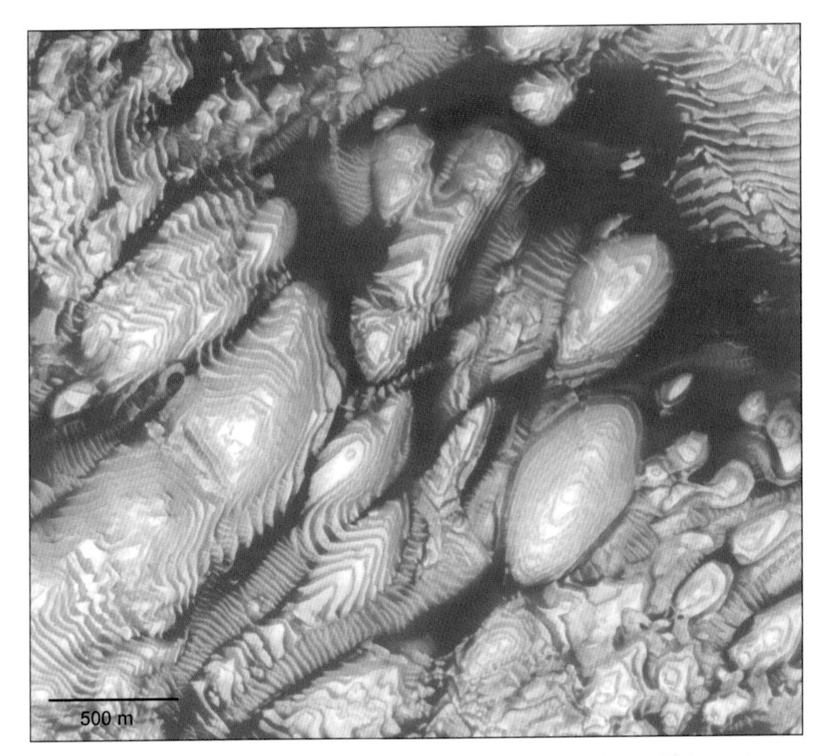

Layered sediments form Martian abstract art. This image reveals a field of light-colored sedimentary layers apparently being eroded down to a darker-colored base layer. The exact nature of the light and dark materials is unknown. This deposit, similar to the layers in Crommelin, is on the floor of a large crater about 200 km to the northeast. (7W 8N, M14-01647.)

borne sediments, the existence of distinct layers plus the different thicknesses of different layers suggests that the Martian environment has been neither constant nor even smoothly varying—as with a model of long, slow loss of atmosphere (or long, slow cooling or long, slow heating). Rather, the implication is that conditions have changed over the whole planet from time to time, favoring deposition during one period, followed by zero or low deposition in other periods.

Malin and Edgett stressed that these layers are not simply late additions plastered on top of preexisting cratered landscapes. In a 2001 review of the MGS images, they pointed out that some authors treat Noachian Mars as if it had formed in two distinct

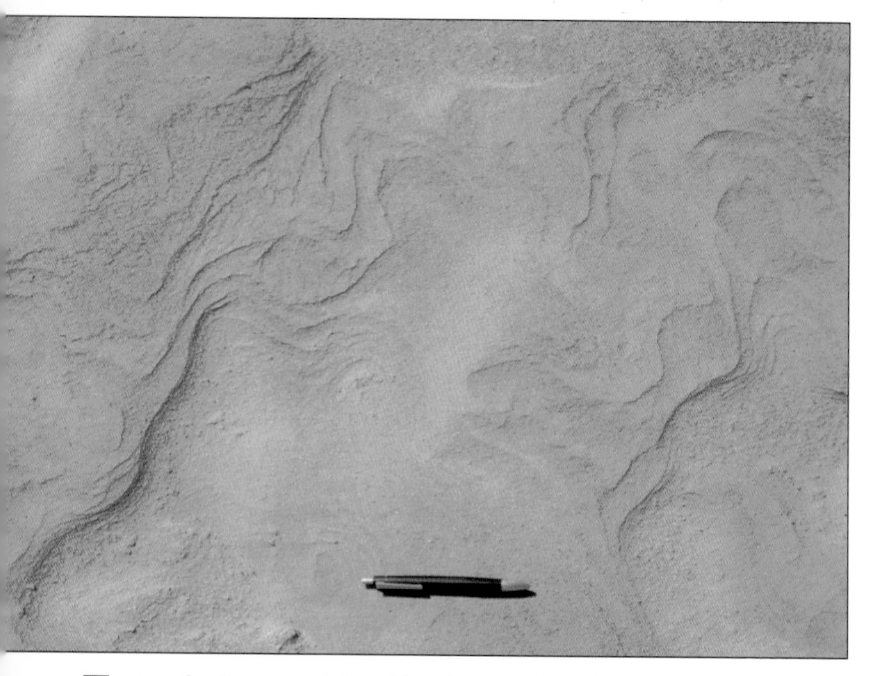

Erosion of sedimentary layers by the wind produces similar features at all scales. This seemingly Mars-like scene, less than 1 meter (1 yard) across, shows erosion of sand layers in coastal dunes at El Golfo, in Sonora, Mexico. Pen shows scale. (Photo by author.)

phases: first, a cratering process that created a Moon-like landscape; and, second, the draping of that landscape with layers of lavas and sediments from water or wind. But in the real geological world, the time of the thicker atmosphere was also the time of maximum cratering, meaning that craters, layers, lakes, more craters, and more layers were probably continually forming, one on top of another. There is probably no simple primordial surface beneath the sediments, but only a complicated stratigraphic column of interbedded, impact-fractured geological features.

As on Earth, each Martian sedimentary layer has its own story to tell. Nature is kindly peeling back these layers like generations of old wallpaper, revealing the story of days gone by. The stories about Martian history that these layers both hide and reveal will occupy geologists for decades to come.

CHAPTER 14

"WHITE ROCK"

An Enigma Explained

As the Mariner 9 mission completed the first mapping of Mars in 1972, camera team members noticed odd formations here and there in the photos. These formations weren't grand cosmic mysteries, but merely natural curiosities—like the Matterhorn, the rock of Gibraltar, or Half Dome on Earth—to which researchers gave whimsical, unofficial names like White Rock and Inca City. Each feature was some quirky combination of ordinary geological processes—the kind that made you stop and ask, "How did nature manage to do *that*?"

In the Mariner 9 discovery pictures, White Rock appeared to be a strange, bone-white mass plunked down in the middle of an ordinary 90-km (56-mile) crater named Pollack, located only 8° S

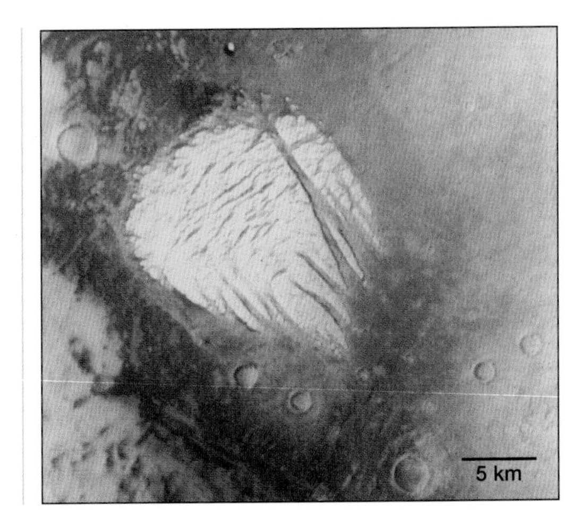

White Rock as it was recorded in a Mariner 9 discovery photo. (NASA, Planetary Science Institute Mariner 9 archive.)

White salt deposits formed by evaporation of salty water in a desert playa near San Luis, Sonora, Mexico. Investigators originally thought White Rock might be a similar deposit. (Photo by author.)

of the equator. Except for the ice deposits in the polar caps, such bright, white-looking material had not shown up anywhere else on Mars. Closer examination revealed that White Rock isn't really a rock, but a plateau-like mass on the crater floor, about 12 by 15 km (7 by 9 miles) in size. What was it and how did it get there? Viking photos in the late '70s did not fully clear up the mystery, because their resolution was not sharp enough to show any diagnostic details.

Many scientists suspected that the White Rock formation might be an example of the long-sought deposits of salt or calcium carbonate, created by evaporation of an ancient crater-filling lake and thus offering "proof" of ancient Martian seas. When MGS got to Mars in 1997, we all held our breath, waiting for the spectrometer to get a look at the infamous white rock. Finally, the slowly evolving orbit took the spacecraft over that formation. The spectrometer measured . . . nothing unusual. Mysteriously, the formation showed basically the same composition as the typical Martian dust that could be found in so many other deposits.

How could the formation look so distinctly bright and yet be the same as the other material? The answer was a classic example of relative contrast. It turned out that the background crater floor was relatively dark in tone, so that when the exposure was correctly set to photograph the crater background, the lighter-toned "white rock" appeared overexposed. In reality, its colors were similar to other, light-toned, reddish brown sediments on Mars. White Rock turned out to be nothing more than a light rock in a dark place.

In retrospect, the White Rock story illustrates the vicissitudes of science. As early as 1976, a team of Mariner 9 researchers headed by James Cutts prepared a manuscript based on Mariner 9 data to show that White Rock was not unusually bright and that it might be made up of sediments, but for one reason or another, the manuscript never got published. Then, in 1979, N. Evans used Viking data to show—once again—that White Rock was not unusually bright; but this paper was only "semipublished"—released as a NASA Technical Memo, a venue for various NASA studies, but not a fully peer-reviewed journal. Not until 2001 did a team of ten Mars scientists headed by Arizona State University researcher Steven Ruff publish a more exhaustive study, using MGS data, in the prestigious *Journal of Geophysical Research.* This confirmed that White Rock is no brighter than normal, lighter-toned Martian dust and that it is probably a mass of weakly consolidated sediment.

One data set utilized in that study was an exhaustive set of temperature measurements that reveal the type of material. (Remember that solid rock conducts daytime heat into its interior and then stays warm at night, while dust acts as an insulator and remains cold at night.) The MGS data showed that White Rock is not covered by loose dust, nor is it solid bedrock, but something in between, like a giant mass of consolidated dirt. Another data set came from the spectrometer, confirming that the supposedly white material was not salt, carbonate, or other typical playa deposit from evaporating water (as many had expected), but rather the same composition of typical Martian dust and soil.

So the mystery of White Rock was reduced to a more down-to-earth (or down-to-Mars) geological riddle: How did such a large, distinct mass of bright Martian sedimentary dust get deposited on the floor of this dark crater? Stunning telephoto MGS images of White Rock show hundreds of thin, flat-lying sedimentary layers. Writing in the *Journal of Geophysical Research* in

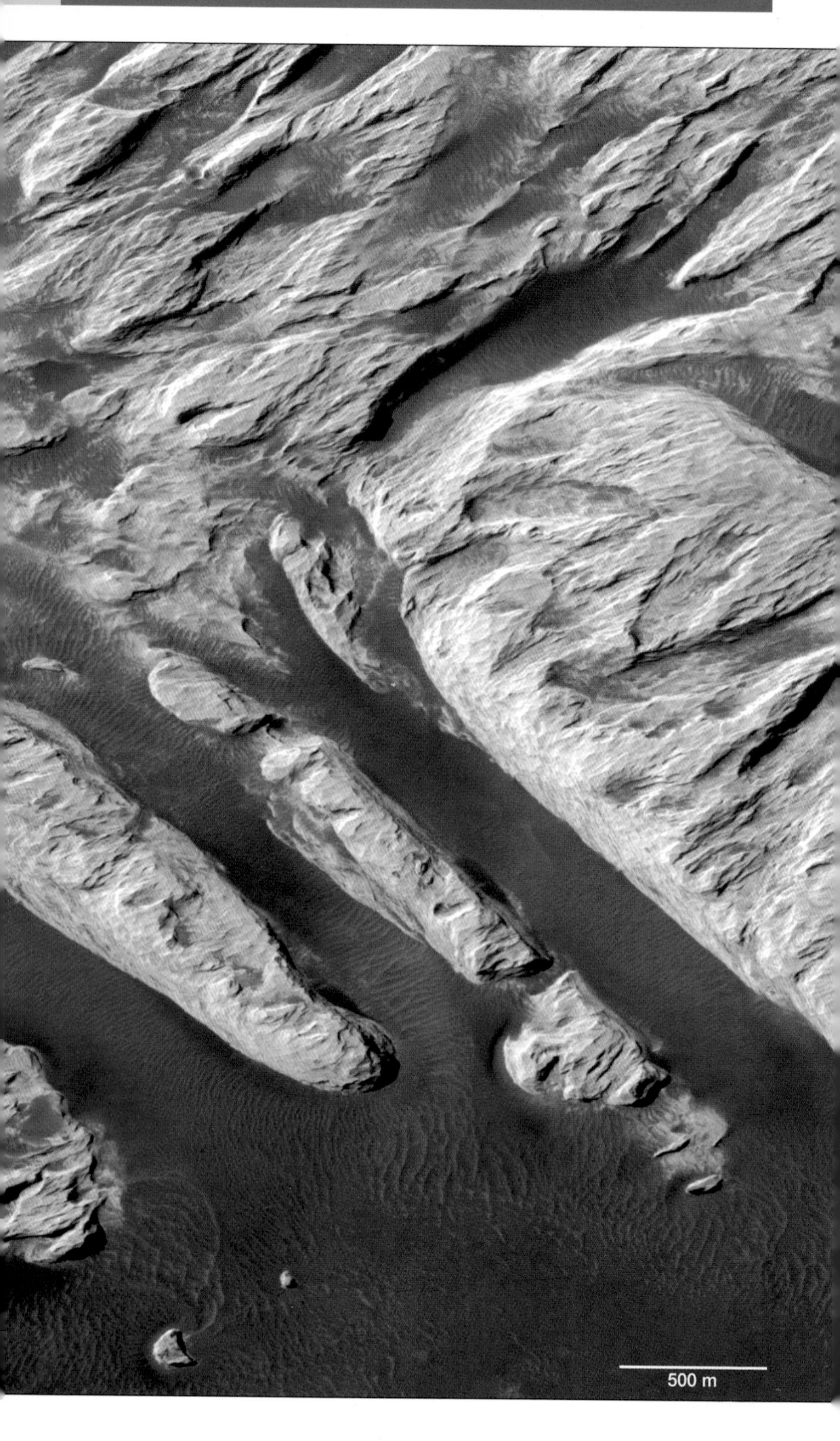
500 m

MGS view of White Rock (facing page) shows a highly striated surface, probably eroded and shaped by wind. Lack of impact craters on the surface suggests that the eroded surface is very young, implying rapid erosion and weak material. (335W 8S, E03-01939.)

2001, Michael Malin and Ken Edgett pointed out an MGS photo of a faint 500-meter crater on the larger floor of Pollack crater, covered by dark sand and all overlaid by the eroded White Rock deposit—which shows that the floor of Pollack crater was exposed and cratered for some time before the "white" sediment layers began to fill it. Again, the issue of dust-storm deposits versus ancient lake-bed deposits arises. Could some of the layers be remnants from the water-rich Noachian period? This part of the mystery hasn't been fully solved and may require human explorers to scale the stair-step cliffs, taking samples of the ancient soils.

The most striking thing about such "white-rock deposits" is that the stacked-up edges of the layers are so well exposed, as if cut away to expose the layers in cross section. Furthermore, the exposed faces of the layer have few if any impact craters, telling us that the erosion process must be recent and geologically rapid—the layers are wearing away faster than they are being cratered. We see badlands country full of gullies and hills, where layer after layer is being cross-sectioned by erosion. The implication is that these are very weak sediments, and that winds, perhaps associated with

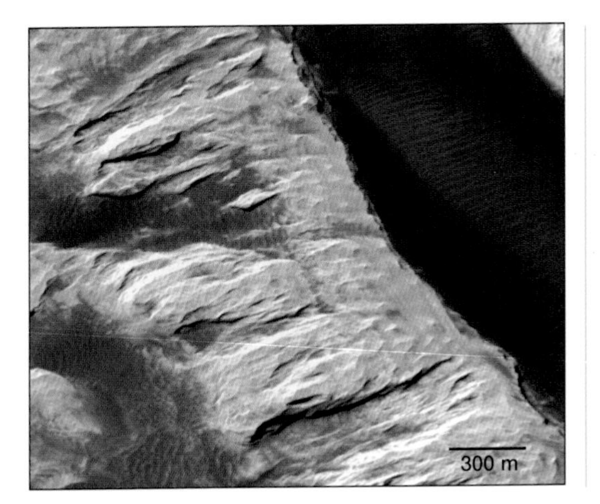

Close-up view near northeast edge of White Rock reveals faint bedding parallel to the edge. Beds are around 5 to 10 meters thick. (335W 8S, M19-00309.)

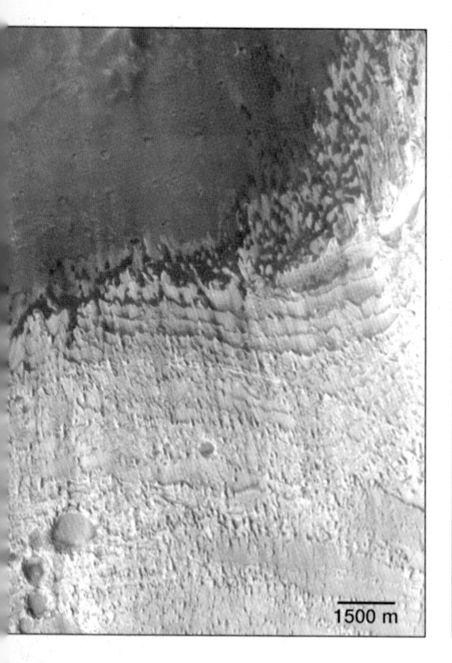

More dramatic layering is visible in a deposit very similar to White Rock in Becquerel crater, northwest of Pollack crater, which contains White Rock. This image was made by the Themis camera of the Mars Odyssey spacecraft. (9W 21N, V020528.)

Above right: A close-up of bright sedimentary deposits in Becquerel crater, reveals extraordinary bedding, with layers only a few meters thick. The layers suggest cycles of climate variation with rapid deposition in some eras and less deposition in other eras. (8W 21N, M08-02650.)

some recent climate change or shift in circulation patterns, are currently eroding and stripping away sediments that were deposited much earlier.

White Rock is to Mars as Ayers Rock, that red desert monolith, is to Australia. It might be a place at which to look for the past. Drill cores into this mass might bring up beautiful records of early conditions, in the same way that drill cores in the Greenland ice pack or in seafloor sediments bring up records of Earth's past climate. Imagine landing on top of White Rock and descending those strange "steps," scrambling downhill, farther and farther back in time. The first explorers on that staircase could have some exciting tales to tell.

PART III

Interlude: Landing on Mars

CHAPTER 15

HEARTBREAK IN HELLES-PONTUS

The Lost Site of the First Lander

So far, we've been trying to reconstruct the earliest history of Mars by visiting Noachian sites and interpreting them through data from orbiting spacecraft and Earth-based telescopes. But there is a parallel stream in Martian exploration—the landing of robotic space probes on the surface. Reaching the surface has been a long and difficult struggle, involving many tales of triumph and tragedy. During the Cold War, a great race was on to see which country could get to the planets first. The ability to mobilize technology and public interest to explore space somehow became *the* test of national prowess. One of the early targets, after the Moon, was Mars.

The roots of our attempts to land instruments on Mars go back to the early twentieth century, when Konstantin Tsiolkovsky in Russia and Robert Goddard in the United States showed that rocket engines would allow travel into space. Rockets with solid gunpowder–like fuels were already common (remember the "rockets' red glare" in Francis Scott Key's poem from the War of 1812), but Tsiolkovsky wrote visionary articles at the turn of the century about designing rockets for space travel; and in the 1920s, Goddard constructed the first crude liquid-fueled rockets that would lead to space vehicles. In the late '40s, German rocket scientists working in both the United States and the Soviet Union popularized plausible designs for such rockets. In the '50s,

President Eisenhower announced that the United States would launch the first artificial satellite into orbit around Earth, but the Russians beat us to it in 1957 with Sputnik, causing national soul-searching throughout the West—and raising fears of cosmic Communist domination. In 1959, Russian engineers crash-landed a probe with Soviet pennants onto the Moon, and in 1961 the Russians triumphed again when Yuri Gagarin became the first human to orbit Earth. Within weeks, President Kennedy recognized that the space race could be a framework for bringing out the best in American technology; he announced that Americans would try to land a human on the Moon within that decade. The Soviet Union feigned indifference but secretly began a program to develop huge lunar booster rockets that would land a Russian on the Moon first. This time, the Americans crossed the finish line first, putting a man on the Moon in 1969.

During the same period, the Russians began an aggressive program to send probes to the two nearest planets, Mars and

Martian terrain near the Mars 2 landing site shows a bland plain with a round-rimmed, eroded crater laced with dark lines that are the tracks of dust devils. (307W 43S, north at right, M12-02673.)

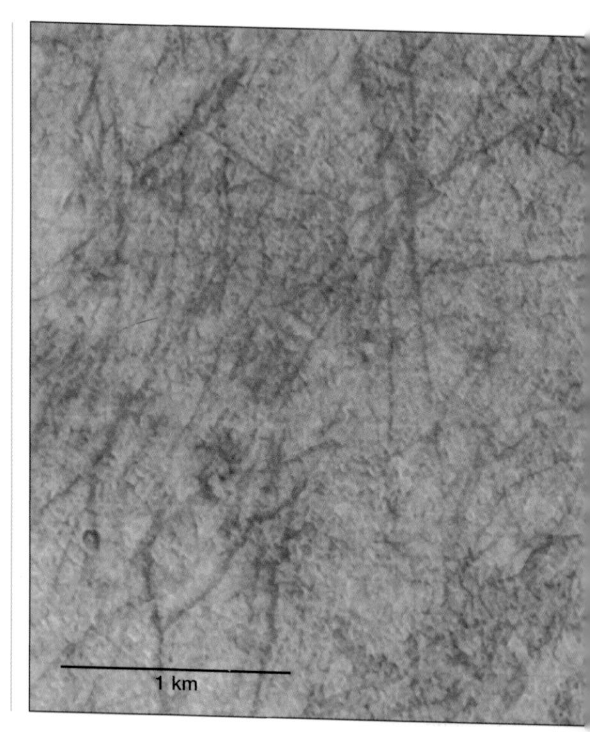

Venus. Their early probes in the '60s failed, but the Soviets persevered, succeeding finally in a brilliant series of robotic soft landings on Venus, beginning with a temperature-measuring lander in 1970 and continuing in the following years with a whole squadron of successful Venus probes, producing what are still the only surface photos of that desolate, cloud-shrouded planet.

It's hard to reconstruct the early stages of the Soviet Mars program because it was less successful and the Soviets were masters of spin-doctoring. If probes designed for Mars had an engine failure that left them stranded in Earth's orbit, they would be described only as routine Earth satellites. If they got partway to Mars and conked out, they would be described as devices to measure interplanetary conditions. These first attempts to reach Mars were not landers, but simpler flyby probes designed to zip past Mars and take the first close-up photos. According to an authoritative log published by British space historian Andrew Wilson in 1986, the Soviets made at least five unsuccessful attempts to send

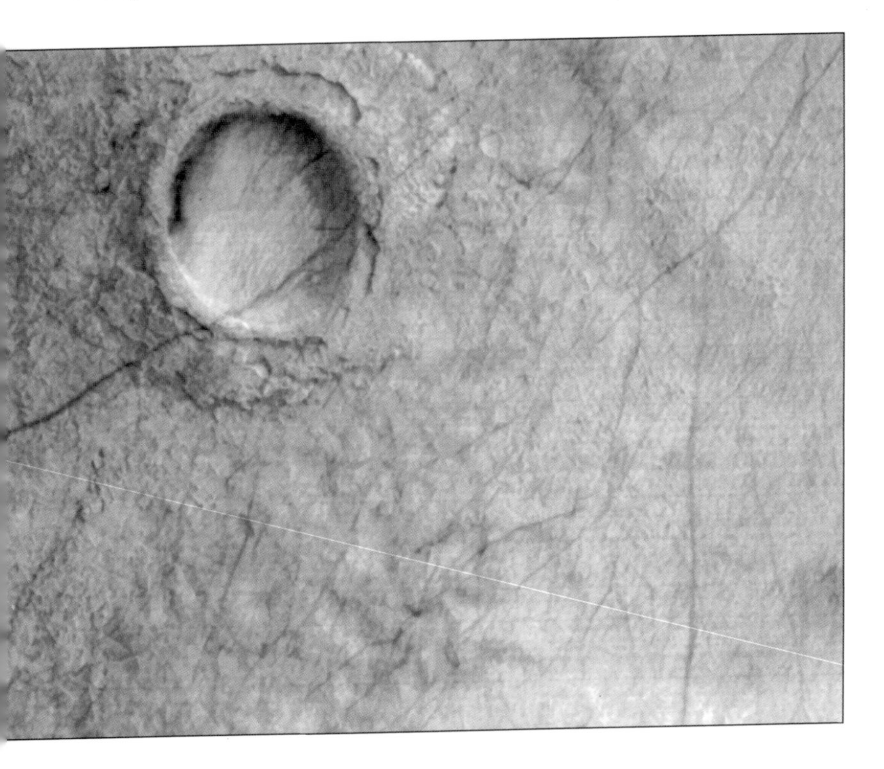

such flybys to Mars between 1960 and 1963. Only one of these, in 1962, was officially declared to be a Mars probe; it was dubbed Mars 1. It beat the American Mariner 2 Venus probe's record for long-distance radio contact, but then it was lost partway to Mars.

The ideal opportunities for launching probes to Mars are dictated by the orbits of Mars and Earth, which come close—a mere 60 million kilometers (40 million miles) apart—about once every two years. When the next close approach came, in 1964, the Americans made their first two shots at Mars. One failed on launch, but the second, Mariner 4, zoomed by Mars in 1965 and became humanity's first ambassador to send back close-up pictures of the red planet, as described in the introduction. In the same year, the Russians upped the ante, apparently trying to *hit* Mars with a lander, but this mission was lost en route.

The next opportunity, 1969, saw two more successful American flybys and two more failed Russian lander launches. But the Soviets were determined; since the Americans had won the flyby race, the Russian scientists set their sights on the next unclaimed Mars prize: the first robotic soft landing on Mars. In 1970, they launched two robotic landers, Mars 2 and Mars 3, which arrived at Mars in 1971. These carried a 3-ton orbiter and a roughly 1-ton, kitchen-stove-size landing probe designed to parachute toward the surface with final rocket braking. Each probe was roughly spheroidal at touchdown but boasted a clever design of four metal petals that would spring open and turn the probe upright on any reasonably smooth surface.

These two landers approached Mars a couple of weeks after the American Mariner 9 orbiter, which was already photographing a huge dust storm that was raging all over the planet. Unfortunately, the dust storm was beyond the control of even Soviet planners. The probes had been designed to go into automatic landing sequence when they arrived, so both probes plunged into the maelstrom. Mars 2 appears to have failed on the way down, a victim of the storm or of mechanical problems, and crashed into Mars on November 27, 1971. The Soviets reported the position as 44.2° S latitude and 313.2° W longitude, in the general vicinity of the Hellespontus mountains on the west rim of the Hellas impact basin. Though Mars 2 failed to send back data, it had the extraordinary and now forgotten distinction of being the first human artifact on the surface of Mars. Its debris still lies somewhere in the Martian southern uplands, west of the Hellas basin. The accuracy of Soviet tracking at that time is uncertain, so the position could be off by some kilometers.

It's fun to look at the modern photographs of the area and think that somewhere down there is the first human artifact on the red planet. Perhaps in some high-resolution orbital photos, future explorers will find a tangled parachute, which may be the most prominent artifact associated with the landing. Images of the area show that much of the surface is bland, with softened or muted craters and a fluffy-looking surface texture. The fluffy-looking material may be a layer of dust that has mantled other features, such as older craters.

This mantle of loose dust might explain the fact that the whole Hellespontus-Noachis area is famous for spawning huge dust storms, some of which grow into planet-encircling global storms like the one in 1971. The pictures reveal something else about the dust. The ground in some photos is crisscrossed by strange dark lines. These are tracks of dust devils—the whirlwind twisters that pick up the Martian dust and inject it into the air. The tracks around the Mars 2 landing site may be a hint that the spacecraft was caught in turbulent local winds, which prevented a successful landing.

Mars 2 still lies somewhere in those dusty hills. I like to imagine that future explorers of Mars will one day hunt down this historic object and place it in a museum, just as I envisioned in my 1997 novel, *Mars Underground.* And perhaps we can turn the site into a historic monument—the first place where humanity managed to make physical contact with the distant red sister-world of Mars.

My Martian Chronicles

PART 6: FACE-TO-FACE WITH THE RACE TO SPACE

In terms of ultimate motivations, early Martian and lunar exploration was as much about global politics and ideology as about an urge to understand our cosmic environment. Nonetheless, to those of us involved at the time, the space race was all about the joy of exploration for its own sake. To us, Kennedy's call to a new frontier meant not just social progress but the opening of the space frontier around us.

On clear evenings in the '60s, when I was a graduate student working on lunar mapping in preparation for the Apollo mission, I would walk across our campus and look up at the Moon, realizing that it would go around Earth only a few dozen more times before we tried to land on it. During those same years, the first Mars close-up photos started coming in. Amid the upheaval of civil rights progress and Kennedy assassinations, engineering flowered. Research grant money flowed. It was a golden age of science. In one grand decade, humanity went from debating the nature of the lunar surface to walking on it. In one generation, we went from squinting at a fuzzy, shimmering ruddy disk in a telescope to analyzing rocks from Mars. Science moved from arguing whether meteorites could cause even a single crater on Earth to recognizing that a horrendous meteorite impact wiped out the dinosaurs. We went from debating continental drift to charting volcanoes on the satellites of Jupiter.

Climax of the space race. Apollo 11 leaves for the Moon.

Much of that excitement has been forgotten, and the Cold War now seems like an outmoded struggle between quaint armies of capitalists and anti-capitalists. But the day will come when people will look back at that period as one of the most dramatic expansions of consciousness in human history—the decades when we took our first steps off Earth and learned the basic nature of other planets. As Tsiolkovsky said, "Earth is the cradle of humanity, but one cannot live in the cradle forever."

Years later, in 1992, just after the fall of the Soviet government, I had a poignant look at how international cooperation could change the world. I had been invited to participate in a Russian mission—the ill-fated Mars 96 that failed during launch and ended up in the Pacific. In a new spirit of cooperation, a group of Americans had been invited to Moscow to participate in the design of a spectrometer for the spacecraft. Russian instruments on previous missions had had calibration problems that made it hard to read their data, and the Russians had asked U.S. scientists to help solve the problem.

We were taken out of Moscow on a bus to a small farm town where Soviet planners had decreed the construction of a space facility. It was a marvelous trip in a time machine. The muddy dirt roads, modest farmhouses, and free-ranging chickens and goats reminded me of boyhood visits to my grandparents' Illinois farm in the America of 1949. Nestled in the trees on the edge of the village was an installation like a small college campus. During the day, the Russians showed us how they were building their spectrometer, and we discussed how a standard color plate could be built at UCLA in Los Angeles, shipped to Moscow, and inserted in the Russian spectrometer so that its readings could be calibrated. Before turning in at night, we ate in the institution cafeteria. After we consumed sufficient vodka, wine, and Pepsi (American capitalists had arrived well before we did), conversation would flow freely. One evening, a Russian engineer leaned forward conspiratorially, with a curious smile that seemed to suggest relief at the end of some great trauma.

On the way to Mars—an eyewitness view of the launch of Mars Observer from Kennedy Space Center. (Painting by author.)

"You know," he said, "a year ago I was working in a weapons factory building atomic bombs, and they were all aimed at you people. I would much rather be doing this."

His casual comment cut through the vodka haze to remind me that we can either keep trying to triumph over each other and then contend with the hatreds that follow, or have the fun of working together toward more transcendent goals.

CHAPTER 16

SORROW IN SIRENUM:

The Second Lander

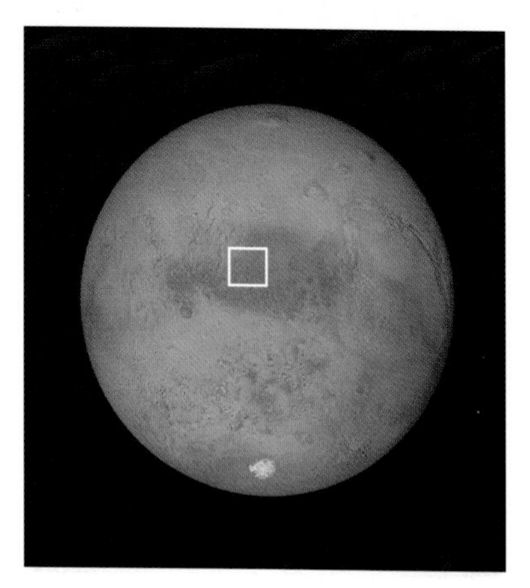

The Soviet Mars 3 probe had a stranger fate than the Mars 2 mission. It also sailed into the 1971 dust storm on December 2, 1971, just a few days after Mars 2, but farther west. This time, the telemetry indicated that the descent was normal and successful. Apparently, a drogue parachute deployed properly and pulled out the large main parachute, which lowered the craft until small solid rockets were ignited to cushion the final landing, three minutes after atmospheric entry. The craft apparently made a bumpy touchdown at about 75 km per hour (45 miles per hour)—seat belts required. But it survived. Ninety seconds later, the automatic timer commenced a photographic panorama of the landscape. TV signals began arriving in the Russian mission control room.

The first lines of the TV image were blank. Was it sky? An image of murky blowing dust? Then, after 20 seconds, all the signals suddenly went dead. After years of effort, the Russian engineers had sent a probe all the way to Mars, achieved a successful landing, and started a photographic scan, only to have the probe die before any part of the landscape could be seen! At almost the very moment of victory, their dreams were

Rugged country near the Mars 3 landing area is seen in this view of a hill (lower left) and a large crater with softened rim, and a floor where the wind may be plucking sediments, leaving a strange pattern of irregular depressions. (155W 46S, E05-00978.)

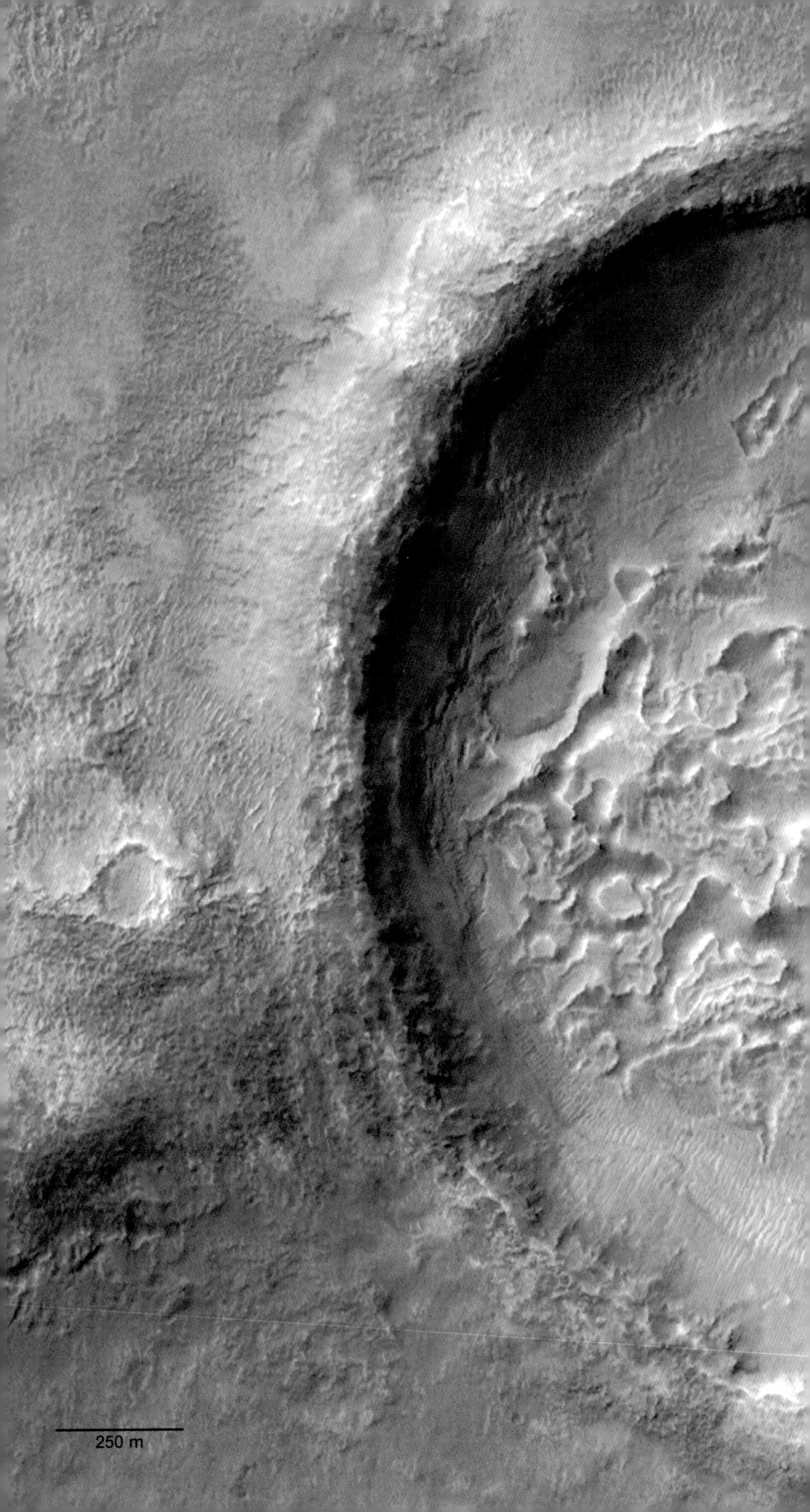
250 m

THE BEST-LAID PLANS FOR MARS AND MEN: TALKING WITH A RUSSIAN SPACE PIONEER

During my work with planetary scientists from different countries, I've had many chances to work with Russians, especially after the thaw between our countries. I enjoy their warmth, generosity, and dark sense of fatalistic humor. Their stories of adversity and perseverance fascinate me.

One such scientist, Mikhail Marov, participated in the attempt to put the Mars 3 lander on Mars, and he has been a leading figure in the Russian space program. He works at the Keldysh Institute of Applied Mathematics in Moscow, and is an internationally respected scientist and now a member of the Russian Academy of Sciences. Misha, as he is known to his friends, kindly provided a foreword for this book, and when I asked him about the Mars 3 landing, he also provided the following account of his activities at that time.

"I perfectly remember that day, December 2, 1971. It was rainy, windy weather in Eupatoria, Crimea, where our main radio telescope dish of the Deep Space Network was located. Weather was not our main concern; my thoughts were millions of miles away, on Mars, where astronomical observations showed that the weather was even worse. There was a severe dust storm raging on the planet, just at the time when humanity's first lander, Mars 3, was trying to touch down on the planet's surface.

"Time went slowly. My colleagues and I were in a state of strained expectation by the time the signal from the lander came in at the predicted moment, about 4:47 P.M. local time. It came in simultaneously over two independent radio channels, with a good signal strength and no interference, so we were sure we had a successful landing. It was clear that a TV transmission from Mars was beginning, although the luminosity detected by the panoramic camera was very low. The signal reception was met by great excitement among us, and shouts of delight! But only 20 seconds later, the

torn from them. The TV screen remained as mute as the sphinx. No further signal ever came from Mars 3, and no one is sure what happened. Perhaps the electronics were damaged by the landing. An even more heartrending suggestion has been made—that the complex lander performed flawlessly and took its pictures of the surface, but that the mother ship, the orbiter that was supposed to relay the signals, had some glitch that prevented the transmission

transmission unexpectedly was broken, and our original joy changed to deep disappointment and frustration. Despite desperate efforts undertaken by the network staff during that day, the signal never reappeared.

"Follow-up analysis brought us evidence that the Martian dust storm, which had concerned me so much earlier that day—especially wind-shear acting on the parachute during descent, not to mention possible rough terrain at the landing site—could have caused irreversible damage to our lander. Meanwhile, high above, the 'buses' that carried the two landers decelerated by igniting their retro rocket engines, and joined the American Mariner 9 orbiter [which had arrived some days before] as the first artificial satellites of Mars. All three orbiters returned data on atmospheric properties and surface features.

"As for the Mars 3 lander, I have no doubt that future astronauts, exploring Mars with long-range rovers, will have a good chance of finding its remnants and even those of its less successful counterpart, Mars 2, which failed soon after atmospheric entry a few days earlier. It's worth remembering that this was a period, during the height of the Cold War, when the first agreement on cooperation in space between NASA and the Soviet Academy of Sciences was signed in Moscow, and the first joint working groups were set up. Agreement was reached to exchange real-time information during those missions to Mars. Based on our current knowledge of Mars, those first attempts at planetary exploration may seem not so impressive, but their importance and contribution to newer, more ambitious projects is difficult to overestimate. We can look back now and say, as I wrote with George Petrov, former director of our Space Research Institute, in the American journal *Icarus* in 1973, 'The year 1971 was marked by an event of great moment, represented by the first artificial satellites being placed around Mars, and by a unique scientific and technical experiment, the first soft-landing on the planet.' "

to Earth. Only a visit to the Mars 3 site will tell the final story.

The reported landing site for Mars 3 was at the same latitude as for Mars 2 but halfway around the planet, measured as 44.9° S and 160.1° W, in the old cratered uplands of Terra Sirenum, between large craters that were named for two giants of ancient science, Ptolemy and Newton. How many years will it be until human hands touch the Mars 3 probe once again?

CHAPTER 17

CHRYSE PLANITIA:

The First Successful Landing

On the cold morning of July 20, 1976, the sun rose as usual over the deserted plains of Chryse on Mars. Dawn temperatures were typical at about –80° C (–112° F), but by midday the morning haze had burned off and the temperature had risen to –29° C (–20° F). It was the seventh anniversary of humanity's first footsteps on the Moon and only a few weeks after the two hundredth birthday of the United States of America. Just after four o'clock in the afternoon, a tiny bright speck appeared in the sky over Mars. It blossomed into a parachute, beneath which dangled a strange three-legged machine. The apparition descended rapidly toward the ground. At the last minute, a braking rocket engine, centered amid the three legs beneath the craft, roared to life. A mighty cloud of tan dust arose as the exhaust blasted the surface, and the craft settled onto the surface. The Martian soil was firm, and the landing pads sank in only about 3 to 4 centimeters—an inch and a half.

The **Viking 1** lander had arrived. The Americans had won the race to put the first working device on the surface of Mars, and humans now had their first working cameras and instruments on the red planet. It was one small step for a machine, but a great leap forward in our understanding of other worlds.

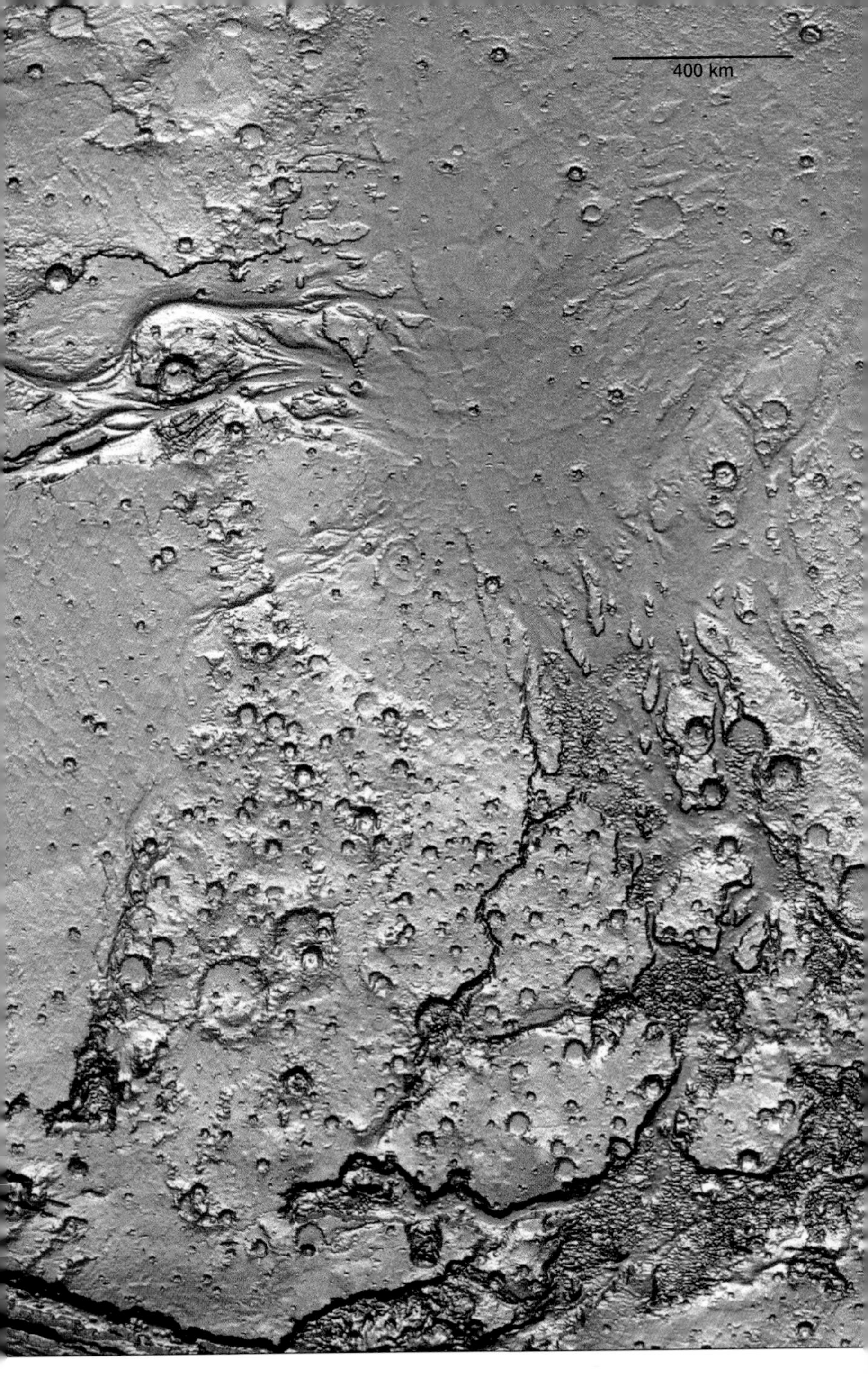

MGS laser altimeter map shows pattern of river drainage from southern uplands (red and green, bottom) toward the Viking I landing site in the Chryse basin (darkest blue, upper center). (MGS MOLA team, processing of data courtesy Nick Hoffman, Victorian Institute of Earth and Planetary Science.)

PLAINS OF GOLD

The lander parachuted to the surface at 22.5° N and 48.0° W. From Earth, this area can be seen in telescopes as a bright desert. Schiaparelli named it Chryse in 1877. To the ancients, Chryse was a rumored land of gold to the east of the Indian Ocean, identified with the lands of southeast Asia. The modern topographic name, Chryse Planitia, can be thought of as meaning "plains of the golden land of Chryse."

The landing was a tricky thing. In the weeks leading up to it, scientists on the Viking team obsessed over pictures from the Viking 1 orbiter, which had gone into orbit around Mars four weeks earlier. The original plan called for releasing the lander from the orbiter for a touchdown on July 4, the bicentennial of the United

Riverbeds of Vedra Vallis system drain eastward toward Chryse basin, where Viking I landed. (56W 18N, Viking mission.)

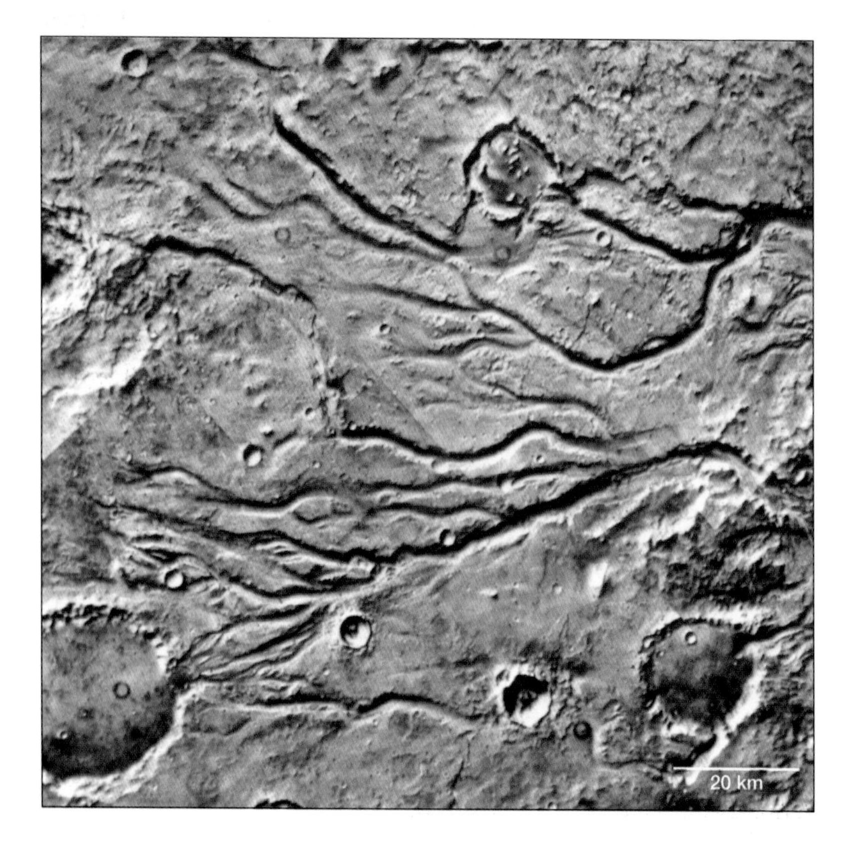

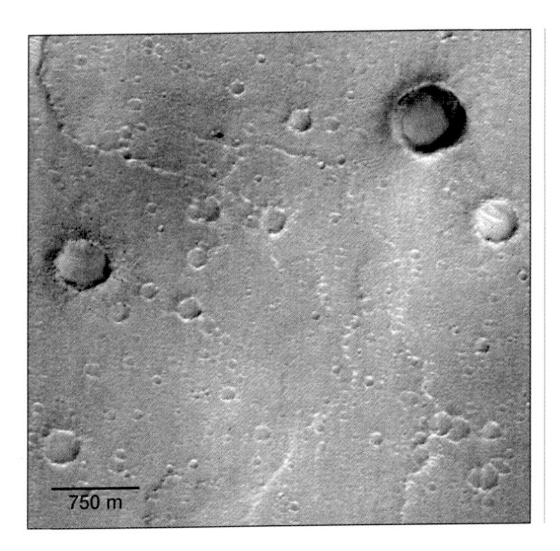

MGS orbital view of the Viking 1 landing site shows a moderately cratered volcanic plain with eroded ridges. (48W 22N, SP1-23503.)

States. Erring on the side of caution, the Viking team decided to wait and assemble more photos from the orbiter in order to determine the safest landing spot for the precious lander. From these pictures, they could see details that were only the size of a football field, but at that scale the Chryse plain appeared level and smooth. The landing on the anniversary of Apollo 11's lunar touchdown was an inspired choice, and since then July 20 has been informally called Space Day in the United States.

Chryse offered another attraction besides smoothness. It appeared to be a region into which huge masses of water had emptied after flowing out of the great waterway and adjacent rivers of the southern highlands. The flooded surfaces were mapped as dating from the Hesperian era, suggesting the flooding happened perhaps 3,500 to 2,500 MY ago. The waters carved many winding channels for 1,200 km (700 miles) through the hills and ridges south and west of the landing site, where streamlined islands testified to the erosive force of the water. The channels emptied onto the plain where Viking 1 landed. Would the lander pick up direct, "ground truth" evidence that water had once flooded the plains of Chryse? And what about life? Would this first look at the surface of Mars reveal lichens, mosses, trees, animals, fossils, or artifacts? Until that afternoon, no one knew.

The first press release picture from the Viking I project was incorrectly color-calibrated and showed the Martian sky with a bluish cast. (NASA.)

Facing page: *The rock-littered landscape of the* Viking I *landing site. The footpad of the three-legged Viking lander is in the lower right. The scene includes some of the same area as the previous view but was made later. Soil in the mid-distance (left center) has been disturbed by the* Viking *trenching tool. (NASA.)*

The first photos of the Martian surface showed no signs of life, only a forbiddingly beautiful desert—an immense vista of boulders and scattered sand dunes untrod by earthly creatures. Some meters northeast was a large boulder as big as a banquet table, three meters long and half as high as a person. Viking team members called it Big Joe. If the lander had come down on top of it, the mission would have been a failure. But Viking 1 was lucky, and the Chryse desert turned out to be a treasure trove of scientific data.

That very first day of the landing, excited scientists pored over the information from the surface of Mars. They were surprised to learn that the Martian sky was not blue, as had been expected, but pinkish tan, due to the great amount of reddish dust kicked up into the atmosphere by the sporadic winds and dust storms. Typical daily winds that were measured from the lander were not enough to raise dust—they were light breezes, peaking around 11:00 A.M. at 7 meters per second (16 miles per hour). Over many months, the cameras took wide-angle vistas, sunset shots, telephoto shots of the distant horizon, and close-ups of soil around the landing pads—but

the photos never showed any signs of life. The images showed that if Mars has any life, it must be microscopic (like bacteria) or underground, or both. The dunes proved the idea that windblown sand and dust is very mobile on Mars. In one photo, Big Joe, the giant boulder near the lander, was capped by a mantle of red dust

Trenches dug by the Viking 1 sampler arm allow measurement of soil compositions at the site. Somewhat out-of-focus parts of the lander are in the foreground. (NASA.)

that had filtered down from the sky after the previous dust storm. Another picture revealed a slippage of dust on a small dune face at the foot of Big Joe. This change, while the lander sat there, gave proof, if any were needed, that things do happen on Mars.

Two terrestrial views show that Mars-like landscapes can be found on Earth. Above left: A cold, wet volcanic desert in Iceland shows the kind of vista that Viking investigators had hoped to find; green vegetation can be seen in the distance. Foreground contains red and brown weathered basaltic boulders as found on Mars. Right: Dry coastal deserts in Peru also illustrate Mars-like conditions. The area is so dry that the sands preserve buried Inca weavings that are centuries old, and rock surfaces have not weathered to red colors. (Both photos by author.)

Viking 1, along with its sister lander, Viking 2 (which we will visit shortly), was designed to make various measurements, in addition to taking pictures. Both landers made chemical analyses of soil samples that had been scooped up by a long robotic arm. The analyses showed that the soils resemble terrestrial soils derived from basaltic lavas, consistent with evidence that Mars has much volcanic terrain. The analyses also found that the soils were unusually rich in sulfur- and chlorine-bearing compounds, the kinds of chemicals that could be left by the evaporation of salty or mineral-rich water. This supported the idea that water had long ago flowed across the Chryse plains, then had evaporated and left salts, sulfates, and other such compounds in the soil. The same process builds up salts in irrigated soils on Earth and is a problem for farmers in the American West.

Highly processed composite of Viking 1 images gives an unusually sharp view of the large boulder Big Joe, as well as a view of a beautiful dune field northeast of the lander. (NASA.)

My Martian Chronicles

PART 7: GRAVEL-TO-GRAVEL COVERAGE OF THE FIRST MARS LANDING

Since I was not an official member of the Viking team, I decided to try on the hat of fledgling science journalist and offered to cover Viking 1's landing on Mars for *Astronomy* magazine. I went to Caltech's Jet Propulsion Lab, where the mission was directed. Outside the press area, giant network TV trucks—ABC, CBS, NBC—clustered by the curb. Even the dusty overflow parking lots were jammed with cars. Inside, reporters from daily newspapers, news magazines, science magazines, and TV and radio milled about in anticipation. Working science reporters from small technical magazines had a certain contempt for big-name science reporters from major networks, who parachuted in to cover big events, but seemed to have little familiarity with actual scientific research. Rumors circulated about a certain famous TV network science reporter who didn't know how to convert from metric units to English units.

Not even the most jaded journalists could completely repress their excitement about what promised to be humanity's first fully successful landing

The first photo successfully transmitted from the surface of Mars to Earth was programmed to show the soil around the footpad of the Viking 1 lander. In case the lander sank into soupy dust and failed, it was hoped that this first photo would give an indication of soil texture and hypothetical small plants, such as mosses or lichens. Instead, the image showed a firm surface of gravel and dust. (NASA.)

on another planet. In those days, no one knew whether the first cameras on the surface would reveal bizarre Martian vegetation, tufts of moss and lichen, or weird fossils in the rocks. Carl Sagan had been advancing the idea that the cameras might do even better than that, revealing living creatures—alien beasts lumbering across the frozen landscape.

The landing went smoothly and there were cheers the moment we knew that humans at last had placed sensors on Mars. But the greatest anticipation was reserved for the first images. The programming called for the first image to be not a landscape panorama, but a shot of the ground immediately adjacent to the lander footpad. There was a good reason for this: in case anything went wrong in the following minutes, engineers would at least have gotten a sense of the soil qualities, the depth of penetration of the pie-pan-shaped footpad, and perhaps the best opportunity to detect the tiny, lichen-like life-forms that some scientists thought might be clinging to the Martian rocks.

When the cameras turned on, a hush fell across the pressroom. The anticipation was heightened because the image, scanned in one line at a time, materialized so slowly on the pressroom's TV monitors. There they were, the first few scan lines. Everyone in the room was peering intently, looking for some ferny frond or fossil shell. The silence intensified as the image finally formed.

Gravel. The pie-pan-shaped landing pad sat atop the dusty soil strewn with gravel and rock chips. The assembled reporters peered intently at the monitor, looking for that ferny frond or fossil shell. Nobody spoke.

Then a voice from the press corps piped up with a response to humanity's first view of the surface of Mars: "Well, hell, that looks just like the parking lot where they made me leave my car."

DID VIKING I DISCOVER LIFE ON MARS?

Viking also carried experiments designed to look for life, and the answer to the above question was not as simple as you might think. For a start, the chemical analyses revealed that the soil was completely sterile—meaning that there were no large organic molecules. This measurement was incredibly sensitive, to an accuracy of a few parts per billion in terms of detecting organic materials. It means that the Chryse desert has less than 1 percent of the amount of organic material found in soils of typical barren deserts of Earth! This result makes it unlikely that life could be prospering anywhere on the current surface of Mars, since viable life has a way of spreading organic residue hither and thither. (A few scientists disagree. At a 1997 NASA meeting on possible Martian biology, during a discussion on certain organic polymers that might not have been detected by Viking experiments, geochemist Fraser Fanale offered the wonderful opinion "If a dog had shit on the ground one meter from a Viking lander, it would never have detected it.")

The case was hardly closed. Viking 1 contained three other experimental packages specifically built to look for Martian life by scooping up soil into test chambers, adding nutrients, and then looking for chemical activity that might indicate microbial life-forms happily engaging in metabolic activity. Despite the fact that the soil tested sterile, these experiments, oddly enough, did spot chemical activity! Here was an enigma! Does Mars really have life after all?

At first, scientists were mystified. If the chemical reactions weren't biology, what were they? The eventual explanation, accepted by the majority of researchers, starts with the fact that Mars lacks ozone, the gas that screens most of the sun's ultraviolet, or UV, radiation from Earth's surface. Photons of UV light are so energetic that they break up molecules in your skin, which is why too much UV exposure and suntanning cause damage. On ozone-free Mars, the surface is exposed to solar ultraviolet radiation, and the energetic UV photons alter soil chemistry, producing highly reactive chemicals such as peroxides. The theory is that these materials—and not microbial life—caused the unexpected reactions in the Viking soil chambers.

Scientists today recognize that the ozone-free exposure to solar ultraviolet is an important factor in understanding the possible biological history of the red planet. Think about the combination of loose dust and UV irradiation. Virtually all mobile dust and sand grains in the top few centimeters have been picked up by the wind and exposed on all sides to UV radiation, which would tend to break up the big, complex molecules needed for life. This is why the Martian topsoil is so sterile.

ECHOES FROM ANTARCTICA: RETHINKING LIFE

On the other side of this question is a minority of scientists who still think that life could exist on Mars and even believe that the Viking measurements detected it! They point to an astonishing discovery in the dry plains in Antarctica. Measurements with Viking-like instruments show more or less sterile soils in this region, but microbes live happily nearby, inside fractures in rocks! In that harsh environment, the rock interiors are sheltered from extremes of wind and cold, and seem to be a more hospitable home for microbes than the exposed soils. So, argue these scientists, Martian microbes might be concentrated inside rocks and pebbles.

Although these scientists think the Viking biology experiments really did detect life, the soil sterility result should probably take precedence. If the soil really is sterile, then the chemical activity in the scooped-up *soil* was probably not biological. Still, the Antarctic findings revitalized thinking about how to look for life on Mars. Mars researchers are realizing that most life on Earth is microbial.

In a way, we have been victims of common sense when it comes to concepts of life. Our cultures and even our religions tell us that we humans, along with the plants and animals around us, are the main story when it comes to the kingdom of life on Earth. But from a different point of view, we are the aberrations! The total mass of living material on the surface of Earth is less than the total mass of living microbial material underground! Even "simple plants," which post-Lowellians thought might exist on Mars, are not so simple. It took about half of the total history of Earth before plants gained any serious foothold on land surfaces of Earth, not to mention the fact that substantial land-dwelling animals appeared only in the last 15 percent of planetary history! A realistic view of total living organisms must take into account that most living creatures are microscopic life-forms underground and we are merely the lumbering multicellular anomalies, produced by evolution run amok on our planet's rare, moist, ozone-shielded, habitable surface.

These ideas change the whole debate about looking for life on Mars. They increase our probability estimate that Mars could have life after all—not on the harsh, exposed surface where Lowell and the Victorians and the Viking landers looked, but rather in more hidden environments. Such life would have evolved in the waters of Noachian Mars and adapted to survive in aquifers at depth, protected from the strongest solar ultraviolet radiation! Perhaps the Viking measurements merely mean that the top few inches are sterile. Conceivably, single-celled Martian microbes lie dormant in the subsurface ice layers or thrive in energy-giving, deep-seated geothermal sites, analogous to the weird and unique life-forms that cluster around dark seafloor geothermal vents on Earth. In hindsight, we can say that Viking 1 never should have expected life amid the ultraviolet-irradiated surface dust of Mars, but in science as in life, it's always better to look and be sure than to sit around theorizing.

PART IV

Hesperian Mars: A Time of Transition

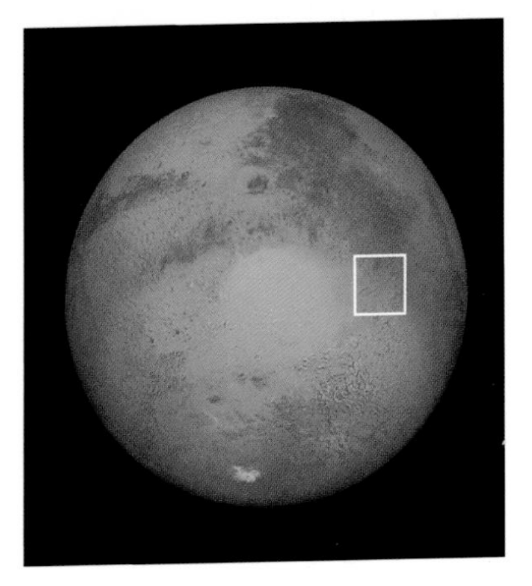

CHAPTER 18

ANCIENT FIRES OF HESPERIA

Off the east rim of the mighty Hellas impact basin lies a moderately cratered plain called Hesperia. The name Hesperia originally referred to an intermediate-toned area between two darker "seas." Schiaparelli assigned the name in 1877, borrowing a classical term for the western frontier. To the Greeks, it referred to Italy; to the Romans, it referred to Spain. For us, it can refer to our future Martian frontier.

Hesperia gives its name to the whole Hesperian era, the transitional interval between the Noachian era and modern, or

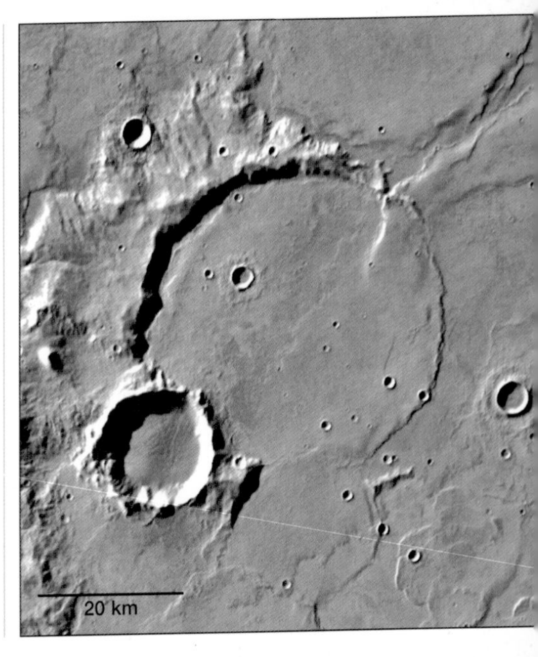

Moderately cratered lava plains of Hesperia Planum (right two-thirds of image) have partly engulfed older upland hills (left edge). The central ring structure is an ancient impact crater whose SE wall has been almost obliterated by lavas. The process by which crater walls are broken down by lavas, leaving only a faint circular ridge—a process that happens in an identical manner on the moon—is unknown. A smaller circular crater remnant is at bottom center. Two still smaller rampart craters can be seen. (239W 37S, Viking mosaic.)

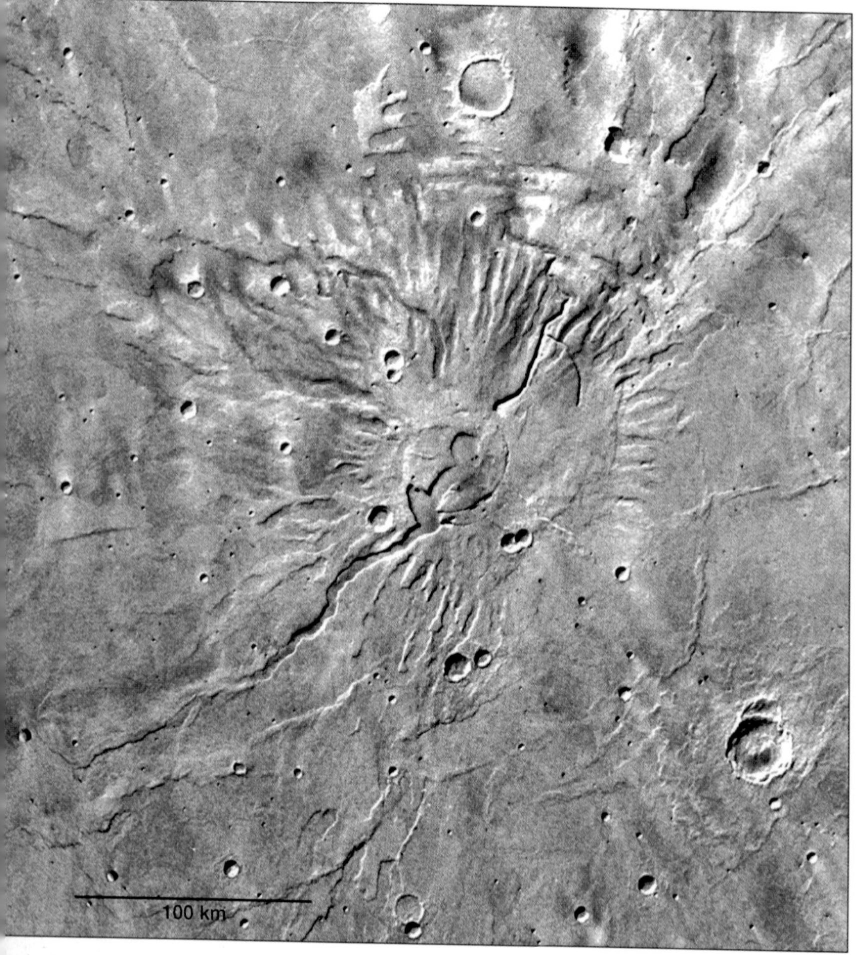

The low, eroded volcanic cone of Tyrrhena Patera is marked by radiating gullies on its flanks and surrounded by Hesperian lava plains. (254W 22S, Viking orbiter mosaic.)

Amazonian, Mars. Hesperian features, from outflow river channels to volcanoes, are more diverse and better preserved than the crater-battered, sediment-covered Noachian features, and we can get a clearer idea of what was happening in Hesperian times. The relatively smooth plain, known today as Hesperia Planum, displays scattered impact craters that typify the modest crater densities used to define the Hesperian era. The plain is composed of early to mid-Hesperian lava flows and dominated in the middle by a large, somewhat younger volcanic mountain, Hadriaca Patera. (Patera is

The flank of Tyrrhena Patera (below), SW of the caldera, has a dissected surface with radial furrows cut by water or fluid lavas. Box shows location of the next image. (255W 23S, E10-04204.)

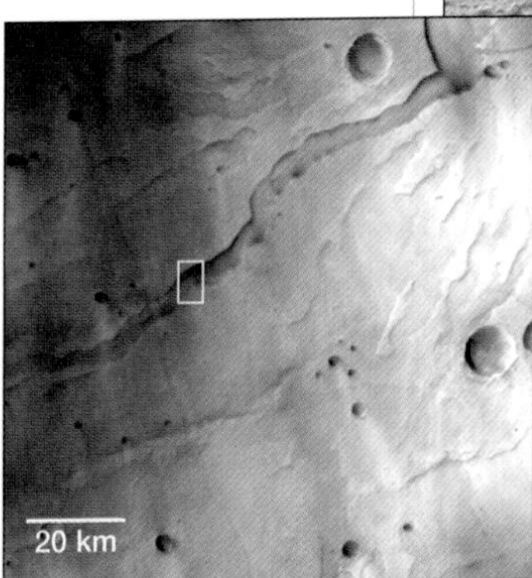

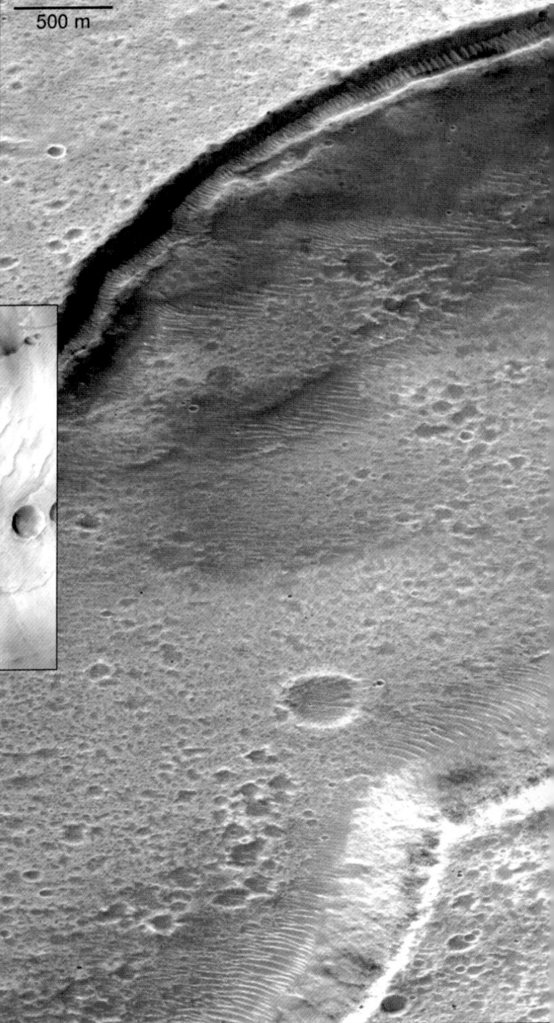

Close-up of Tyrrhena Patera (right) shows one of the radial furrows. The whole surface has been modified by cratering and covered here and there by dunes. (255W 23S, E10-04203.)

the Mars geologist's Latin-derived term for a volcanic caldera, or volcanic collapsed crater.) The cone is about 200 km wide, comparable in size to West Virginia. Its slope is gentle, and it rises only a few hundred meters, but it is striking in the radial pattern of grooves or gullies running down its slopes. About 900 km away, on the southwest corner of the plain and the northeast rim of the Hellas basin, lies a second volcano, Tyrrhena Patera, of similar size and appearance. Its lavas ran to the southwest into the Hellas basin, paralleling the drainage of water in a nearby, 900-km-long

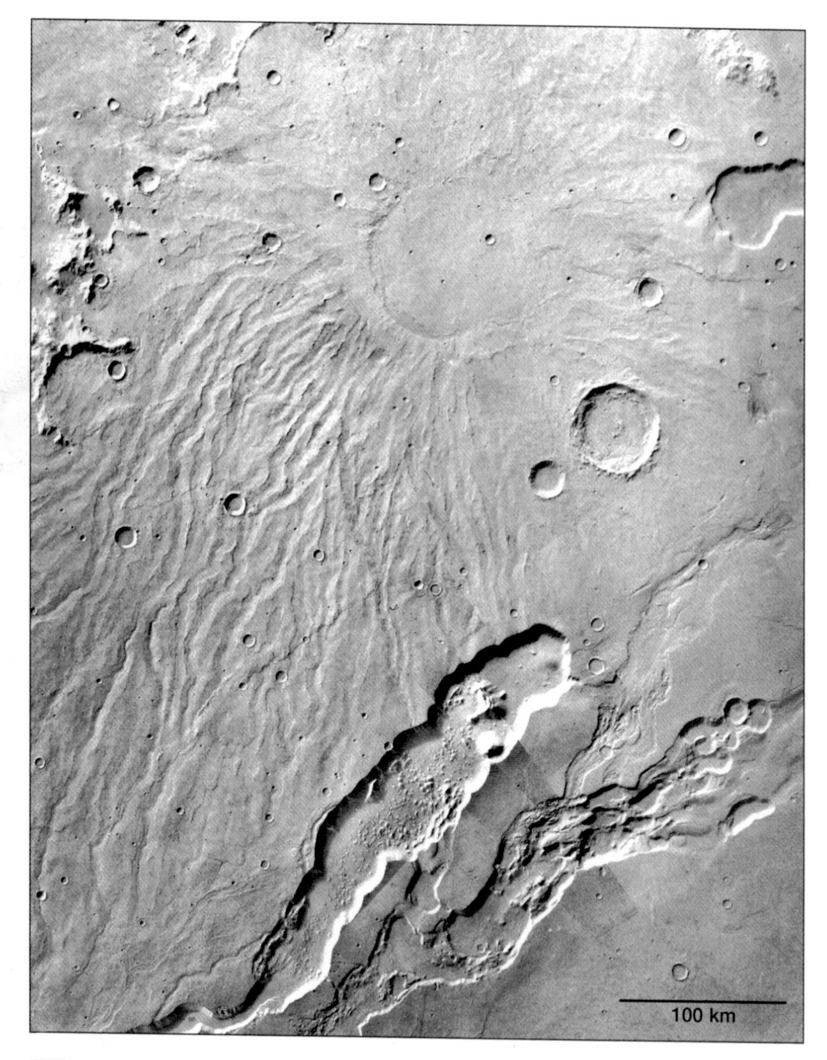

The radially furrowed volcanic cone of Hadriaca Patera resembles that of Tyrrhena Patera. At its southern base are collapsed regions from which issue the water-eroded channel of Reull Vallis. (267W 32S, Viking mosaic.)

outflow channel. Both volcanoes are dated as forming in mid- to late-Hesperian times, with some activity extending into the early Amazonian period. They were likely active around 2,500 to 3,000 MY ago and are among the oldest prominent volcanoes on Mars. They give mute testimony to spectacular ancient vistas in which fiery fountains of molten lava thundered in the thin Martian air.

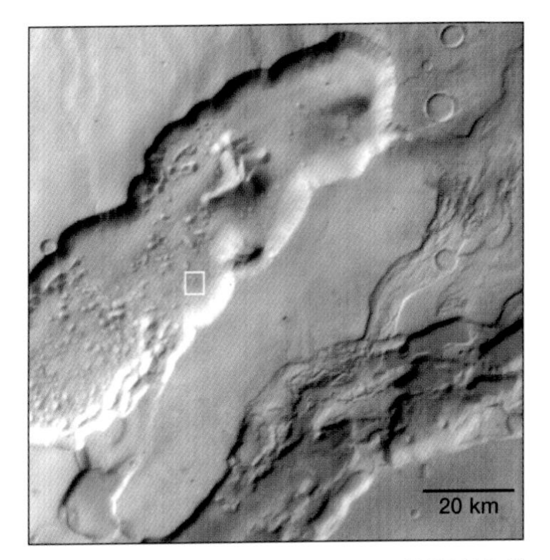

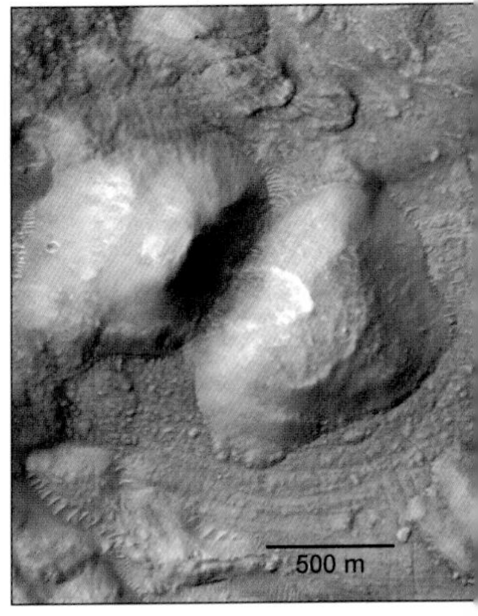

Close-up of the collapsed areas at the base of Hadriaca Patera shows water drainage channels (lower right) among the depressions. Box shows region of next image. (267W 34S, M01-02149.)

Above right: *The floor of the collapsed region at the source of Reull Vallis contains striations possibly associated with water flowing out of the area as the collapse occurred. Scattered hills have slopes where eroded debris overlaps onto the striated surface. (267W 34S, M01-02148.)*

Why did we encounter no volcanoes among the Noachian sites we visited? Did volcanism begin only in the Hesperian era? Hardly. Careful mapping indicates that there was abundant Noachian volcanism, probably tens of times more intense than later volcanism in terms of volume of lava produced per millennium. We don't see such prominent Noachian volcanic peaks because, first, the style of volcanism may have changed; Noachian volcanism in the presence of liquid water tended to produce cinders and ash that are easily eroded away. Second, Noachian volcanism occurred during the early, intense cratering of Mars, so that the volcanic landforms are torn up by overlapping craters. Third, if lava piles up to form a tall volcanic mountain, the weight of the mountain tends to make the crust of the planet sag, so that the mountain gradually sinks unless fresh lavas renew it; for this reason, the famous large volcanoes of Mars date from the most recent Amazonian era, as we'll see in part VI. A fourth reason is that Mars lacks enough geological energy to have much continental drift. Since the crust does not shift positions, a region located over a "hot spot" just keeps accumulating lava flows and the earliest features tend to be covered by more recent lavas.

RAIN ON TYRRHENA PATERA AND HADRIACA PATERA?

These two volcanoes are noteworthy because their slopes have been shaped not just by the piling up of volcanic overlapping lava flows, but by dramatic erosive gullies. The entire southwest slope of Hadriaca Patera is cut by these furrows. They look like they were cut by water—but is that possible? It's one thing to talk about underground sources of water erupting onto the surface, but how would water appear on slopes near the peak of a mountain? And was there enough water to cut into solid lava rock?

The area was studied by Arizona geologists Dave Crown and Ron Greeley, who used knowledge of terrestrial volcanoes to conclude that Hadriaca Patera must have been built not from hard lava rock but rather from another common volcanic product: ash. As mentioned above, water affects volcanism. When magmas ascend underground and encounter water or wet soils, steam forms and may blast the lava into the air as fine, molten particles, producing massive clouds of ash and cinders instead of a lava flow. The near-molten cinders that fall close to the volcano are loosely cemented by heat and weakly weld to each other. Additionally, fine grains of ash may be lofted by the wind, eventually falling to the ground at some distance to produce beds of weak, ashy soil. The upshot of all this is that the volcanic cone can thus be easily eroded, and in the case of Hadriaca Patera, the slopes were easily cut into furrows.

But what fluid caused the eroding? Once again, as with the valley networks, the geological patterns raise the specter of Martian rainfall. Could these features be proof of an earlier, very different climate? In this case, the answer is probably no, because rainfall near a volcano can be a local product of the eruption, not a manifestation of a different climate. The billowing steam clouds

The main constituents of gas coming from volcanic eruption on Earth are water vapor and carbon dioxide. This is believed to be true of Martian volcanism, too, thus explaining the CO_2 atmosphere and the apparent abundance of frozen water. (Photo of Kilauea volcano, Hawaii, by author.)

produced by the eruption through water-rich soils may condense and form water, which falls out as rain. This is commonly observed in terrestrial volcanoes.

A completely different hypothesis argues that the eroding agent was not water, but very fluid lava. Lavas of high temperature and certain compositions can be very fluid and flow long distances, and the eruption of such lavas from the mountain's summit caldera could have cut furrows on the sides. However, this idea is somewhat contradicted by the existence of another feature . . .

OUTFLOW CHANNELS

Nestled against the southeast base of Hadriaca Patera is a system of two parallel outflow channels known as Dao Vallis and Niger Vallis. Dao Vallis, the one in contact with the base of Hadriaca Patera, starts in a box canyon at its northeast end. This is a closed oval depression of collapsed terrain out of which water apparently flowed, cutting a narrower channel downstream, emptying into the Hellas basin. The close association of a collapsed terrain and outflowing water in the midst of several large volcanoes in ancient, high-latitude upland terrain strongly supports the idea that this is an area with soils rich in ice. The maps of ice depth, created by Stephen Squyres and others from studies of rampart craters, show the top of the ground ice to be unusually shallow in these regions, less than 100 meters deep.

A reconstruction of volcanism in the Hesperian era. A distant volcanic eruption is seen just after sunset, when the sun still illuminates the high volcanic plume from below the horizon. The atmosphere, somewhat thicker than it is today, may have had a bluer cast than the current Martian sky. Also, just after sunset, only the high, relatively dust-free atmosphere is illuminated by the sun, and like the high atmosphere of Earth at twilight, it scatters more blue light than the current Martian daytime sky. (Painting by author.)

Putting all these lines of evidence together, we can sketch a dramatic picture: The magma that rose up through this material encountered massive ice, formed steam clouds, and created local rain, which fell on the volcano slopes and cut the furrows. The geothermal heating associated with the magma melted the ice in ice-rich upland soils. The overlying landforms collapsed as the water flowed out, and the debris-laden waters erupted onto the surface, carving the channels of Dao Vallis and Niger Vallis. We will see these kinds of processes echoed at many locations in different areas of Mars.

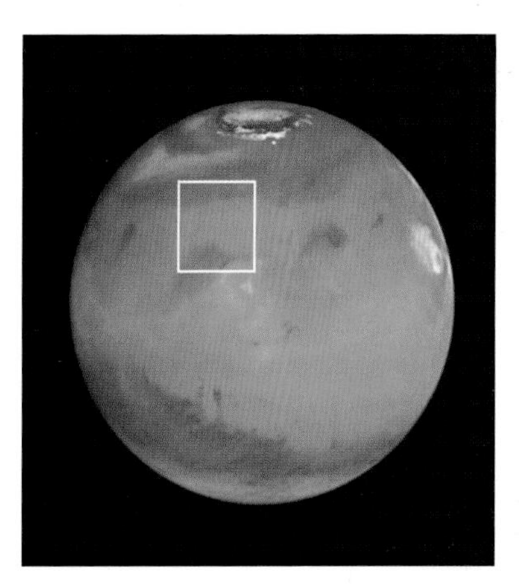

CHAPTER 19

UTOPIA PLANITIA

Viking 2 in the Northern Plains

NASA's 1976 Viking mission involved two duplicate spacecraft, Vikings 1 and 2, on the theory that if one failed, the other might make it to Mars. After Viking 1 landed successfully on Mars in July of 1976, Viking 2 waited in orbit around Mars. Passionate debates ensued about where to land the second probe. In the case of Viking 1, the research teams had picked the flattest, smoothest plain they could find—Chryse Planitia—to maximize the chances for a safe landing. Now they argued whether to use the same philosophy to maximize the chance for a second safe landing.

Two factions emerged. The engineering-oriented members of the team argued in favor of caution, to ensure a safe landing. After all, the main goal was to get the instruments on Mars. Even in the case of Viking 1, the cautious approach had yielded a flat plain strewn with dangerously large rocks, and if the "smooth" plain of Chryse had rocks big enough hang up a Viking lander, who knew what would happen at a more complex site like a volcano or a river channel? The most exciting place on Mars would yield no science if it produced a crash landing.

The exploration-oriented faction argued in favor of trying something different. Their feeling was that a second landing in a flat plain would merely duplicate Viking 1's science, not to mention disappoint the public. Shouldn't Viking 2 go to some radically

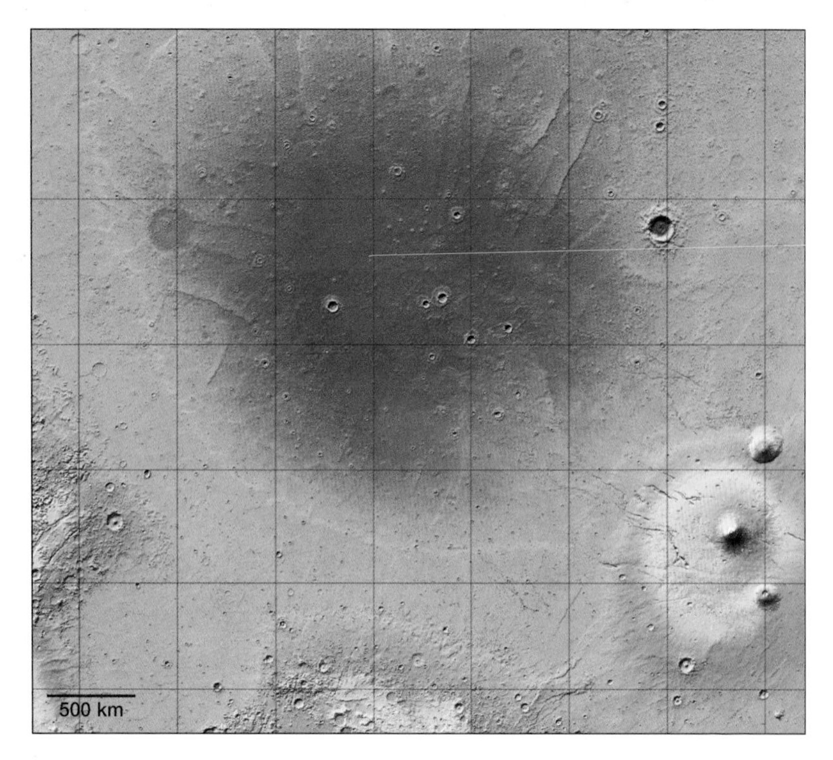

MGS laser altimeter map shows the low expansive plain of Utopia Planitia (blue) with bits of surrounding cratered uplands (green), and the high cone of Elysium Mons (red, lower right).

different topography of Mars in hopes of sampling different geology and perhaps an environment that might harbor life? What about landing on the biggest volcano in the solar system, Olympus Mons, or perhaps in the enormous canyon Valles Marineris?

In the end, prudence prevailed and Viking 2 was parachuted toward another bright, flat plain several thousand kilometers east-northeast of the Viking 1 site. Like Chryse, this plain had been mapped as dating from the Hesperian era. The original name, Utopia, was bestowed by Schiaparelli in his 1881–1882 map, taken from the title of Thomas More's famous sixteenth-century book that described an imaginary perfect society. More, in turn, had chosen the name from a Greek root meaning "nowhere"—the only likely location for a Utopia. Viking 2 landed in the Plains of Nowhere on September 3, 1976.

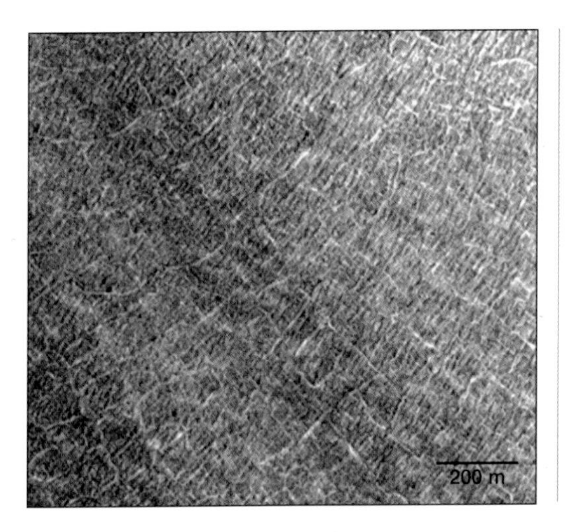

Viking orbiter view of patterned ground at 65° latitude, a possible sign of underground ice deposits. (8W 65N, M21-01650.)

Judging from the first lander photos, Utopia looked no different from the Chryse plains of Viking 1. Boulders of basaltic rock were scattered carelessly across a rolling landscape to the distant horizon. But further study showed that Nowhere was somewhere of its own. While Viking 1 landed at latitude 22° N, Viking 2 landed more than twice as far from the equator, at 48° N. The longitude was 226° W. Remember that a latitude of 48° is well up into the zone in which terrain softening starts and where photos show abundant signs of ice-soil interactions. Even before Viking 2 landed, orbiter photos revealed rampart craters and patterned ground—evidence for underground ice in Utopia and the surrounding plains.

THE DEPTH TO THE ICE

As discussed in chapter 7, rampart craters are deep enough to tap into buried ice-rich soils. Published studies of such craters suggest that depths to ice are less than 100 to 150 meters in Utopia and the surrounding northern plains.

The notion of shallow ice under the northern plains is supported by another line of evidence that we already mentioned in chapter 4. Not too far from Utopia, the dark northern plain of Acidalium has a type of basaltic lava that matches what would be formed if magma erupted through ground ice and interacted with the water.

The view from Viking 2. With antennae, color calibration charts, and an American flag on the foreground parts of the lander, the Viking 2 cameras looked out across a rocky plain similar to that at the Viking 1 site. (NASA.)

THE VIKING 2 LANDSCAPE

Study of the landscape photos and soil data showed other subtle differences from the Viking 1 site, in addition to the probable shallow ground ice. First, Viking 2 images showed faint, rivulet-like channels, only a meter or so wide, crossing the field of view. These might be related to the larger polygonal patterns that

Summer and winter in Utopia. The first view shows part of the flat, rocky landscape under normal conditions. The second view shows the same scene when morning frost covered much of the surface, throwing the basaltic rocks into sharp contrast with the bright deposit. (NASA Viking 2.)

could be seen from above—either fine-scale expression of patterned ground or possibly even some sort of small-scale water runoff channels (although liquid water running across the surface in such small-scale channels seems unlikely in present climatic conditions because of the high evaporation rate of water).

Second, the Viking 1 site appeared to feature more exposures of cemented soil called **duricrust.** At both sites, exhaust from the landing rockets blew away loose dust and exposed areas of duricrust underneath. The duricrust seemed weak, breaking up into platy clods when disturbed by the landing pads or the sampler arm that scooped up soil samples. It resembled the dried, adobe-like soils of the American Southwest. As at the Viking 1 site, the soils tested high in chlorine and sulfur, interpreted to indicate a high

A small dune deposit is seen in the mid-distance, beyond a trench dug by the Viking 2 soil sampling tool. Note that although the rocks look reddish brown, the color on many of them comes from a mantle of air-deposited dust, and some of them (as at upper left) can be seen to be dark gray under the dust, a color more typical of unweathered lava. The metal object is not an alien artifact, but a protective cover designed to pop off one of the cameras after landing! (NASA Viking 2.)

concentration of salts (such as ordinary table salt, sodium chloride) and sulfates. The weak "glue" that cements the soil into duricrust is thus probably the salts left behind in the soil as briny water evaporated. Duricrust existed at both landing sites, but the Viking 2 site had more striking examples. Third, there are fewer sand dunes in the Viking 2 pictures. The Viking 2 site is a rolling plain of rocks, and many rocks are perched on the surface, as if winds have eaten away at the sand beneath them. The rock textures resemble those of basaltic lavas on Earth, and many have vesicles, or small bubbles, characteristic of lavas that are charged with gas when they erupt. At the time of the Viking 2 landing, many Viking team members felt that the scattered rocks had been thrown into place as debris ejected from a large, young crater named Mie, located 200 km east of the lander. With a diameter of 90 km (56 miles), its impact formation would have involved a colossal explosion that threw abundant boulders hundreds of kilometers. However, we now know that three out of three of the first Martian landing sites are rock-littered plains, so that rock-strewn vistas may be a common product of Martian geology rather than a specific product of a nearby big crater. My own idea is that on surfaces of Hesperian age or older, repeated 10-meter-scale cratering has plowed the upper few meters of Martian ground, as has been observed directly on lunar surfaces of similar age. Then the Martian winds would have blown away the fine material, which would collect in sand dunes and sediment layers. As the fine dust and sand are removed, the boulders would remain behind. Places that aren't covered by dunes and sediments may commonly become rock-strewn landscapes.

VIKING 2 LOOKS FOR LIFE

The Viking 2 lander had the same life-detecting equipment as Viking 1, and it got the same results. The composition tests showed that the soils were sterile. The soil had almost the same basic composition as at the Viking 1 site, confirming the idea that the wind mixes dust from all parts of the planet. In hopes of finding soil shielded from sterilizing ultraviolet radiation—soil that might thus be able to harbor organisms—scientists programmed the Viking 2 sampler arm to scoop gravel from under a small rock. But even that soil was sterile. As with Viking 1, the Viking 2 biology experiments showed chemical reactions when nutrients were added, but the interpretation was the same: The reactions were probably nonbiological.

The bottom-line conclusion from Vikings 1 and 2 is that loose dust in the upper centimeters or meters of Mars have been blown around by wind and irradiated by the solar ultraviolet radiation, and that there is no sign of life in the surface material.

My Martian Chronicles

PART 8: YESTERDAY'S NEWS

On the day when Viking 1 landed, and again when Viking 2 landed, there was a feeding frenzy at the Jet Propulsion Lab newsroom in Pasadena. In those pre-Web days of 1976, newly arrived images from our Martian emissaries could not be instantly disseminated by e-mail. Rather, the best images were sent to the JPL darkroom, where hundreds of prints were developed as fast as possible and then handed out to clamoring reporters at press briefings.

Today, NASA's new images are spread around the world in real time by the electronic media, and each day's new pictures are posted on the Web for armchair explorers and e-surfers to enjoy in their living rooms. The feeding frenzies of the '70s, however, taught me a lesson about planetary exploration and the press that still applies today.

During Viking missions, reporters massed to get the first lander image. A few days later, however, the press lines shortened. The ABC, NBC, and CBS vans disappeared from their privileged parking spots in front of JPL. Even though Viking 1 and then Viking 2 were on Mars, producing new pictures and snooping around in the soil, the media lost interest. The first pictures had been published and everything was unfolding according to plan, and there was no new *story.* The irony—and this applies to almost every space mission (perhaps every major news event)—is that the first pictures, the ones that reporters lust after, are never the best or most spectacular. In fact, as the days pass, engineers gain better control over the cameras, and even the earliest pictures get cleaned up. More spectacular pictures are tried. Wide-angle vistas. Sunsets. Full color instead of black and white.

In this context, the Viking 1 story includes a wonderful example. Before the Viking 1 landing, Mars scientists around the world reasoned that because the low air pressure on Mars matched that at altitudes of about 37 km (120,000 feet) above Earth, the Martian sky would be a deep gray-blue. When the first Viking 1 landscape images arrived, mission controllers noticed the sky was unexpectedly bright, so they dutifully tinkered with their color adjustments until the sky was rendered blue. This was the first color picture handed out to the press—Mars with a blue sky! A day or two later, Viking scientists, among them the late Jim Pollock, noticed that more light was coming through the red filters than the blue filters, and they realized that the

The Martian sky. When the sun is seen through dust or haze, it is surrounded by a wide bright glow, but when it is seen though clear air, the sky coloration approaches much closer to the sun's disk. The former case is seen here, with a wide part of the Martian sky in the upper right washed out by light scattered from the fine airborne dust. The sun itself is out of the frame, to the upper right. (NASA Viking 2.)

normal Martian sky color has a pale, apricot-pinkish tone. The pictures were corrected, and scientists began to report their first results. But Viking had been on Mars for days, and the media had folded their tents and moved on. Although the first pictures—a dull black-and-white view of the footpad sitting on gravel, and the blue-sky picture—had been published on front pages around the world, most of the public never got to see the best color vistas of the Martian prairies. A week or so after the first landing, I asked one of the few reporters still hanging around JPL why they weren't getting the really good stuff out to the public. "You know it's fantastic and I know it's fantastic," he said, "but to the editors back home, it's yesterday's news."

During different space missions over the years, I've seen the media's short attention span operate again and again. It's a fundamental flaw in our society's conception of news. The problem is that science and exploration, by their very nature, need time to assemble results. A space-probe mission to another world begins to pay off with its most spectacular color imagery and new discoveries only in the weeks, months, and even years after arrival. During today's space missions, Web sites such as that of Malin Space Science Systems site on MGS (www.msss.com) present a steady stream of updated images. Nonetheless, TV networks, newspapers, and the public are conditioned to overvalue the first images and first results. Hapless scientists at mandatory press conferences an hour after "arrival" are prodded by rooms full of reporters, who clamor for an explanation as to what has been discovered before any appreciable data have accumulated. Fickle front pages soon return to the latest murder, fire, earthquake, or sex scandal, and the public is cheated once again.

I once heard Ray Bradbury say that media aren't reporting what matters, because what they report doesn't have to do with our future. We need more science journalists who understand that the real stories—the events that affect our future—unfold slowly. Scientists often don't know how to communicate these stories. We need journalists who have learned how to knock on university doors and ferret them out.

POST-VIKING MARS: DRY, LIFELESS, AND GEOLOGICALLY DEAD?

The final result of the two Martian landers of 1976 was a dramatic swing of the scientific pendulum away from previous optimism about biology and habitable conditions on Mars. Such a paradigm shift is never just an arbitrary scientific fad, as might be suggested by superficial newspaper coverage. It's forced by new measurements. From 1972 to 1976, post–Mariner 9 Mars had been an enigma where cameras had spotted dry river valleys, and where future lander cameras might even show strange alien plant life; but post-Viking Mars was a planet where ground truth showed only sterile volcanic soil bathed in life-damaging ultraviolet radiation. After Viking, many Mars scientists assumed that virtually all geologic activity had been concentrated at the beginning of Martian history and that most Martian volcanoes and rivers had been inactive for billions of years. One prominent model proposed a scenario in which volcanism had essentially stopped by 2,500 MY ago. In contrast, my own crater-count studies suggested that some relatively crater-free lava flows were no older than 200 to 300 MY, but not everyone agreed. The pervading sense of a relatively dead Mars persisted for two decades, until the '90s, when new data altered the paradigm once again, and, as we will soon see, a modern, more intriguing picture of Mars began to emerge.

CHAPTER 20

MARTIAN MOUNTAINS

A Bit of Continental Drift?

A planetary rule of thumb, going back to Percival Lowell, says that the larger the planet, the longer it retains its internal heat and the more geologically active it is. Our world has lots of internal heat and energy that drives movements of Earth's crust. Earth's megafeatures are controlled by large-scale movements of the crust, called **plate tectonics**, or, more popularly, **continental drift**. Continent-size regions of the crust, both on land and under the sea, move relative to each other. Sometimes they collide, scrunching the edges of the plates into massive folds and contortions—mountain ranges. The Rockies and Andes are examples. The Himalayas may be the best example, where the Indian plate ponderously crashed into southern Asia 10 to 50 MY ago.

Mars and the Moon, both smaller than Earth, lack this fundamental type of terrestrial geology. Their topography is dominated by ancient impacts and lava plains that fill lowlands. The main mountains are curved rims of giant basins and craters, and from this we infer that Mars and the Moon have almost no plate tectonics.

There is one possible exception on Mars—a mass of mountains that jumps out of the laser-altimeter topographic map as being unconnected to any major impact feature. It is sometimes called the Thaumasia mountain range, west of the great waterway and bounding the southeast side of the Tharsis volcanic province.

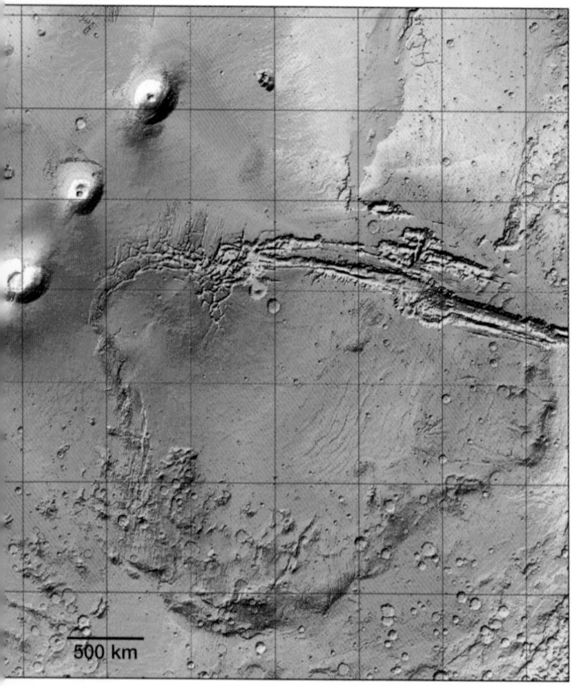

An arc-shaped mountain range on Mars (bottom, brown) rises above the high volcanic plains of Tharsis (brown, upper left) and Solis Lacus (red, bottom center). Three of the huge volcanoes of Tharsis are at upper left (white), and the Valles Marineris canyon fractures are at upper right (blue). The mountain range's form resembles so-called island-arc forms, such as the Himalayas and Japan, caused by continental drift on Earth. (MGS laser altimeter map.)

While most investigators agree that Mars has not had very significant plate tectonics, some believe that this mountain mass may be an indicator of short-lived incipient tectonic activity at a global or plate-size scale. James Dohm and his colleagues at the University of Arizona have studied the history of the whole Tharsis region and interpret the building of these mountains as a tectonic response to the forces that produced the long-lived volcanic center there. Perhaps it is significant that these mountains lie just southwest of the great canyon system of Valles Marineris, which bears many similarities in size and structure to certain rift valleys on Earth, such as the Gulf of California, a fracture caused by terrestrial plate-tectonic motions.

Mars falls somewhere between the Moon and Earth on a tectonic-activity scale, but the bottom line remains the same: It is a planet without full-fledged plate tectonics or continental drift, although it did have localized tectonic stresses and some limited mountain-building activity.

CHAPTER 21

ARES VALLIS

A Riverbed on Arid Mars

"There must be rivers on Mars," wrote British astronomer R. A. Proctor in 1871. That was the high-water mark, so to speak, of belief in an Earth-like Mars. "The mere existence of continents and oceans on Mars proves the action of [Earth-like] forces," Proctor wrote. "There must be mountains and hills, valleys and ravines, watersheds and watercourses."

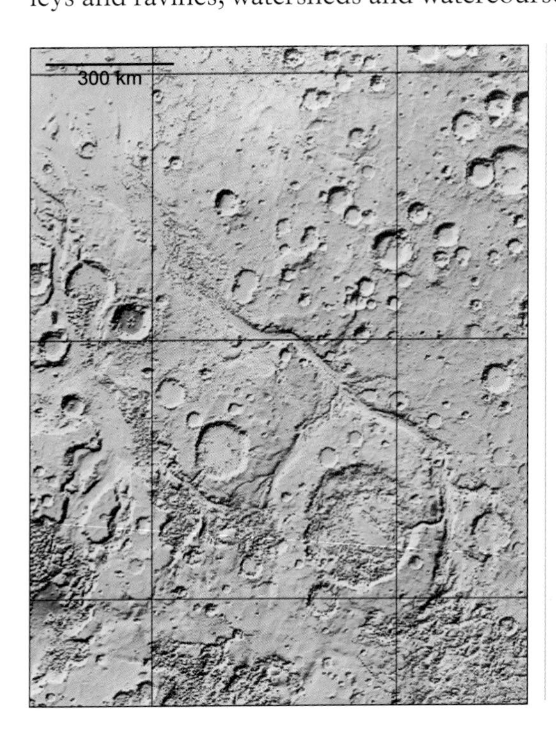

Ares Vallis (center and lower right) is a large outflow channel, or riverbed, seen here in MGS laser altimeter map. Its origin lies in collapsed, or chaotic, terrain in lower right corner. Tributaries can be seen flowing into the main riverbed. The large circular crater with the collapsed floor is Aram Chaos, discussed in the next chapter. (MGS laser altimeter team.)

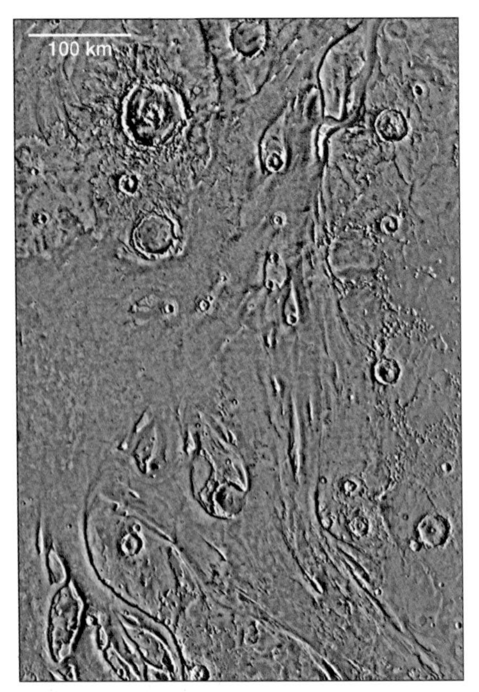

Mouth of Ares Vallis is a region where the outflow channel system (bottom) emptied onto a plain. Craters and other obstacles diverted the water flow, creating beautifully formed, teardrop-shaped islands (center and upper center). The central part of this area can be seen in upper left corner of preceding radar image. (Viking mosaic, NASA.)

Above right: *The main channel of Ares Vallis shows terraced banks. The terraces may represent layers in the surrounding landforms, or could represent shorelines eroded by the river as water flow declined to different depths. (Mars Odyssey Themis camera thermal infrared image, where bright regions are warm [heated by sun] and dark regions are cold; 23W 9N, I01-24901I.)*

Proctor was on the right track more than he could know, but just three generations later scientific opinion had shifted dramatically toward a dry Mars. "Spectrographic results . . . agree in pointing to a much higher degree of dryness . . . than in the most arid terrestrial deserts," concluded the French astronomer Gerard de Vaucouleurs in a 1954 book. "This extreme desiccation is in excellent agreement with the whole body of observations." De Vaucouleurs's generation marked the low tide of belief in any fluvial activity on Mars. But barely one generaton after that came the stunning surprise that this dry, frozen planet—which cannot now sustain liquid water—is laced with ancient riverbeds. De Vaucouleurs was right about contemporary Mars. Proctor was right about its past.

Gigantic ancient riverbed systems, or channels, snake for hundreds of kilometers over the Martian landscape. Many formed in massive floods characteristic of Hesperian Mars, possibly in a matter of weeks or months. Although the first Martian rivers flowed on Noachian Mars, most riverbeds appear to have witnessed their final major floodwaters during the Hesperian transition from wet Mars to dry Mars.

Ares Vallis is a good example. The floor of the channel itself is mapped as late Hesperian in age, it cuts through Noachian and early Hesperian plains, and we can't tell the date of the earliest flow. It curves about 1,600 km (1,000 miles) from a Hesperian-age region of collapsed terrain on the south to its mouth on the north, where it empties onto a plain covered by early Amazonian lavas. This is comparable to the length of the Missouri and Mississippi Rivers from Minnesota to New Orleans.

Ares Vallis and the other large Martian riverbeds, or **outflow channels** as they are called, are different from the valley network systems discussed in chapter 7. Outflow channels are larger and deeper, and involve larger amounts of water. Many originated in restricted, collapsed areas and have few tributaries. Water apparently gushed from these areas as they collapsed. The collapsed source areas are often Hesperian in age.

Ares Vallis is such a channel. Its name, granted in 1973, a year after its discovery via Mariner 9 photos, comes from the Greek name for Mars. This naming follows a decision of the International Astronomical Union to name all the major Martian riverbeds after names for Mars in various languages.

ARE CHANNELS REALLY RIVERBEDS?

After the outflow channels were discovered, an argument broke out among Martian researchers. Were these features really carved by water? What about some other fluid? Could lava have done the job, or liquid carbon dioxide? Or horrific winds, whistling through dusty canyons? One by one, the alternatives were dismissed. Lava creates modest channels but tends to solidify as it flows downstream, so that lava channels often dwindle to nothing, whereas the Martian channels get wider and deeper downstream, matching the properties of large terrestrial rivers. Carbon dioxide is plentiful on Mars, comprising 96 percent of Mars's scanty atmosphere, but its liquid form requires five times the pressure of air found on Earth's surface, so it is not a good candidate to flow hundreds of kilometers under the extremely thin air of Mars. Wind has never been seen to carve the kinds of valleys found on Mars. Many Martian channels have tributary systems, streamlined islands, and other features of earthly river channels, especially those associated with flood systems. The same features can be seen in terrestrial water-flow systems, from the Mississippi to the dribble of water flow from your garden hose across a patch of bare dirt. Most researchers finally agreed that most Martian channels must have come from outflows of water.

SLOW FLOW OR SUDDEN FLOOD?

Did the Hesperian rivers flow constantly for millions of years, or was each river a one-shot thing? Or were they sporadic, coming and going as Hesperian Mars wavered back and forth between a clement climate and an arctic climate? These are difficult questions, but a Hesperian Martian river probably did not look like a bucolic vista along the Mississippi. Application of terrestrial hydrologic theories suggests that many of the large Martian channels were carved not during millennia of leisurely flow, but during catastrophic, sporadic flood events.

Ares Vallis is one of the most dramatic of the river channels and a good example of how the above interpretations are made. By terrestrial standards, Ares Vallis represents what must have been an enormous river. It not only winds its way 1,600 km among ancient craters but is also part of a larger system of channels that are fed both by the great waterway on the south (chapter 10) and by outflow from the enormous canyon complex of Valles Marineris on the west. Typical sections of the main channel of Ares Vallis are 2.5 km (1.5 miles) wide or more, with a depth of 100 meters (100 yards) to as much as a kilometer (1,000 yards).

If you filled such a channel with water and allowed for the slope, how fast would water flow out the lower end? Terrestrial hydrologists have formulas to estimate very rough answers, allowing for the lower gravity that pulls the water downhill. Arizona hydrologist Vic Baker, an expert on both terrestrial and Martian channels, made such calculations in his 1982 book on Martian channels; he concluded that Ares Vallis, in its heyday, disgorged 2 million to 70 million cubic meters of water per second. This is about 50 to 2,000 times the flow rate of the Mississippi! Michael Carr, the U.S. geological scientist who is one of the leading experts on Martian fluvial history, cites similar figures in his 1996 book, *Water on Mars.* Carr estimates the total volume of material removed from the source area as 2,000 cubic kilometers, and estimates that if the river typically ran 100 to 200 meters deep, the flood would have lasted on the order of 9 to 50 days. The figures are why Mars researchers believe that Ares Vallis and many other outflow channels were not million-year rivers, but more like catastrophic floods.

The estimated Martian flow rate is not totally beyond our own terrestrial experience. In the American Northwest, a flood channel of similar size carved its way through basaltic lava rocks in several catastrophic floods between 13,000 and 30,000 years ago. What happened was that glaciers during the last Ice Age dammed large lakes in Idaho and Montana. At the end of the Ice Age, the glacial ice melted, and the dam got thinner until the lake water broke through, pouring enormous floods across Idaho and Washington, like a man-made dam bursting upon a hapless village. As pointed out by Baker, Carr, and other Mars geologists, the resulting channels strongly resemble the channels of Mars.

Where did the Martian water come from? The Martian floods probably originated not in surface lakes but in underground ice. We've seen evidence that Noachian water sank into the soils and froze. Later, in certain regions, large masses of ground ice must have melted. The weight of the overlying rock and soil squeezed the water onto the surface, and the rock layers broke up and collapsed into the underground.

Fine, but *why* did the ice melt? Geologists who argue for short-lived, catastrophic floods require some mechanism to melt large volumes of ice rapidly. Before we tackle the cause of the floods in the next chapter, let's pause to look further at properties of the channels themselves and the water that flowed in them.

KEEPING WATER IN MARTIAN RIVERS

Why didn't the water freeze or evaporate in the harsh, dry Martian environment? Would a river freeze promptly when the temperature dropped below 0° C (32° F)? The answers require consideration of the behavior of water on Mars as well as careful geologic detective work on each Martian riverbed. Two environmental conditions must be met to produce and maintain liquid water: temperatures above freezing and pressures above a certain minimum. Let's consider temperature first.

Pure water freezes at 0° C (32° F), but salty brines stay liquid down to lower temperatures. This is why cities spread salt on icy roads. The saltiest brines freeze at roughly the point at which the Centigrade and Fahrenheit scales happen to be equal, –40° C (–40° F), or even –50° C (–58° F) for the right kinds of salty mixtures. Note that the equality at –40° C is a happy coincidence that helps you to remember the freezing point of brines. That temperature defines an important benchmark temperature for Martian water. If Mars is warmer than –40° C, then salty water could stay liquid, assuming adequate air pressure. This suggestion is not new. Scientists such as the Swedish chemist Svante Arrhenius (1859–1927) suggested as early as a century ago that

Martian water might be salty and therefore could avoid freezing even at Martian temperatures. The soils at the Viking landing sites already gave us evidence that Martian water was salty and that it contained ordinary salt (sodium chloride, NaCl), as well as other chlorides, sulfates, and perhaps carbonates.

Martian surface soil temperatures exceed the briny –40° C freezing point on warm afternoons. The typical mean daytime surface soil temperature at Martian mid-latitudes rises to as much as 10° C (50° F) in southern summer afternoons, though it might barely reach –40° C in winter. The upshot is that, from a temperature point of view, salty rivers would have been plausible on an early Hesperian Mars, when traces of an early, thick atmosphere may have helped maintain slightly higher Martian temperatures than exist today.

The second condition to allow liquid water is adequate pressure. If the pressure is too low, water won't be stable in a liquid state, as I've remarked in earlier chapters. The critical pressure for water is 6.1 millibars. (Remember that air pressure at sea level on Earth is defined as 1 bar, or 1,000 millibars.) If the pressure is below that value, ice will not melt into liquid water as the temperature increases, but rather ice will transform directly into gas, a process chemists call **sublimation.** We are familiar with sublimation from the example of dry ice (frozen carbon dioxide), which transforms directly from ice to a vapor without ever producing a puddle of liquid.

By remarkable coincidence, the average surface pressure at an average elevation on Mars today hovers around this critical value of 6 millibars! As on Earth, the air pressure on Mars is lower on mountaintops and higher in the low regions. So from the pressure point of view, salty water could be stable even today during summer in the low plains, deeper canyons, and crater floors. It would not be stable in the uplands or on high volcanic slopes. If it were suddenly exposed in those areas, it would bubble away into vapor. Even in the lowest areas, like the Hellas or Utopia basins, "stable" does not mean the water would last very long. The air pressure rarely gets above 10 to 11 millibars in those areas, as shown by analyses with MGS data in 2001 by NASA researcher Robert Haberle and his colleagues. Even at those pressures, water would evaporate much more rapidly than it would in a desert on Earth.

The real question, then, is about the timescales: If a massive melting of underground ice dumped salty water onto the surface of present Mars or past Mars, how long would it last before it evaporated or froze? Could it run long distances in a river?

Keeping water in a liquid state is still a challenge, even if it is a brine and even if the pressure and daytime temperature are right. At night, typical temperatures drop to roughly –100° C (–47° F), well below the freezing point of brines. Thus, a Martian river is likely to have developed a surface layer of ice within a few hours, like a Canadian arctic river in winter. Actually, the ice would have helped to maintain the river, because the blanket of solid ice would protect the underlying liquid water from evaporation.

Nonetheless, a river on Mars would be a short-lived, unstable affair as long as the air pressure was at today's near-vacuum level. The key, then, may be the higher air pressure on early Mars. We've already seen that it may have been anywhere from 100 millibars to 1,000 millibars. These higher pressures, working with the ice blanket, would have slowed evaporation and helped sustain early rivers.

To summarize, the water that surged through Ares Vallis and many other channel systems was probably a surging flood that cut deeply into the old, fragmented and cratered soils. Its main surge may have lasted less than a year, although lesser flows may have dribbled out sporadically over longer periods of time. The river itself was probably capped by ice or choked with ice chunks during much of its life. The water likely had a high salt content and flowed under cloudy carbon dioxide skies with higher air pressure.

EXPLORING ARES VALLIS ON THE GROUND

Near the mouth of Ares Vallis, the rushing water cut into the plains, swerving around obstacles such as mountains and crater rims in its path. The water carved the sides of these obstacles,

The area where the Pathfinder probe landed in the mouth of Ares Vallis was a smooth, cratered plain, but still had more topographic features near the lander (circle) than at the Viking 1 and 2 sites. Names indicate some features seen on the horizon in various Pathfinder surface photos. Twin Peaks in particular, a kilometer (1,000 yards) WSW of the lander, was prominently shown in many lander photos. (NASA.)

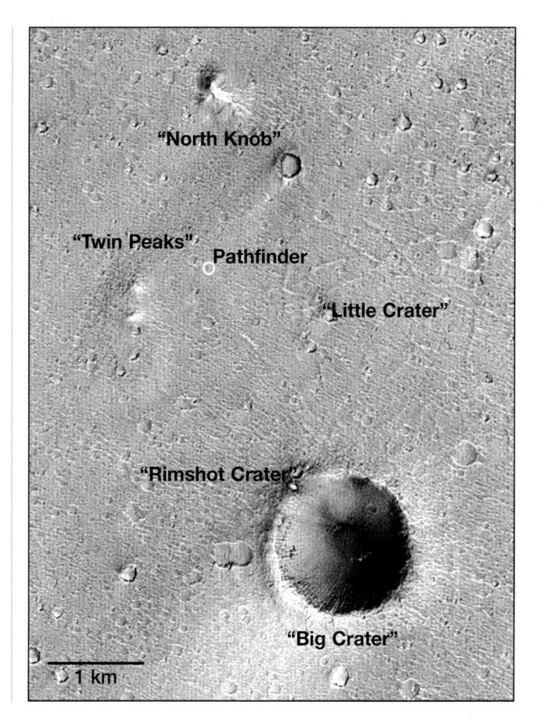

Twin Peaks, as seen across the rock-strewn landscape from the Pathfinder lander. Bright patches on the flank of the larger hill appear to be dust drifts. (NASA.)

Sand atop mountains. Above left: *High-resolution view of bright markings on one of the Twin Peaks. The material is probably drifts of sand piled up by the wind. (Super-resolution view courtesy Peter and Sarah Smith, Lunar and Planetary Lab, University of Arizona.)* Above right: *Similar drifts of sand high on mountain slopes in the Gran Desierto, NW Sonora, Mexico; aerial photo by author.)*

creating beautifully sculpted, teardrop-shaped islands. The broad, round end of each island is upstream; the downstream end of the island is drawn out into a streamlined tail. Each island looks like a giant wind tunnel experiment. The islands are dotted far into the desert plains, as much as 300 km (190 miles) beyond the ill-defined mouth of the river—proof that a massive flood emptied out of the channel's mouth, eroding everything in its path.

We've actually seen the mouth of Ares Vallis from the ground, thanks to Earth's third successful robotic ambassador to Mars. When the Pathfinder probe landed there in 1997, it took beautiful color photos of the surroundings and sent out the small, plucky rover, Sojourner, that roamed from rock to rock, measuring compositions. Remarkably similar to the Vikings 1 and 2 landscapes, the vista at the mouth of Ares Vallis is another stark landscape of tumbled rocks and windblown dunes. As at the Viking sites, soils are salty. Unlike the Viking site, the vista includes nearby hills on the horizon, banked up with dunes, similar to formations in windblown terrestrial deserts.

The landscape held teasing clues about the Ares Vallis flood. Examination of images in stereo pairs gave a view of the landscape in 3-D, revealing shallow grooves or ditches running parallel to the flood direction. These may be remnants of the original channels that

An example of rocks piled against each other (above), interpreted by some Pathfinder scientists as a configuration typical of debris left by a flood. (NASA.)

Below: *Rocks and sand. Extreme processing of the Mars Pathfinder images brings out varied tonalities of different materials. The weathered dust and sand are the reddest materials, although drifts of fine gray dust are concentrated here and there by the wind. The rocks generally have a grayer color, typical of basaltic lavas. (NASA.)*

were scoured out by the flood. Most of the Pathfinder science team concluded that certain groups of rocks show an imbricated structure, in which slabby rocks lean against one another; it is a form typical on Earth when rocks have been tumbled into place by chaotic floods. Team leader Mike Golombek and others concluded that the rocks seen at this site are lying just where they were

Image from the crude camera on the Sojourner rover shows the Pathfinder lander atop the deflated, balloon-like gasbags that cushioned the probe's landing on Mars. (NASA.)

A view from Sojourner's camera as it approached a pitted rock informally dubbed Half Dome. (NASA.)

deposited by the floodwaters, perhaps 3,000 or 3,500 MY ago. My own calculations, however, suggest that during such a long period of time, meteorite impacts would have tilled the surface like a drunken plowman, churning things to a depth of a meter or so, in which case most rocks would have been disturbed by impact effects. We haven't resolved this discrepancy. Perhaps the imbricated rock formations left by the flood were covered half the time by drifting dunes, protecting them from impact-gardening. Perhaps also, by the luck of statistics, here and there among the scattered rocky debris, clusters of rocks were spared from impact battering, and survived in their original flood-deposited configurations.

The Sojourner rover explored the landscape near the Pathfinder lander. It nuzzled against various rocks with a measuring device that was able to determine the composition of the rocks. (NASA photo from the Pathfinder lander.)

THE MYSTERY OF THE UPLANDS

Investigators carefully studied not only the rock composition measurements from the Sojourner rover, but also the detailed photos of the various rocks, to look for clues about the rock structure and origin. There was a clever strategy at work here. The Pathfinder landing site had been chosen at the mouth of Ares Vallis not only to study the channel itself, but also because (in theory) the deposits at the mouth should contain rock fragments eroded from hundreds of kilometers upstream, in the uplands, and carried downstream by the surging floods. Thus, even though Pathfinder landed in a plain the way Vikings 1 and 2 did, it would give us our first sample of upland materials.

The results of this plan were ambiguous. On the one hand, some investigators felt they were seeing unusual rocks at the Pathfinder site. One group of Pathfinder image analysts concluded from photos that one of the boulders near Pathfinder was not an igneous or lava rock, but a conglomerate made of pebbles cemented together—which would likely be a sign of sedimentary rocks upstream in the highlands. Also, some studies of Sojourner's composition data led to an idea that the ancient uplands might be rich in andesites, in contrast to the younger basaltic lowlands. However, other groups criticize the conglomerate interpretation as unconvincing—generally, it's hard to tell a rock by its appearance in a photo, and one person's conglomerate is another's lumpy basalt—and many workers downplay the significance of the andesitic tendencies.

The basic conclusion to draw from all this is that exploration of the red planet is still in its infancy. At the time of this writing, we've landed on three different plains of Mars but the ancient uplands are still unvisited and still mysterious in terms of rock types, geologic diversity, origin of layered sediments, numbers of lake-bed deposits, and abundance of underground ice.

CHAPTER 22

ARAM CHAOS

Melting the Ground Ice

Now let's return to the question of why the underground ice melted in some areas. As mentioned previously, the water in many major outflow channels, such as Ares Vallis, drained from huge tracts of collapsed landforms known officially as chaotic terrain. These areas hold the key to at least one mechanism for flowing water on Mars. One area of special interest is a circular region, partially collapsed, called Aram Chaos.

In 2000, only a few months after announcing the discovery of a hematite deposit near zero latitude and zero longitude, the MGS spectrometer team announced a second, smaller deposit about 900 km to the west. While the argument raged about whether the first site had to do with ancient water, the evidence was clearer in the second case.

The name Aram Chaos was an ancient term for Syria and was originally assigned to a small bright area of Mars by the French observer Antoniadi in 1929. Later, topographic mappers assigned it to the region of chaotic terrain. Closer examination showed that the chaotic terrain actually fills a huge ancient crater, about 300 km across. As a result, it's convenient to refer to the crater itself as Aram, although craters are named officially after deceased scientists.

The 300-km-diameter crater containing Aram Chaos (bottom center) is framed on the E and N by the dry river channel of Ares Vallis. Comparison with the laser topographic map on page 219 shows that the lowest, collapsed part of Aram, as well as the low floor of Ares Vallis, tend to be dark. The NW part of Aram's floor, along with the light triangular area near the center of Aram, are uncollapsed higher parts of the original sediments that filled the crater. Between Aram and Ares Vallis is a narrow E-W gorge where water emptied from the crater into the river channel. (Mars Global Surveyor Atlas, Malin Space Science Systems.)

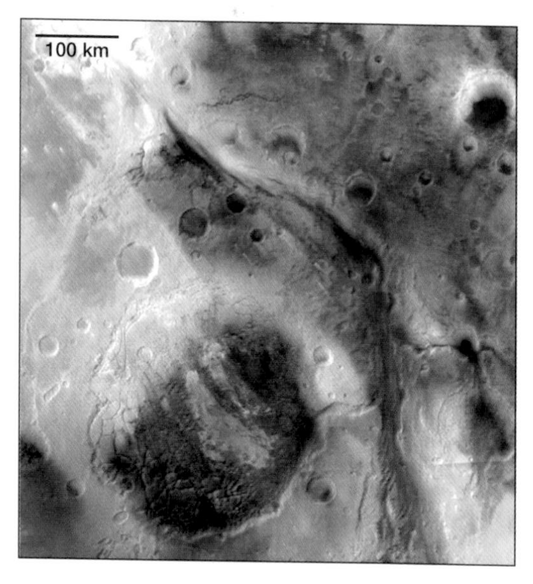

THE GREAT COLLAPSE

The history of Aram can be read in the photos. In Noachian times, the floor of the crater, and many nearby craters as well, was filled in and covered with sediments, hundreds of meters deep. When the process was done, the craters had smooth, shallow floors. Parts of that filled-in surface are still visible and are mapped as mid- to upper Noachian plains. The sediments must not have been dry dust, but wet soil, such as lake-floor sediments, saturated with water that later froze into massive ground ice deposits.

A profound change happened during the Hesperian era: the ice in those sediments melted. The overlying sediments sagged and broke into blocks separated by water-filled fissures. Water was trapped in Aram crater and accumulated as the ground subsided, ponding in the lower, eastern part of the crater's interior. As more water seeped out of the ground, more collapse occurred. Many craters near Aram show the same effects, so the Hesperian melting must have been widespread in this area.

The fractured landscapes inside the Aram crater. Regional context view shows heavily fractured sedimentary fill (dark material, lower left) with smoother, brighter units of uncertain nature. Box shows location of next image (22W 2N, M19-00255.)

Telephoto view of the fractured dark material. Many of the fractures may have once drained water out of the system, but their floors are now covered with sand dune deposits. (22W 2N, M19-00254.)

What caused it is one of the great Martian mysteries. Did it occur during a temporary period of warm climatic conditions? Was ancient Mars cooling due to a loss of the greenhouse effect as the air thinned, or was it warming due to an increase in solar luminosity? Or were there irregular fluctuations in climate, so that the warm periods were merely "bumps" on a wiggly curve of rising or falling temperature? We must recall that Earth's history is laced with much greater fluctuations, such as glacial epochs that covered the northern United States and Europe with ice on timescales of tens of thousands of years. The influences that drive these long-term climate fluctuations are poorly understood. In Earth's case, they include cyclic changes in the planet's orbit, massive emissions of carbon dioxide and haze during the largest volcanic eruptions, and changes in ocean currents due to movement of continental landmasses during continental drift. On Mars, we know from calculations that the axial tilt varies. With the pole tipped more toward the sun in some eras, summer heating would have been stronger and could have transformed all the polar ice at the summer pole into vapor, thus adding to greenhouse effects. Or episodes of CO_2 enhancement from volcanic eruptions might have created short-lived periods of global warming. Who knows what long-term heating variations acted on Mars's underground ice? Researchers are still pursuing these questions. Answers may shed light on climate stability and life's variability on planets throughout the universe.

What we do know is that not all sediment-filled craters collapsed equally, which suggests the possibility of regional, not global, heating episodes. Regional heating of the underground layers could come from underground geothermal fluctuations (for example, those associated with movements of magma in the crust). As mentioned earlier, calculations based on total radioactivity content of rocks show clearly that the heat flow on ancient Mars was greater than on current Mars (because radioactive elements gradually decay and become depleted). Underground heat flow from inside Mars thus declined from the Noachian into the Hesperian era, and surely this decline saw variations and temporary surges of geothermal heating due to local volcanism.

Regardless of whether heat came from above or below, a temperature increase of only a few tens of degrees could heat the underground ice layers from their mean temperature, around –70° C, up to the magic range of –40° C to 0° C, transforming the once-solid ice into liquid, briny water. The water broke out and left a tumbled-down chaos of low valleys and mesa-like remnants of old, overlying surface.

Some of this activity could have been explosive. As pointed out by Australian researcher Nick Hoffman, the underground pore spaces in the soil would be filled by carbon dioxide on Mars, because that is the main gas in the atmosphere. At a depth of a few hundred meters, if the soils were sealed by overlying ice, the gas pressure would build to above five bars (five times Earth's sea-level atmosphere pressure), due to the weight of the overlying soils, and the CO_2 in the soil's pore spaces would be liquefied. To put it in other words, the sediment masses in craters such as Aram, sealed by frozen ice in the upper layers, may have contained high-pressure liquid CO_2. Thus, if local water ice melted and the overhead layers of soil suddenly cracked and collapsed into the water, massive amounts of underground CO_2 fluid could be exposed at a pressure of hundreds of times the local air pressure. As Hoffman has theorized, an explosive release of this underground fluid could drive water out of the ground in violent eruptions, perhaps explaining the catastrophic floodlike volumes.

Hoffman goes a step further and sees the liquid CO_2 itself as the fluid that carved gullies in chaotic terrain, and even the major outflow channels themselves. But liquid CO_2 is so unstable that I would favor an alternate scenario in which the pressurized liquid merely aided in blowing out the subsurface water, enhancing the process of rapid collapse and aiding the formation of chaotic terrain.

A GORGE LEADING EAST

As the floor of Aram crater disintegrated, the water accumulated in the low areas against the east wall until it broke through a low spot in the eastern rim of the crater. Torrents of water poured through this "break in the dike," cutting it deeper and deeper. This crater rim material seems to have been easy to erode, probably because it was mere rubble blasted out of the crater itself. The surging water soon carved a dramatic gorge, 80 km (50 miles) long and 13 km (8 miles)

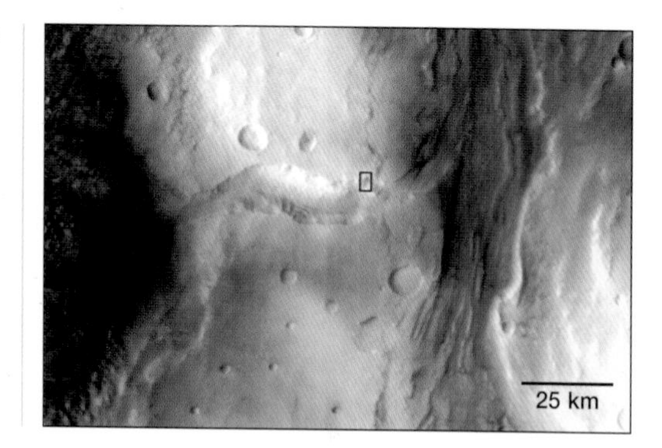

The gorge eroded by the drainage of water from Aram Chaos (out of frame, left) into Ares Vallis (right center). Box gives location of close-up of the canyon wall. (Mosaic of MGS M07-01690 and M07-00263.)

wide, clear through the wall of Aram. The water pouring out of the crater flowed directly into the riverbed of Ares Vallis, which lies immediately east of the crater rim. The gorge is lined by magnificent cliffs. We can still see the fluvial striations on the floor as they swerve left into Ares Vallis, to flow north, downstream, and eventually empty onto the plain where Pathfinder landed in 1997.

HEMATITE DEPOSITS IN ARAM

The hematite deposits, which called our attention to this area in the first place, are smaller and patchier than those of the Terra Meridiani site. They exist mostly in the east half of the crater, where the water ponded, supporting the idea of hematite's connection to aqueous activity. It's difficult to correlate the hematite patches directly with the chaotic topography in Aram, but as pointed out by spectrometer team analyst Melissa Lane, they may be weathering out of the sides of the blocks of ancient sediments. The overall association of hematite with the water-rich Aram sediments reinforces the idea that water may have been the agent that transported the dissolved iron into the crater in the first place, leaving concentrations of iron-rich minerals as it evaporated.

The labyrinthine chaos of badlands in Aram crater is one of those special Mars sites that may, like Earth's Grand Canyon, expose keys to the stratigraphic record and geologic processes of the planet's past. And the great gorge, leading out to the chaos to the east, is one of the erosional wonders of the planet.

A mighty cliff cut by running water. This cliff is the north wall of the gorge shown in the preceding image, through which water drained out of Aram Chaos. Bedded layers (lava? wind-borne soil layers? water-deposited strata?) are clear partway down from the top of the cliff. The floor of the gorge is covered by hilly debris and dunes at the foot of the cliff. (18W 3N, M07-01689.)

My Martian Chronicles

PART 9: WATER RELEASE

The very nature of chaotic terrain makes it hard to interpret photos of the landscape. Instead of geometric simplicity—a circular crater, a long canyon, a conical volcano—we have the almost fractal intricacy of ground masses broken into ever smaller plateaus, fragments, and wedge-shaped blocks.

One day, while looking through frustratingly messy images of Aram Chaos for this book, I chanced on an important feature in MGS frame 22-00311, taken in 2000. The larger-scale context frame of this region was especially vague, showing mottled surfaces, some bright and some dark, with complex fractures at the north and northeast. The telephoto view shows what seems to be a broken trough cutting between smoother remnant plateaus of Noachian sediments on the north and south.

For some reason, my eye caught on a distinctive triangular block, or plateau, on the southwest side of the trough. On its northeast face, descending into the trough, were a series of amazing features—gullies, carved by water emanating from the triangular plateau itself. The whole northeast face of this plateau is cut by these gullies. To make the image more sensible, I enlarged the triangular plateau and turned it with south at the top, giving the viewer the sensation of looking obliquely down toward the gullied cliff face. To help you interpret the picture, note that the light in this view comes from the right. The top of the triangular block and the base of the image are dark gray and are laced by ridges or dunes. The slope facing us is completely carved by the gullies. The gullies originate at the cliff top in 100-meter-wide alcoves, which narrow into roughly 50-meter-wide gullies that run a couple of hundred meters downslope. At the end of each gully is a long, smooth gray triangular fan of debris, running another 200 meters downslope. The gullies are so numerous and closely spaced that the debris fans coalesce into a continuous smooth apron that covers the bottom half of the cliff.

An extraordinary aspect of these features is that they look young. Crater densities are very low on the gullied slope, and the debris fans appear to have been emplaced on top of the darker, dune-streaked valley floor. The gullies of Aram Chaos thus seem not to be Noachian or Hesperian but, rather, have formed fairly recently and deposited their debris on top of valley floor dunes—a wholly unexpected

Smoking-gun evidence of water release. Large triangular block (center) is raised or tilted above dark dunes (bottom third of image). The eroded face of the block (upper left center to lower right) is composed of bright-walled canyons or gullies with amphitheater-like upper ends. Bottom half of this slope is covered by light gray delta-like deposits of mud or gravel eroded out of the canyons, and coalescing into a talus slope. Water (or some other fluid) must have issued out of the canyons and carried debris downslope to make the deposits, implying that the whole block was probably ice-rich. (South up; 21W 4N, M22-00311.)

finding, since creation of chaotic terrain was supposed to be ancient.

These gullies are similar to ones discovered throughout the near-polar latitudes in 2000 by Mike Malin and Ken Edgett, which we will visit in chapter 34. Those gullies, too, originate in upslope alcoves, run down cliff faces, and leave fans of debris. The big difference is that the gullies of Aram are located only 4 degrees off the equator, where conditions are supposed to be too warm for near-surface ice to serve as a source of water. Then how did the water survive within these plateau-like blocks? To my knowledge, this is the first evidence of near-equatorial gullies and the first case of their association with chaotic terrain and hematite deposits.

The fact that water-eroded gullies completely dominate the face of a cliff on one of the blocks of collapsed, chaotic terrain in Aram craters seems to offer smoking-gun proof that the crater-filling sediments were water- or ice-rich materials that collapsed as water drained out of them. In that sense, the image is very satisfying. It seems to establish that Noachian-era craters, at least in some areas, filled with water-rich sediments that froze into ice-rich masses. At the same time, it raises new questions about whether some of the sediment blocks in Aram and other filled-in craters still contain ice, and how ice melting, along with local-scale emission of water, can continue to occur in more relatively recent geologic time.

CHAPTER 23

NANEDI VALLIS

Climate Change on Another World

Another class of Martian riverbed is exemplified by a major outflow channel named Nanedi Vallis. Its claim to fame is its indication of multiple episodes of flows. Nanedi Vallis originates in mid-Noachian era plains and flows about 400 km (250 miles) north across late Noachian plains. It would stretch across New Mexico. Like Ares Vallis and the Aram outflow, Nanedi Vallis is mapped as Hesperian in age, and it may have been active some time between 2,500 and 3,500 MY ago.

Unlike the gently curving Ares Vallis, Nanedi Vallis winds in tight sinuosities, like a sidewinder on a sandy playa. In terrestrial rivers, such meanders are usually the mark not of a sudden flood but of a mature river like the Mississippi, which, by virtue of a long history of flooding and overflowing its banks, has deposited sediments and built a flat plain. In the absence of a strong slope, the water meanders here and there.

DISTINCT EPISODES OF WATER FLOW

The striking thing is that among the meanders of Nanedi Vallis there are indicators of different stages of flow or perhaps even distinct water-release episodes. The best example is in an image obtained at the very beginning of the MGS mission and widely reproduced in technical papers. Here, we see a 2-km-wide

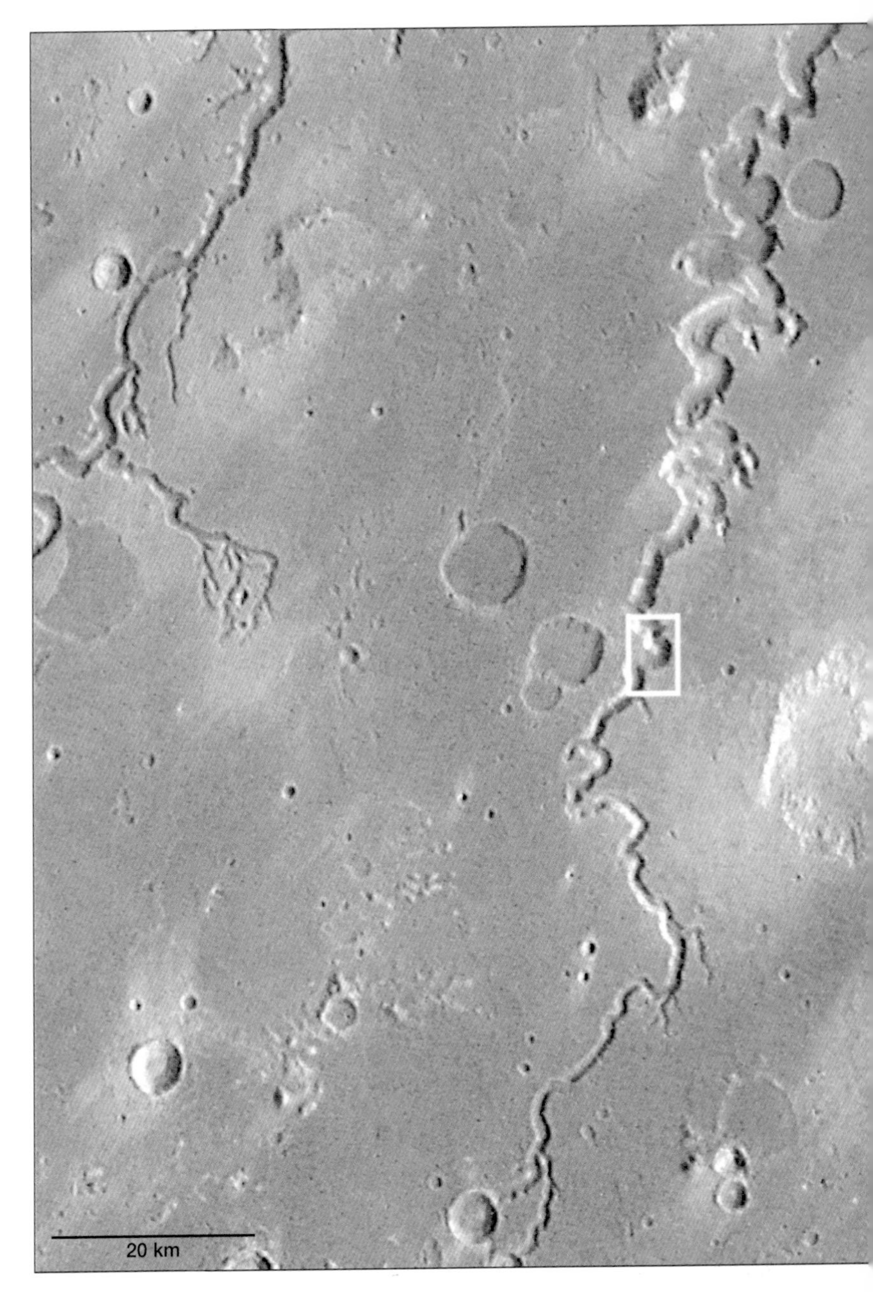

Regional view shows two branches of Nanedi Vallis draining north across moderately cratered Martian plains. The snake-like winding patterns are typical of a river established on a relatively flat surface. Note stubby tributaries at the source of the western branch and along the eastern branch. Box shows location of close-up image on page 244. (Viking 1 orbiter, 897A32.)

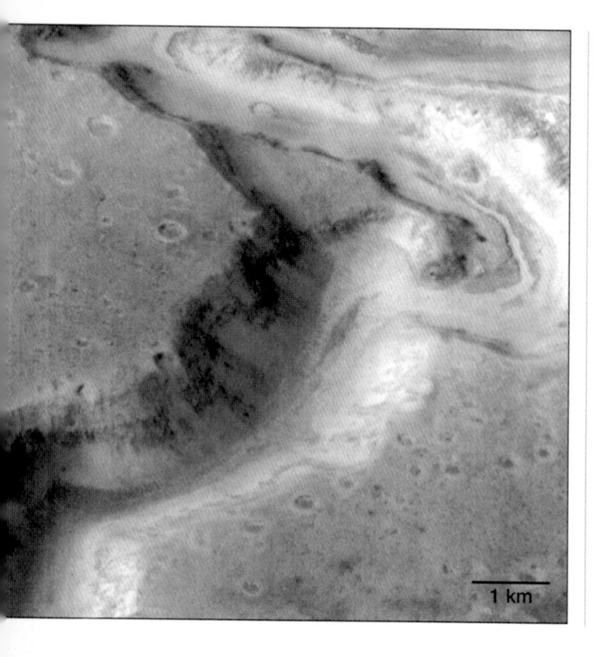

The MGS discovery image that first showed evidence of late flow down the center of Nanedi Vallis some time after the main channel was created. The small, final flow created a gully in the center of the channel, visible in the bend at upper right, but covered by dust drifts and dunes in most of the channel floor. (48W 5N, AB1-08704).

valley snaking across a cratered plain in its meandering pattern. Second, some meanders have almost created oxbow lakes, classic features seen in the Mississippi and other rivers in which the gradually changing course of the river cut off an original bend, leaving a crescent-shaped lake, while the river itself shifted to a new position. Third, on the west side of the river at the most prominent bend, a triangular flat terrace or shelf is exposed a few hundred meters down from the plain but a few hundred meters *above* the main channel. This terrace is a stranded bit of an earlier, higher floor of the river. The river first flowed there, cutting down to that level, and then meandered to the east, cutting deeper and stranding that bit of floor as a terrace. Fourth, the most prominent bend in the river contains a striking feature: a small arroyo only 200 meters wide, with sharply defined banks, winding down the middle of the broader channel. This must represent a final "trickle" of water, the last bit of water to flow down the channel. The arroyo itself is mostly buried under later dust deposits in the channel floor, probably windblown dust and sand that have partly filled in the channel and buried most of the floor. The central arroyo can be glimpsed in other parts of Nanedi Vallis, but nowhere as clearly as in this image.

Notice another feature of the image. The banks of the broad channel reveal stratified resistant layers about halfway down to the main channel floor. The stranded, triangular terrace represents part of the riverbed that cut only as far as these layers, which can be seen in a craggy bluff immediately below the terrace. We don't know if these layers are hard rock, like an ancient lava flow, or merely layers of crumbly mud. The fact that the early stage of flow eroded just down to one of these layers and no farther, as seen at the terrace, would favor the idea that the layers are fairly strong material, resistant to erosion. The stronger those layers, the longer it would have taken to erode through them, suggesting that Nanedi Vallis was not caused by a single rapid flood but may have persisted for some time.

To summarize, a geologist could speak of three phases of activity in this river. In the first phase, the river cut down to the level of the terrace and during its winding nearly created some of the oxbow lakes. In the next phase, it shifted to the east and cut farther downward, stranding the terrace and creating a deeper adjacent floor. Finally, a last stream of water flowed down the middle of the broad channel, cutting the central arroyo.

The big mystery here is that we don't know whether these three stages were telescoped into a single week or stretched out in distinct flow episodes over, say, 100 million years. In dry riverbeds of the Southwest, for example, a massive storm can send torrents of water down a channel in a single day. During the week after the storm, the declining flow of water can first leave terraces stranded along the riverbank, then meander along the floor, creating streamlined islands and a braided network of separate, twisting flows, and then finally settle into one last narrow channel, like the arroyo on Nanedi Vallis. Still, the meanders and erosion of resistant layers suggest extended, sustained water flow. MGS imaging team leader Mike Malin and his associate Ken Edgett, writing in

2001, used the above features to call Nanedi Vallis "perhaps the best evidence of sustained fluid flow on Mars."

WAS ALL WATER FLOW ON THE SURFACE?

I've already remarked that most ancient Martian rivers may have had a solid ice cover, like a Canadian river frozen over in winter. Nothing surprising there because Mars is a cold planet. There has been much controversy, however, over whether some channel features were carved underneath still more massive glacier-like ice packs, or even underground, by underground aquifers. It may yet turn out that many major channels were formed under such a thick ice pack that most of the erosive action was essentially underground, but Malin and Edgett, in their 2001 review, regarded Nanedi Vallis as a good candidate for having formed on the surface. They also noted that Nanedi Vallis appears with no visible source in the south and terminates abruptly at its northern mouth without much evidence of a delta or any other deposition of flood materials. A possible explanation of the missing features at the source and terminus is that those regions have since been covered by massive sheets of wind-drift sediments, which seem to be common on Mars.

LONG-TERM CLIMATE VARIATIONS?

The possibility that the last flows coursing down the channel of Nanedi Vallis, cutting that last arroyo, might have been truly distinct flow episodes, millions of years after the first ones, brings us back to questions of Mars's climate stability. Over and over we've encountered subtle hints that something strange has been happening in the Martian environment: a melted-down crater rim, ghostlike crater remnants, exhumed ancient surfaces, isotopic evidence of thicker atmospheres, collapsed areas from which water gushed, and enigmatic riverbeds. Could the Martian environment fluctuate in ways so as to create sporadic warmings, releasing water in different episodes, separated by many millennia?

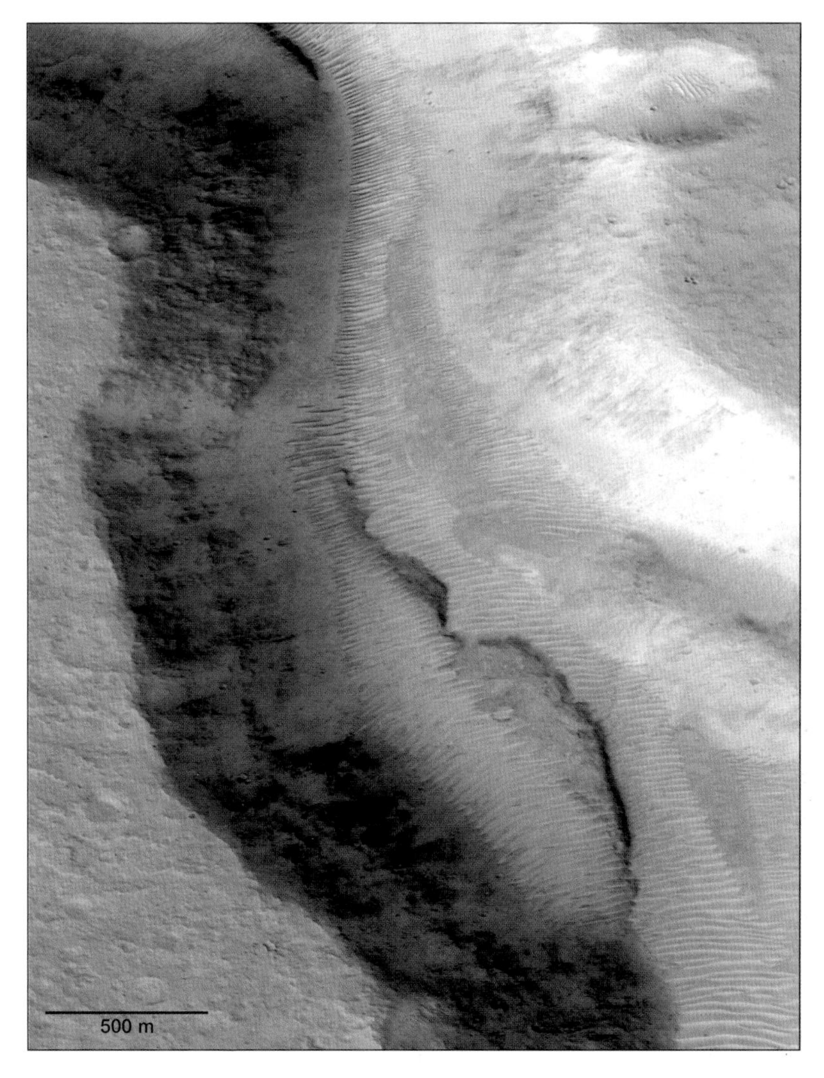

Detailed view of Nanedi Vallis and the gully eroded down its central axis. This view shows the dunes that fill that gully and other parts of the floor of the larger channel. (49W 4N, E10-02339.)

After all, if an Earth-like planet were located a little farther from the sun, it would be colder, and the off-and-on-again cycles that we call Ice Ages might be seen instead as "normal" frozen periods interspersed every few hundred thousand or million years with warm spells in which the ice melted and rivers run. That's a scenario suggested by some Martian features. Here is a quick review of some mechanisms we've discussed that might bring about such episodic climate fluctuations:

- During gradual atmospheric loss, sporadic volcanic eruptions could add enough gas to push the total atmospheric density past some critical minimum required for greenhouse heating and ice melting. This idea remains unproven because we don't know if any volcanic eruptions could produce enough carbon dioxide and water vapor to do the job.

- Internal volcanic activity and heat flow from the interior have been declining, as established both by direct tabulations of volcanic surface areas with different ages and by calculations of the changes in heat production by radioactivity. However, we can see many areas of recent volcanism and lava production, which confirm that local crustal regions experience changes in heat flow. Thus, underground layers may experience sporadic heating. Even mild heating by tens of degrees could melt ground ice without producing full-fledged volcanic eruptions, just as is the case in geothermal areas of Earth.

- Mars's axial tilt is not restricted to its present value of 25 degrees. Earth's axial tilt (by a very similar angle of 23.5 degrees) is maintained by forces associated with the Moon. Lacking a large moon, Mars experiences changes in its axial tilt down to zero and up to much higher values, over timescales of a few million years. This would cause extreme changes in seasonal behavior from one period to another. Colorado dynamicist William Ward and others have made the arduous calculations necessary to understand these effects. They estimate that under current conditions, Mars's axial tilt varies from as low as 13° to as high as 40°. The variations are somewhat aimless, and the typical timescale for a cycle of changes is about 10 to 20 MY. The calculations are good for only the last 10 MY; before that, the uncertainties caused by many competing effects become too great. In current parlance, this is called chaotic behavior. According to a 1992 summary of the effect, the last period of 40° axial tilt may have been about 4 to 5 MY ago, but after new data were received from the Pathfinder mission, the estimate was revised so that the last period of tilt approaching 40° would have been more like 3 to 4 MY ago. The theoretical work is still evolving, but the general 10-MY timescale for cycles of change seems secure. At the times when the tilt is high, the summer polar cap could point more directly at the sun, receive increased heat, and add all its water and carbon dioxide to the atmosphere, in the form of gas. In estimates cited by the French researcher F. M. Costard in 2001, the total amount of precipitable water in the atmosphere could be fifty times higher under those conditions (though the total amount would still be small). These effects are a major area of current research and may lead to radical new views of the past Martian climate.

- Even higher axial tilts may have been possible in the ancient past than at present, due to effects of lava concentration in a large dome on one side of the planet, as we will see when we visit the Tharsis area.

- Any of the above effects might produce an episode of ice melting and water production, but, in addition, combinations of all the above cycles of changes might produce unique circumstances that favor ice melting in certain regions or latitudes.

One day, if all goes well with Earthly society, astronauts will visit Nanedi Vallis and walk down the central arroyo, looking at strata in the wall, measuring the strength of the blocky layers in the main channel walls, climbing onto the 1,000-meter-wide terrace, and taking drill cores that may reveal the history of this river system and tell us, perhaps for the first time, whether Mars had distinct episodes of more Earth-like conditions when water flowed along its winding valleys.

CHAPTER 24

NIRGAL VALLIS

Sand in the Tracks

Because outflow channels are so dramatic in influencing our modern picture of Mars, we'll visit a third major example, which reveals yet a different style of water erosion. Nirgal Vallis was one of the first channels to be recognized, when Mariner 9 carried out the first orbital mapping in 1972. Mapped as Hesperian in age, it winds across Noachian-age cratered highlands just a few hundred kilometers west of the great waterway. It has a striking appearance, different from most other outflow channels. The sharply defined main

Regional view of Nirgal Vallis shows the riverbed winding across a Noachian-era cratered desert. Short, stubby tributaries may be caused by sapping from underground water, instead of surface water runoff. (NASA, mosaic of Viking images.)

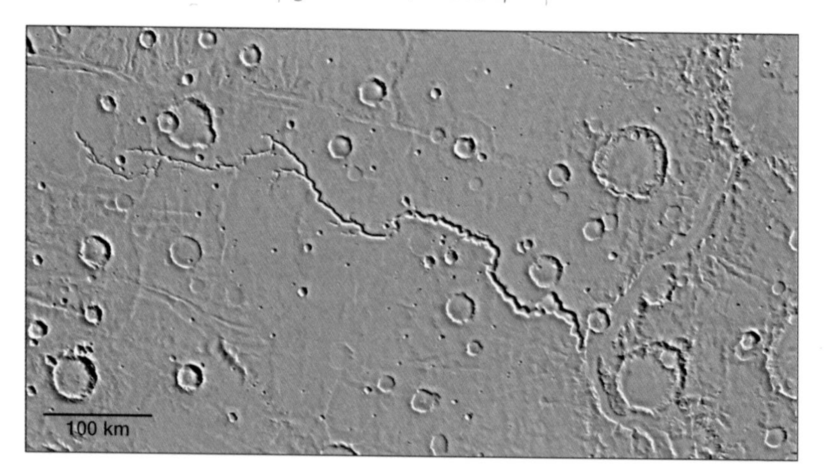

channel has short, stubby tributaries extending into the desert on either side and ending abruptly. The tributary system is not dendritic, in the sense of looking like a spreading tree; rather, the tributaries look like little twigs stuck onto a main stem. This pattern of stubby side valleys is not typical of water runoff drainage across a broad surface, as in valley networks, nor does it fit the pattern of a channel such as Ares Vallis, where the water seems to originate primarily in a fixed area of collapsed, chaotic terrain. What water source, then, explains the channel pattern that eroded Nirgal Vallis?

GROUNDWATER SAPPING

The pattern seen in Nirgal Vallis is recognizable on Earth in valleys that have grown by a process called **groundwater sapping**. Remember that just as *ground ice* refers to underground ice, *groundwater* refers to underground water. It is the opposite of *surface water*, or runoff from rainstorms or melting snow. As used

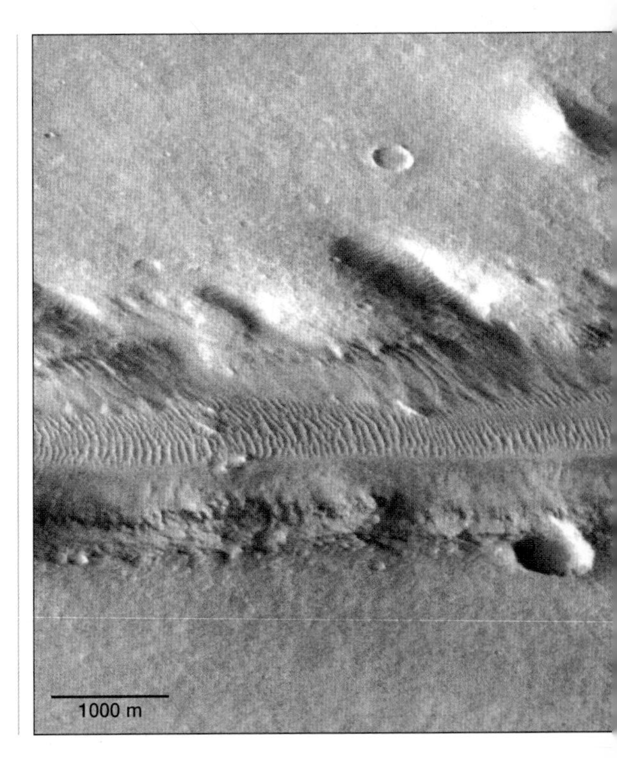

A portion of Nirgal Vallis, showing the stubby tributary channels (upper center) that may be a result of groundwater sapping. Note also the very small, narrow gullies running down the north wall of the channel (center)—more evidence of underground water release. One of the earliest MGS telephoto images, this view showed that the channel floor is silted in with sand dunes. (42W 29S, AB1-00605.)

on Earth, *sapping* refers to a process of gradual collapse of a riverbank or hillside as groundwater leaks out. Imagine a spring in the side of a hill. Water trickles out of the spring and erodes a small gully. Gradually, the water carries away soil particles, eating back into the slope and removing support for the hillside above the

Ground view of a winding, 3-mm-wide dry channel in a windswept Mexican desert playa that mimics a Mars-like pattern: dunes cover the channel floor, and small obstacles leave bright wind streaks across the channel, due to prevailing winds from left to right. (Photo by author, near El Golfo, NW Sonora, Mexico.)

spring. That part of the hill can collapse. The spring is buried at first, but gradually the water works its way out, forming a new spring and starting the process over again. The water keeps eroding farther back into the hill. Therefore, groundwater sapping can enlarge a valley without either rainstorms or full-fledged floodlike runoff.

In his writings on Martian water, U.S. Geological Survey researcher Mike Carr argues that groundwater sapping may be a major Martian process. He sees it as active even in valley network formation. Every time a cliff exposes a layer of ground ice it could initiate sapping that produces runoff.

A RIVER OF SAND DUNES

Nirgal Vallis was one of the first channels imaged in high-resolution photos from MGS. Scientists were excited about searching these images for diagnostic channel-floor features that

Rapid gully-washer activity from a few major thunderstorms in New Mexico formed this winding channel in a few decades after cattle grazing removed some of the vegetation in the headwater area. This shows the ease of erosion in loosely consolidated desert materials. (Photo by author, near Las Vegas, New Mexico.)

might reveal the history of such valleys, but they were disappointed to find that much of Nirgal Vallis, and other river channels, has been filled in by windblown sand. Sand and dust tend to collect in Martian low spots—crater floors, canyons, and river valleys. Once a sand grain blows over a rim into a depression such as a valley or a crater, it's hard to get it back out. As a result, Nirgal Vallis displays a striking lane of dunes running down the center of the ancient riverbed. When I showed this early picture to friends, many responded that it looked like some sort of massive tire tracks, similar to what we see in the Southwest when four-wheel drive enthusiasts roar down dry desert washes. Even the side valleys don't reveal their original outlines but seem softened, worn down, and sandblasted.

By concealing the features we want to see, sand-dune fill may be the enemy of future Martian astronauts' pursuing research on river channels. That's why Nanedi Vallis, with its exposed "last gasp" central streambed, described in the previous chapter, is such an attractive candidate for exploration.

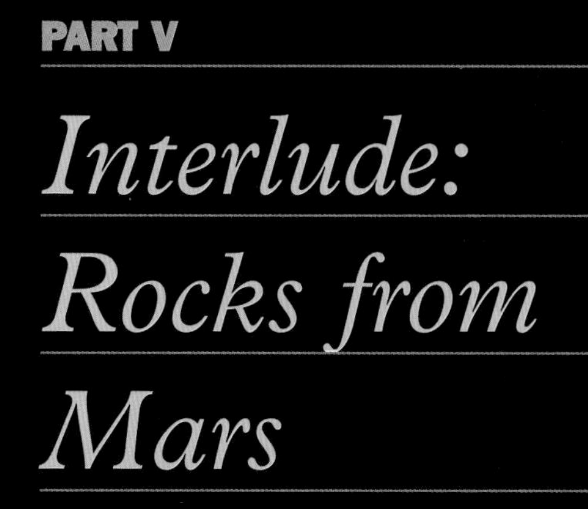

PART V

Interlude: Rocks from Mars

CHAPTER 25

MARTIAN METEORITES

Ground Truth at Last

At 9:00 A.M. on June 28, 1911, a shower of rocks fell out of the sky on El-Nakhla, a town near Alexandria, Egypt. The rocks were pieces of a single meteorite that had broken apart as it slammed into the atmosphere. At least forty pieces fell, totaling about 10 kilograms (22 pounds) of rock. One of them reportedly struck and killed a dog. The largest recovered piece weighed about 1.8 kilograms (4 pounds). Meteorites are named for the place where they are found, so this one went into the database simply as Nakhla. The fall was spectacular, but Nakhla's role in science was only beginning.

Meteorites, or pieces of rock and iron that fall from space, have been called the most complex and mysterious rocks known to geologists. After generations of work, we know that most of these rocks are fragments of asteroids that came from the asteroid belt between Mars and Jupiter. They are created during asteroid collisions in the belt, when rock fragments are blasted out in all directions. Particular gravitational interactions with gigantic Jupiter steer some asteroid trajectories onto orbits that cross those of Mars, Earth, or Venus. Those that hit Earth, if found and recognized, end up in museums, planetariums, and private collections. Because asteroids as well as planets were formed and cooled 4,500 to 4,550 MY ago, during the creation of the solar system, virtually all meteorites are that old.

By the 1970s, however, various laboratories had confirmed through isotopic dating that the Nakhla meteorite and a handful of others were much younger, ranging from a few hundred MY to 1,300 MY old. These rocks were found to have similar compositions, typically containing basaltic lavas. What could they be? Where did they come from? They were clearly extraterrestrial, because their isotope chemistry was different from terrestrial rocks, and some, like Nakhla, had even been seen to fall from space. Even the largest asteroids are so small that they probably cooled off before 4,400 MY ago and should not have had lava flows since then. The youngest rocks on our own Moon are generally older than 3,000 MY. Scientists were left with a puzzle. What body in the solar system could be the source of lavas and igneous rocks less than 1,300 MY old?

Between 1979 and 1981, a few hardy scientists, such as NASA's Larry Nyquist, UCLA's John Wasson, Carnegie Institution's George Wetherill, and NASA's Charles Wood, began to argue that the offending meteorites must have originated on Mars. The idea was that an asteroid impact might blow rocks clear out of Mars's thin atmosphere and launch them into space, from which some of them might eventually reach Earth.

Other experts scoffed. Several impact specialists said that no rock could be blasted off Mars because the pressures needed to accelerate the rock into space would crush the rock. But, as the same experts admitted later, this was wrong. It was based on what I call a "golfball model" of launching rocks into space. Obviously, if you try to whack a rock hard enough to launch it into space, it will break into dust. But a large crater-forming explosion produces an expanding fireball of gas and debris, and is a much more complex process than hitting a ball off a tee. A chaotic mass of material, kilometers wide, is accelerated out of the crater during an enormous explosion. The pressures break kilometer-scale chunks into 100-meter chunks, and 100-meter chunks into 10-meter boulders. Some of these make it through the atmosphere,

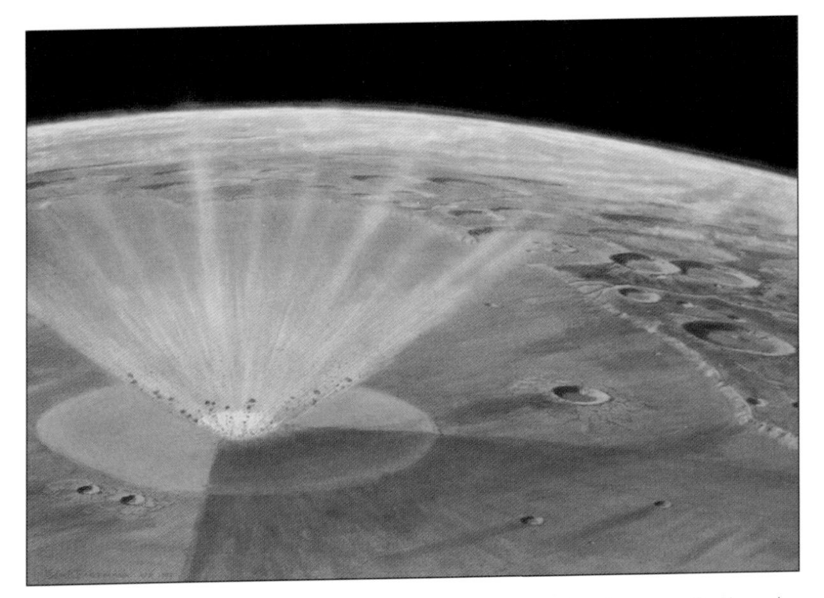

Impact of a modest-size asteroid on Mars creates a crater a few kilometers across and blows rocks off Mars. Some of these rocks will eventually be gravitationally deflected into orbits that intersect Earth and fall on our planet as meteorites. (Painting by author.)

lofted in a cloud of ejecta that reaches heights above much of the Martian atmosphere itself.

The clincher came as early as 1983, when geochemists Don Bogard, P. Johnson, R. H. Becker, Robert Pepin, and others, in various papers, made an astonishing discovery: the meteorites in question contain gases that exactly match the composition of the Martian atmosphere. The rocks were definitely from Mars.

This discovery, in turn, allowed identification of other meteorites that come from Mars. New ones were turning up every year or so, and by 2002 twenty were identified. (A similar number have been proven to come from the Moon, an identification that would not have been possible without the rock samples returned earlier by Apollo astronauts.) Many of the Martian rocks have been found in Antarctica. The frozen continent is a meteorite hunter's treasure chest because meteorites that hit glaciers get embedded in the ice, the ice flows downhill, and the rocks eventually get exposed in a restricted area at the mouth of the glacier. Free delivery.

DATING MARTIAN ROCKS

The system used for dating rocks from Mars, Earth, the Moon, and asteroids is called radioisotopic dating. As you learned in high school chemistry, elements come in various forms, called isotopes. Some isotopes are radioactive, meaning that they are unstable and spontaneously give off particles, transforming them into other, stable elements. The radioactive isotopes, or radioisotopes, are said to decay into daughter isotopes. An example is the element potassium, common in rock-forming minerals. It decays into the gas argon. Thus, radioactive potassium atoms that were originally incorporated in a rock when it formed constantly decay into argon atoms, and the argon atoms accumulate in the mineral. The basis of the potassium-argon dating system, then, is to measure the amount of accumulated daughter atoms of argon as well as the amount of potassium in a rock; the more daughter argon, the older the rock. In any given sample of radioactive potassium, half of it decays to argon in 1,300 MY (we say that the half-life is 1,300 MY), and so the potassium-argon system has a very useful timescale for dating solar system rocks. There are many chemical complexities in accurately measuring the amounts of potassium and accumulated argon in a rock, but several laboratories around the world are equipped to do it.

Strengthening the technique, there are other radioactive systems useful for measuring ages of solar system rocks. For example, one radioactive

UNKNOWN SITES ON MARS

You might think that planetary geologists would be in ecstasy to have free delivery of rocks from Mars, but their joy is mitigated by the frustration that we don't know what specific locations on the planet these rocks came from. Careful study of the Martian rocks shows that all of them are igneous rocks, some having crystallized on the surface as lava flows while others crystallized underground. But which craters launched them? Theorists estimate it would take a crater measuring at least a few kilometers across to launch the rocks into space, but there are hundreds of candidate young craters. Not knowing the specific site of origin is a severe handicap because it means we can't correlate rock composition or age with specific surface units. Are the rocks from the isotope of uranium decays to lead with a half-life of 4,510 MY; another uranium isotope decays into a different lead isotope with a half-life of 713 MY; radioactive rubidium decays to strontium with a half-life of 47 MY; and so on. When a given rock sample is obtained from Mars, the Moon, or a meteorite, different labs around the world can date it by using different isotopic systems. In general, there is good agreement among these estimations, but even discrepancies in ages measured for a given rock can give us information about some unique process that affected that particular rock. It is this same international system of radioisotopic dating that overthrew the outmoded seventeenth century idea that Earth is only a few thousand years old (though this idea is still encountered among religious fundamentalists who want to teach "creation science" in our public schools).

As new Martian meteorites are identified, samples are sent to various labs and scientific meetings are held at which geochemists from the different labs gather to announce and compare their results. Some of these geochemists have been at this business since the first rocks were brought back from the Moon, and they now find themselves managing the labs that measure ages of Martian rocks. This transition from the "dirty work" of lab dating to the white-collar work of administration prompted the German geochemist Elmar Jessberger to quip that you can identify the most experienced geochemists because "they are the ones who are still aging but no longer dating."

lava-covered plains? From uplands? From ice-rich high latitudes? Which types of geological units are *missing* from the collection? It's like trying to characterize the whole planet Earth with a handful of rocks from an unknown number of sites randomly scattered all over the globe.

Nonetheless, clever sleuthing allows us to estimate the number of impact sites that have been sampled, as follows. Thanks to isotopic dating, we know not only the age of each rock but also the amount of time since it was launched into space (determined from changes due to cosmic rays while the rock is in space—but that's another story). These in-space flight times are all "recent," within the last 20 MY. This agrees with orbital studies, which show that a rock blasted off Mars would hit Mars, Earth, or another planet typically within the first 10 to 20 MY after its launch. Some groupings of rocks that have similar composition, the same age, and the same date of launch are assumed to have come from one impact. Thus we can estimate how many "sample sites" on Mars produced our collection of Mars meteorites. Using this logic, an international team of geochemists estimated that the first twenty Martian meteorites found on Earth represent four to eight sample sites on Mars. The rocks tell us stories about the geological events at these unknown spots on our neighboring planet.

A PIECE OF THE ORIGINAL MARTIAN CRUST

The oldest of the Martian meteorites is a unique rock found in the Alan Hills of Antarctica, known by the dull catalog name ALH 84001. A best estimate of its age, based on several different radioisotopic measurements in multiple labs, is 4,510 MY, plus or minus 110 MY, and that it was knocked off Mars only 15 MY ago. As of this writing, we have only one rock from that impact site.

The age of 4,510 MY excited scientists because it's only a few tens of millions of years after the formation of Mars itself. This is a rock from the very beginning of the Noachian era at the very beginning of Mars. The coarse texture of the rock's crystals indicates that the original molten magma cooled slowly. This means that the magma was not on the surface in a lava flow, where it would cool fast, but in a magma chamber as deep as 40 km below the surface. Taken together, these facts mean that ALH 84001 is a piece of the original Martian crust and that the crust formed soon (in geological terms) after the planet itself formed.

This rock is unique. From the whole solar system, it's the only rock we have that samples any planet's original crustal formation. On Earth and Venus, early crustal rocks have been destroyed by our planet's incessant geologic activity, such as continental drift, erosion, and volcanism. On the airless Moon, Mercury, and Pluto, they were pulverized by the accumulated impacts that cratered and broke up the surface layers; any remaining initial crust is buried under kilometers of impact-blasted debris. Gas giant planets like Jupiter probably don't even have conventional rocky surfaces.

Mars, too, must have experienced impact blasting of its early rocks, so how can we have a sample of its original crust? My own suspicion is that Martian erosion has exposed at least one area of primordial crustal rock, where an asteroid impact was able to launch a sample of this material into space. Probably the area of exposed crust is substantial; if it covered only 1 percent of Mars, we would be unlikely to get a sample of it in only four to eight impact samplings. The existence of ALH 84001 tells us that Mars may be the only planet in the solar system where we can walk around on exposures of primeval planetary crust—rock units that formed as the planets themselves were first coming into existence. A complete sampling of such rocks could help us construct the first days of a planet's history.

It's yet another reason to go to Mars.

Scientists have labored over ALH 84001, prying loose its secrets about the earliest conditions on Mars. For example, as mentioned in chapter 5, it experienced some sort of geological disturbance 3,920 MY ago, plus or minus 40 MY. The disturbance possibly involved an impact, exposure to water, and/or mild heat. Carbonates were deposited in fractures in the rock about that time, presumably when mineral-rich water seeped into the rock and then evaporated—bathtub rings in a Martian rock. This affirms the idea that Noachian Mars had abundant mobile water. But the greatest secret of ALH 84001 was something else.

MARTIAN FOSSILS?

In 1996, a shocking new ingredient was thrown into the stew of Martian speculation. A NASA research team headed by David S. McKay and Everett Gibson at the NASA–Johnson Space Center in Houston, announced the identification of possible fossil microbes in ALH 84001. The possible fossils were microscopic structures

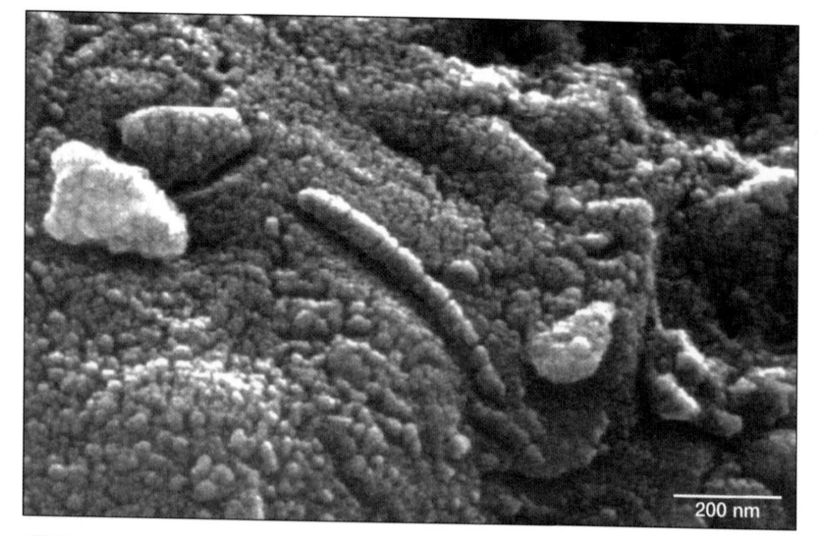

Electron microscope photo of mineral surfaces inside the Martian meteorite ALH 84001, found in Antarctica, shows tiny worm-like structures that some investigators think could be remains of fossil microbial life-forms that evolved on Mars. (Photo courtesy Everett Gibson, Johnson Space Center, NASA.)

embedded in the 3,900-MY-old carbonate globules. This fits the idea that the possible microbes were in the water that got into the rock.

Needless to say, the announcement of fossils in a Martian rock was extremely controversial, but the NASA team cited several lines of evidence, including the existence of organic materials in the carbonates. Some critics claimed that the organics were contaminants that entered the rock from outside while it was lying on Earth, but the NASA team answered that the putative Martian organics were scattered evenly through the rock, not concentrated toward the outside as might be expected from contamination. Other critics suggested that most of the structures are too tiny to be functioning cells because they would not have room for enough genetic material, and that they look more like natural nonbiological mineral formations. Others hypothesized that the structures might be just fragments of larger, highly damaged cells; still others claimed (again, controversially) that terrestrial microorganisms of that small size do exist. The case for biological origin was strengthened a few years later when the Martian structures were shown to contain tiny crystals of the mineral magnetite, matching types of magnetite crystals produced by microbes on Earth.

Does ALH 84001 really contain evidence for ancient Martian life? It may not be an all-or-nothing issue. I have suggested the intermediate possibility that because the origins of life on any planet involve gradual chemical formation of more and more complex molecules and nonliving cell-like globules, there may be "missing link" structures that were never quite alive. It's known, for example, that natural processes in various parts of the universe create amino acids, the building blocks of proteins. Another example is a natural chemical process in which organic molecules in a solution spontaneously aggregate into cell-size coacervate globules of organics, floating in a relatively pure water medium. These processes do not produce full-fledged living material, but the next step beyond such processes might produce the kinds of micro-objects found in ALH 84001. Has the NASA team detected remnants of precursor building blocks strewn along the road to life? If so, it would be a very important discovery, because such precursors on Earth would have been overwhelmed and destroyed by later organisms. My own view is that the chances are fifty-fifty that the NASA group detected such proto-biological material.

THE 1,300-MY-OLD NAKHLITES

Nakhla, the rock from Mars that fell in Egypt, is a member of a group of Martian meteorites named nakhlites. This is the next youngest group of Martian meteorites, which solidified about 1,300 MY ago. They were blasted off Mars 11 MY ago. Like ALH 84001, these igneous rocks have large, coarse crystals, which implies moderately slow cooling underground. The closest analogs on Earth are found in lava masses at least 125 meters underground, masses that either intruded into underground chambers or formed at the bottom of thick lava flows. The nakhlites are generally interpreted as fragments of a volcanic complex formed from

A NASTY CONTROVERSY

Compared to many branches of science, planetary studies have been remarkably congenial. I credit Carl Sagan as an influence who helped set the tone. During early planetary science conferences in the 1970s, when a personal argument threatened to break out, I saw Sagan rise and, in the guise of asking a question, deftly and gently remind us all how privileged we were to be the first generation to study other worlds from space probes, and how we were all in it together.

The argument over the detection of possible evidence of Martian life is the most heated I've seen in our field. Perhaps it's because the related field of biology is rife with contention. Anyway, when an issue has excited interest for generations—like the issue of life on Mars—the stakes are high.

Some critics therefore attacked the McKay-Gibson team on philosophic grounds. They argued that if you want to announce a finding as dramatic as Martian fossils, you need higher standards of proof than for more mundane results. They asserted that the NASA team should never have announced the result at all, in order to avoid any hint of a quest for publicity.

As you can imagine, this sort of situation can get personal fast. A nuance here or an ill-chosen phrase there can instantly change the abstract philosophic debate into an escalating free-for-all: "Your data are inadequate. You published immature results because you wanted to grab headlines. You were unethical . . ." I've been at meetings where nominally fair-minded researchers from the two sides traded accusations of inept science, where participants virtually shook with red-faced rage, and where session chairpersons were compelled to suggest that the discussion was growing, um, unproductive, and that it was time to move on to the next scheduled speaker.

Scientists are constantly dogged by a question from the Gray Zone: When do I have enough interesting new data to publish? Some researchers have published too soon and subsequently had to retract overly enthusiastic interpretations; others have waited too long, trying to get that last bit of

the cooling of volcanic magma that was deeper than 125 meters but probably not deeper than 1,000 meters.

Another Martian rock that is part of the same 1,300-MY-old grouping, but with a somewhat different composition, is a 4-kilogram (9-pound) stone that dropped out of the sky in France in 1815 and is named Chassigny after the place where it fell. It is better proof, only to find that their work has been outmoded by other, less cautious teams. My own view is that the McKay-Gibson team clearly had provocative data and had a responsibility to alert others to what they had found. In general, they were careful to say only that the findings were not a final proof and should be investigated further. Scientific progress involves a responsibility to make rapid and open publication of results that will fertilize the thinking of other teams, and the ALH 84001 proposals clearly inspired a rush of good research from other parties.

These issues arise not just in science. As the world is threatened by ancient antagonisms, we need open discussion of new data, new ideas. An open system based on evidence is the best system invented for arriving at truth. In medieval days, the main system for seeking truth was, instead, the appeal to authority. Fundamentalist ecclesiastics appealed to ancient writings from one source or another for their "proof"; early humanists appealed to Aristotle or Ptolemy; and legal judgments came from kings or potentates. For better or worse, our own society relies on the adversarial system. In the developed world's legal system, one lawyer is paid to defend one side and another is paid to defend the other side, with the presumption that truth will emerge as each side makes its best case. In America's marketing system, one salesman or advertiser is paid to tout the merits of one product and someone else touts the competing product; the presumption is that quality emerges from the free market. Yet, as we see more and more clearly each year, the danger in our modern systems is that many people care less about truth than about winning—the case, the big contract, the election.

Science works differently, by appeal to evidence. In ideal science, the glory goes not to the person who wins an argument but to the person who brings the best data to the table. All the data are spread out, and the best estimate of the truth emerges from it, not from the rhetoric of the person who makes the best case. Scientists, being human, don't always meet the ideal, but good evidence always beats rhetoric in the long run.

dominated by dense greenish crystals of olivine. On Earth, olivine-rich igneous rocks, called dunite, often form at great depth, near the base of the crust. They are often found in complexes of basaltic lavas, brought up by the lavas themselves. At a well-known site in Hawaii, on the volcano Hualalai, you can find fist-size olivine nodules, not unlike this Martian sample, encased in basaltic lava.

Based on the mineralogy and ages, the interpretation of the nakhlites and Chassigny is that they came from a single volcanic complex on Mars, probably formed by different pulses of deep-seated magma, which formed lava flows a few hundred meters thick on the surface of Mars 1,300 MY ago. Then a single asteroid, hitting 11 MY ago, blasted the rocks from these various layers, giving us this grouping of rock fragments.

YOUNG LAVAS FROM MARS

Another grouping of Martian rocks are the youngest we have from the planet and represent basaltic lava flows on the surface of Mars. They are ordinary lavas. If they turned up in a volcanic field, they could easily be passed off as pieces of terrestrial lava flows, and it would take sophisticated laboratory measurements (of the Martian gases, for example) to prove they are Martian. Their very ordinariness reminds us that Mars is not so unlike Earth.

They crystallized from lavas 170 MY ago and were blown off Mars 3 MY ago. At least seven basalt chunks belong to this group, and detailed mineralogical analysis suggests a multistage eruption history for at least some of them. First, they were brought up from deep magmas, more than 7 kilometers below the surface, and cooled somewhat in an underground magma chamber; then they were erupted onto the cold Martian surface in a lava flow at least 10 meters thick (the thickness being needed to explain the cooling rate and texture of small crystals). This pattern is very typical in terrestrial volcanoes, in which magmas come up from the deep crust, collect in magma chambers, and then erupt onto the surface in volcanic outbursts.

The exciting news here is that these rocks prove without doubt that volcanism extended through Martian history, even to the present day. Remember that an age of 170 MY represents only the last 4 percent of the age of the planet. It is unlikely that volcanic eruptions on Mars extended through 96 percent of the planet's history and then suddenly shut off forever in the last 4 percent of time, just before we got there!

A few other rocks are also basalts from surface lava flows on Mars, but they seem to come from different impact, or sampling, sites on Mars. One has a well-determined age of about 327 MY, plus or minus about 19 MY. Another has a less well-determined age, but probably formed about 500 MY ago.

COMPARING WITH THE ROCKS IN ARES VALLIS: ROCK DIVERSITY

Remember that we have one other suite of rock measurements from Mars, the rough compositional measurements of several boulders made by the Sojourner rover at the Pathfinder landing site in the mouth of Ares Vallis. Naturally, researchers rushed to compare the two data sets. Are the Martian meteorites like the Ares Vallis boulders? Broadly, yes—but in detail, no.

The Martian meteorite rocks are essentially basaltic, while the rocks at the Pathfinder site (assuming correct analysis of valid data from the remote X-ray spectrometer in the Sojourner rover) belong more to the andesite or basaltic andesite suite discussed earlier. If Mars really has andesites, why don't the Martian meteorites bring us samples of them? Is it just luck of the draw? Or are the andesite interpretations wrong? It may take a few more Martian meteorites or a few more Mars lander missions to find out.

SALTY ROCKS

In the late 1980s and '90s, geochemists (chiefly Houston's James Gooding and Susan Wentworth) discovered that most Martian meteorites were at some time exposed to water, and the water was salty. As it evaporated from the rock, the water left its salts and alteration products behind in the meteorites: clays, halite (ordinary table salt or sodium chloride), carbonates, gypsum, and others. From the composition of these telltale deposits, British geochemists J.C.Bridges and M. M. Grady and other geochemical teams determined that

A view on the surface of Mars from the Sojourner rover emphasizes that the three areas of Mars seen so far from the ground are strewn with rocks, probably dislodged by impacts. Computer simulations show that rocks launched from Mars by impacts would come from near the surface (top 50 meters?) but may favor fragments from intact lava layers instead of loose rocks like the ones shown here. (NASA.)

the Martian water that wetted the rocks was not only salty but strikingly like terrestrial seawater! This confirms the century of theorizing that we described earlier, that Martian water was salty and stayed liquid at temperatures lower than the 0° C or 32° F.

The interesting news was not in the salt itself but in the tale it had to tell. In one of the meteorites, a 1,300-MY-old nakhlite called Lafayette, the briny water caused enough mineral alteration to be dated by radioisotopic methods. Two groups, one under C.-Y. Shih at Lockheed Martin Space Operations in Houston in 1998 and the other under Tim Swindle at the University of Arizona in 2000, measured the age of iddingsite, a mineral caused by the weathering of olivine grains in the rock. The two groups inferred that the alteration happened near or later than 670 MY ago. Remember that the rock itself formed 1,300 MY ago. So at least 630 MY went by before water got into the rock and altered its minerals. The water was not necessarily present in large amounts or long-lived; a few days of exposure to a trickle of condensed water or steam may have done the job. Still, the water release was not caused by the volcanism that formed the rock but happened 630 MY later. This tells us two things. First, a given spot on Mars may experience sporadic water-exposure events separate from local volcanism. Second, this fluvial activity occurred not just back in the dim Noachian and Hesperian past but in relatively modern times! An age of 670 MY marks an event that happened in the last 15 percent of Martian history.

Thus, the Martian rocks confirm the revolutionary conclusion that both fluvial and volcanic activity happen at least sporadically in modern times. Mars is radically different from the Moon, where there was never any water and virtually all volcanism ended about 2,200 to 3,200 MY ago. Mars is also different from the planet pictured by many Viking-era researchers, who thought that volcanism and water flow essentially ended there by 2,500 MY ago.

WHY NO METEORITES FROM EARLY TO MIDDLE MARTIAN HISTORY?

There is one more mystery lurking in Martian meteorite ages. Let's review the situation. From four to eight places on Mars, we have rocks that formed 170 MY, 330 MY, 500 MY, and 1,300 MY ago, and one additional rock from 4,510 MY ago. Now, remember from chapter 1 that Noachian surfaces are likely 3,500 to 4,500 MY old and Hesperian surfaces might range from as little as 2,000 or 2,500 MY to 3,500 MY old. There are lots of Hesperian and Noachian surfaces mapped on Mars, and according to studies by U.S. Geological Survey researcher Ken Tanaka, two-thirds of Mars was resurfaced in either the Noachian or Hesperian periods. That means two-thirds of the Martian surface should be older than 2,000 MY. So why, out of four to eight sites on Mars, do we have no rocks from 1,300 to 4,500 MY ago? Why no samples from the Hesperian era or most of the Noachian era? And given the scarcity of rocks before 1,300 MY, why do we have one sample from all the way back to the beginning? Is it just the luck of the draw?

The answer may be more interesting than that. Impact studies show that the best kinds of material to launch rocks into space are hard, solid rock surfaces, and that in an impact explosion rocks get ejected primarily from near the surface, perhaps the top 50 meters. Weak, gravelly materials dissipate too much of the energy into shaking up the dust. So the reason we have no rocks from 1,300-to-4,500-MY-old Martian upland surfaces may be that those surfaces are mostly sand, dust, gravels, and weakly consolidated materials that don't efficiently launch rocks into space. Probably the oldest areas were pulverized to depths of a kilometer or more by impacts and then weakly consolidated with salts and ice. We've already seen that many older surfaces on Mars have been covered by layers of sediments, hundreds of meters thick, and the way these sediments are eroding proves that they are not strong. A cosmic impact into such weakly consolidated old materials may produce only a kablooey blast of steam and dust but no rocks shot into space.

If this logic is right, it means the statistics of the Martian meteorites are telling us about the nature of Martian surface materials. Solid-rock lava flows date back to at least 1,300 MY but may be harder to find before that (just as on Earth). If that's true, then the 4,510-MY-old sample may be telling us that erosion in some areas removed loose material all the way down to the original solid rock crust of Mars, exposing the crust in some regions.

There's a lesson here for future explorers. If we want samples of Noachian rocks, such as lake-floor sediments—the materials that contain clues about ancient, wet Mars and its possible biological activity—we may not be able to wait for asteroid impacts' free delivery service. In the first place, it's a poor postal system because it doesn't reveal the point of origin. In the second place, it doesn't deliver the kinds of materials we want. To get them, we'll most likely need to go to Mars ourselves.

PART VI

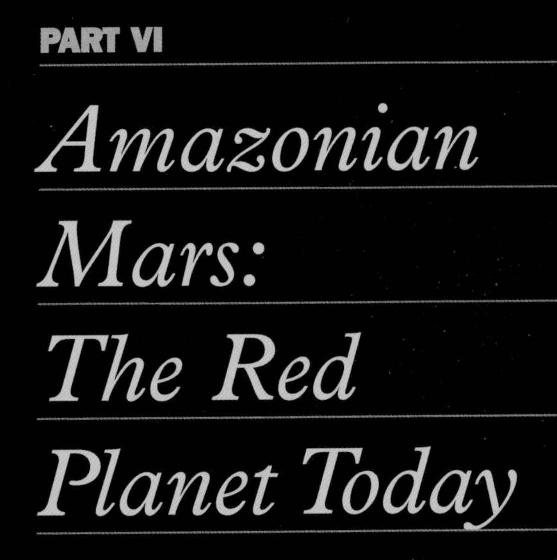

Amazonian Mars: The Red Planet Today

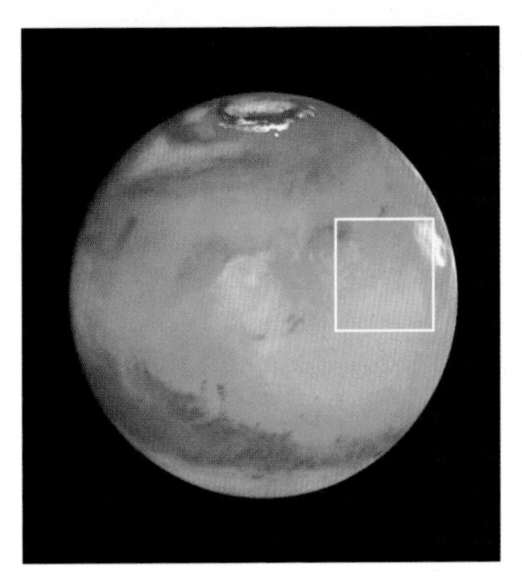

CHAPTER 26

AMAZONIS AND ELYSIUM

Yesterday's Lava Eruptions

Amazonis Planitia, named for the land of the mythical Amazons, gave its name to the whole Amazonian era. It is a bright, dusty volcanic desert crossed by many fresh-looking lava flows. It extends west into another lava-covered plain, Elysium Planitia. Elysium is named for the mythical land of the afterlife, located by Homer at the western ends of the earth. In the center of Elysium Planitia is a notable volcanic mountain called Elysium Mons. According to MGS laser altimeter mapping, Elysium Mons rises 14.1 km (46,000 feet) above the mean Martian surface and about 9 km (30,000 feet) above the surrounding plain. Elysium Mons has the classic features of a Martian or terrestrial shield volcano: it is a shallow-sloped mountain with a collapsed caldera at the summit. Amazonis and Elysium, both named by Schiaparelli in 1877, are notable for having some of the youngest-looking lava flows on Mars. Orbital images show craggy textures that match those in active volcanic sites on Earth and look as if they might have formed yesterday.

Older lava flows on Mars are pocked with craters and shrouded with dust and dunes, but the sparsely cratered, rough, fresh-looking Amazonian flows immediately raise the question of how young they actually are. Can we find sites of volcanic eruptions that were active in very recent geological time? Are some Martian volcanoes merely dormant? Could they erupt again at any time?

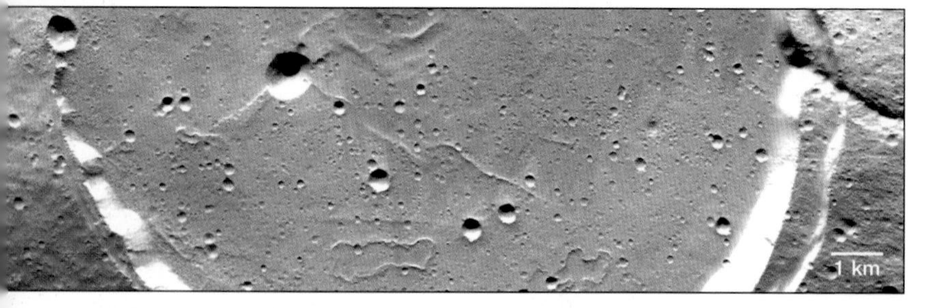

The summit caldera of Elysium Mons volcano (above) has a moderately high density of impact craters, proving that it has not experienced very recent lava flows. (214W 25N, oriented with W at the top, MGS image SP2-40303.)

Facing page: A striking lava flow on the plains of Elysium Planitia gives a clear example of superposition of geologic units. The background is heavily cratered. The lava flow, somewhat darker than the background, has spread over the old landscape like molasses and is much less cratered because it is younger. The youngest craters have created fresh exposures of dark basaltic lava, but the older craters are mantled by bright dust. (200W 14N, M07-01051.)

IN SEARCH OF RECENT LAVAS

As geochemists began to report young ages such as 170 MY and 330 MY for certain Martian meteorites, and as the first data started coming in from MGS in the late 1990s, the first-priority questions about the Amazonian era were: How young is "young" on Mars? Is Mars really an exhausted planet? Has some volcanic activity extended to the present day? Are some volcanoes still active?

In approaching these questions, I was aware that a few maverick geochemists had cautioned that since many of the Martian rocks have been exposed to mineral-bearing waters on Mars, the isotope systems used to date them might thus be messed up by chemicals brought in by the water. In other words, the young ages might be wrong. For those reasons, an independent test of young Martian surface ages would be very important, and as part of my own role in the MGS mission, I was especially interested in seeing whether we could find lava flows and other surfaces that gave ages as young or younger than the reported meteorite ages.

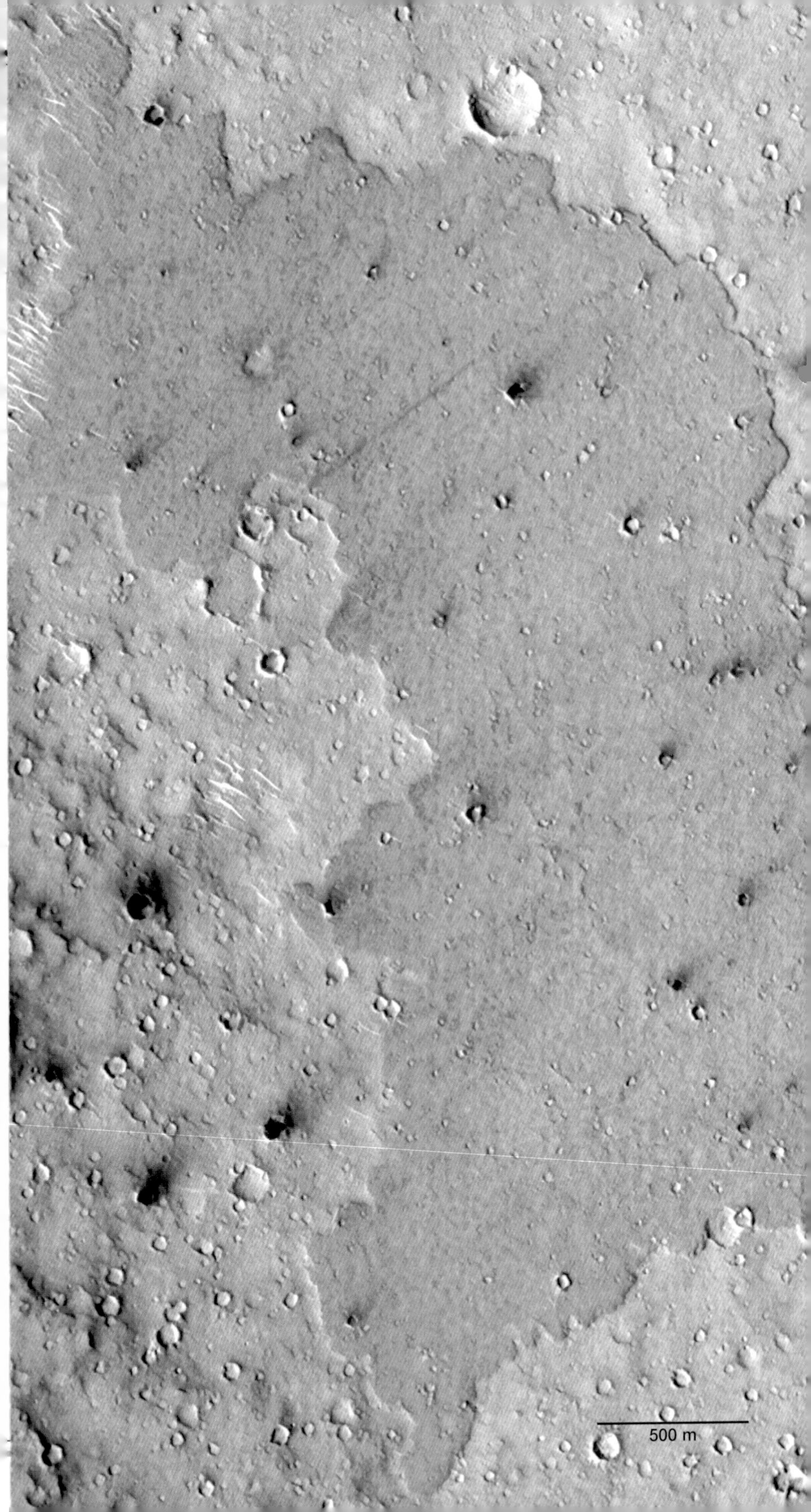
500 m

A rough surface in the Elysium plain shows craggy textures associated with some types of fresh lava. The surface is young, but depressions between the crags are silted in with windblown dust. (197W 33N, M02-04160.)

Our search was centered around the Amazonis-Elysium area. Elysium Mons itself was not a great candidate. It has more impact craters than other big young volcanoes, such as those in the Tharsis region, which we will visit shortly. The conclusion was that the summit and slopes of Elysium Mons are older, on average, than those of Olympus Mons or the other Tharsis volcanoes—a conclusion reached as early as 1988 by Florida researcher Nadine Barlow, who made counts of the larger impact craters visible in Viking images. Using the improved MGS crater statistics in 1999, I estimated that Elysium Mons's surface lavas dated to the second half of Martian time, perhaps 600 to 2,000 MY. This was not out of line with 500- and 1,300-MY-old Martian meteorites, but not spectacularly young, either. So our quest for the youngest lavas continued, and it led us to a nearly uncratered region of dark lava flows south and east of Elysium Mons itself, in the plains of Amazonis and Elysium.

Researchers searching for young lavas in the Viking photos from the 1970s zeroed in on southern Elysium Planitia because of its few impact craters, but couldn't be sure what kind of plain it was. Jeff Plescia, a planetary geologist then with Caltech's Jet Propulsion Lab, studied the area exhaustively in 1990 and again in 1993, concluding that it was covered with unusually young lavas. Viking-era altimetry maps showed a broad depression in this area, and U.S. Geological Survey Mars mappers Scott and Mary Chapman debated in 1991 whether the flat plain in that low spot might be young lavas or might include lake-bed deposits from a body of water that ponded in the low area. Mapping with the MGS laser altimeter proved that the broad basin in the Viking-era maps didn't really exist, casting doubt on the idea of a large ancient lake bed. The area was more of a flat plain, with a small, localized depression toward the south.

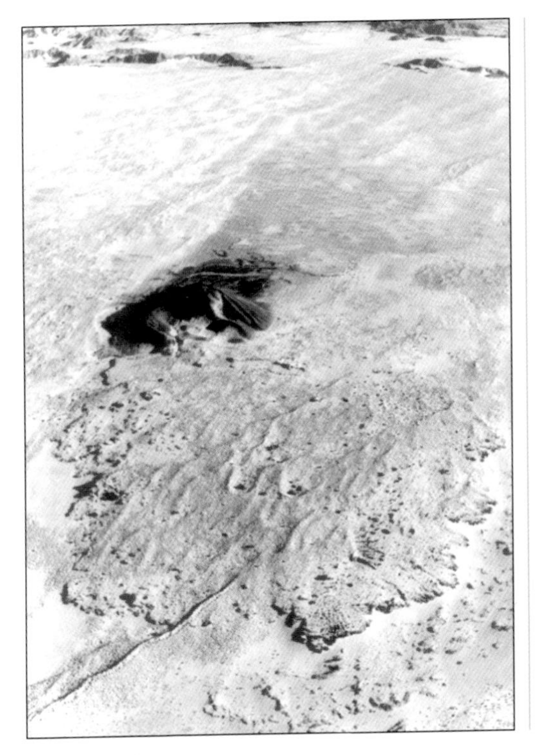

The Pinacate volcanic field of NW Sonora, Mexico, shows many examples of Martian features. The black cinder cone (left center) is associated with lava (lower center) that flowed toward the bottom of the frame. The flow was then mantled by bright, windblown sand, creating lava textures with bright desert colors, similar to the previous Martian lava flow image. Prevailing winds from lower left to upper right have created a dark wind tail of volcanic dust blowing off the cone, extending toward the upper right. (Aerial photo by author.)

10 km
1 km

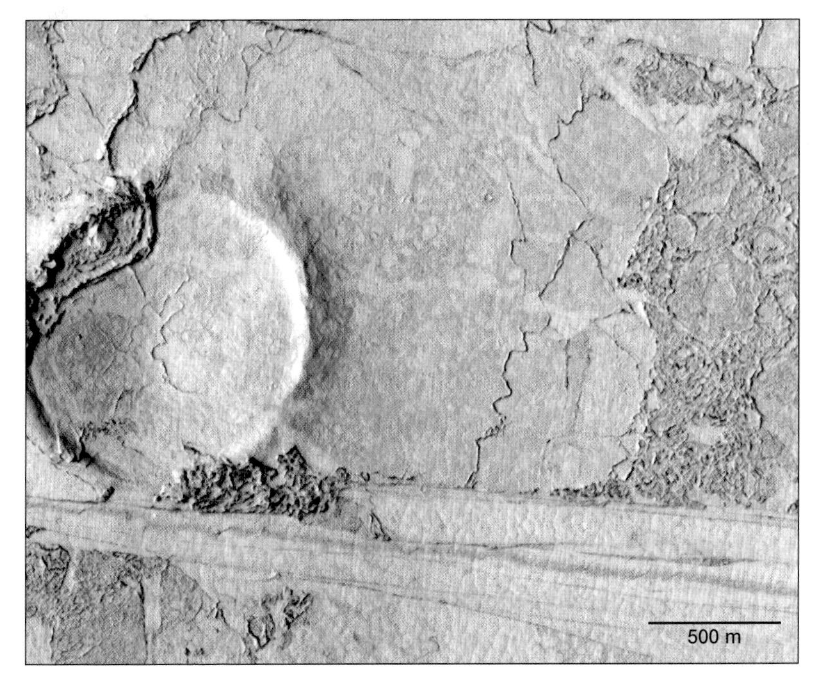

An impact crater almost buried by young lava in Amazonis Planitia (facing page). *In the inset Viking photo, this 5-km impact crater seems fresh, but the telephoto MGS view shows that it has been partly covered by even fresher lavas, probably less than 50 MY old, that lap up against the crater rim. If the lava had been much deeper, it would have broken through the rim at upper left and filled the crater. (166W 26N, M02-00364.)*

Above: *An impact crater partly covered by young lavas in Elysium Planitia. Unlike the previous example, this crater appears to have been partly filled by the lavas. One lava flow poured over the NW rim onto the floor (far left). The flows may be less than 20 MY old, so young that few if any meteorite impact craters scar the surface. This image contrasts rough flows (right) with smooth flows (center), as well as with mystifying patterns of winding ridges, linear streaks, and offsets, typical of very well-preserved young lavas on Mars. (207W 6N, M03-06931.)*

Imaging wizard and geologist Alfred McEwen at the University of Arizona was the first to concentrate on the MGS images of Elysium Planitia, and he made a striking discovery. The whole area was covered by nearly impact-free surfaces with strange textures unlike anything we had seen. Strange cobweb-like networks of ridges and crags divided smoother areas into a pattern something

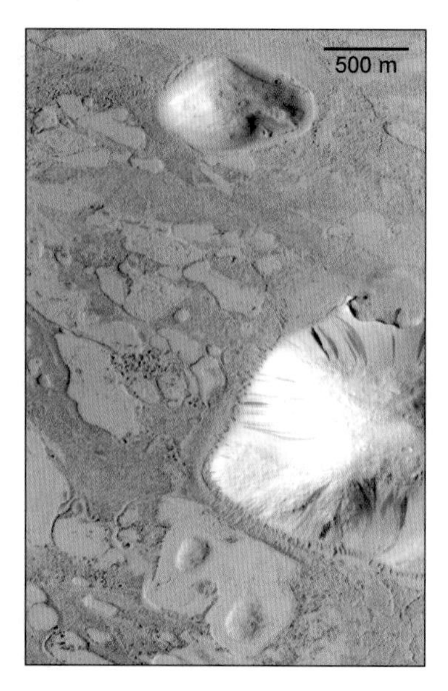

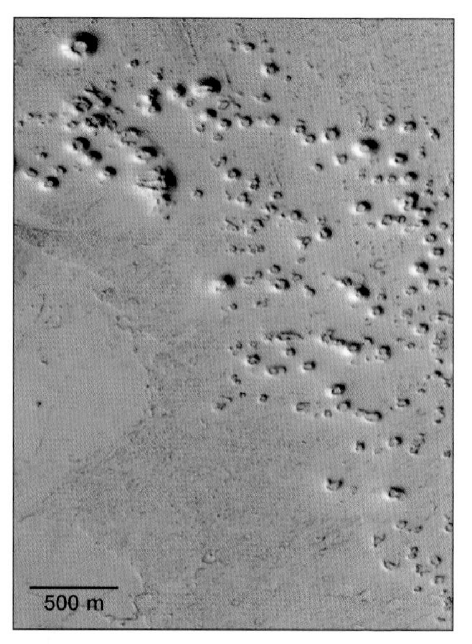

A morass of young lava flows surrounded these two hills in Elysium Planitia (above left). *The intricate lava patterns are a result of the chaotic processes of flow among fluid lavas. The hills display some impact craters, confirming that they are older than the lavas. Dark streaks on the hillsides are landslips where loose material slipped downslope; the darker they are, the younger they are. (203W 3N, M03-07387.)*

Above right: *Volcanic cones in the midst of young lavas on the Elysium plain. Nearly all the cones appear to have a central crater. They may be cinder cones (piles of debris around eruptive vents) or so-called "pingos," which are cones produced by lava interacting with ice. (190W 26N, M08-01962.)*

like fragments of a broken plate. McEwen and his graduate students recognized this as a pattern reminiscent of very fluid lava flows on Earth, especially those in Iceland. On that cold, windswept island, eruptions during the last few centuries poured fluid basaltic lavas from Earth's mantle across a beautiful, gray, mossy landscape. McEwen flew off to Iceland, chartered a plane to take aerial photos that would simulate MGS images, and came back with excellent matches to the new Martian images, giving a good argument that these might be fresh, uneroded Martian lavas.

Meanwhile, Dan Berman, a young research assistant at the Planetary Science Institute, worked with me to assemble crater statistics for this area and to quantify how fresh was "fresh." This was tricky. Among the few craters bigger than a few kilometers across, we could see scattered rims of old craters protruding above the much younger surface lava flows. Their appearance and numbers suggested that they themselves had formed on underlying lava layers hundreds of MY old and then were partly buried by younger flows. We focused on the freshest-looking flows that were superimposed on the older flows, both in Elysium and in Amazonis. On those surfaces, the small number of impact craters made the counting itself easy, but it meant surveying large areas in order to get a decent statistical sample. For many of the surface units with McEwen's well-preserved lava textures, our crater counts gave age estimates that were less than 100 MY. For the youngest, most recent flows, the few craters that we could count suggested ages as low as 10 MY or less. This sounds old by human standards, but we must always remember that 45 MY represents the last 1 percent of the age of Mars. We suspect that Elysium

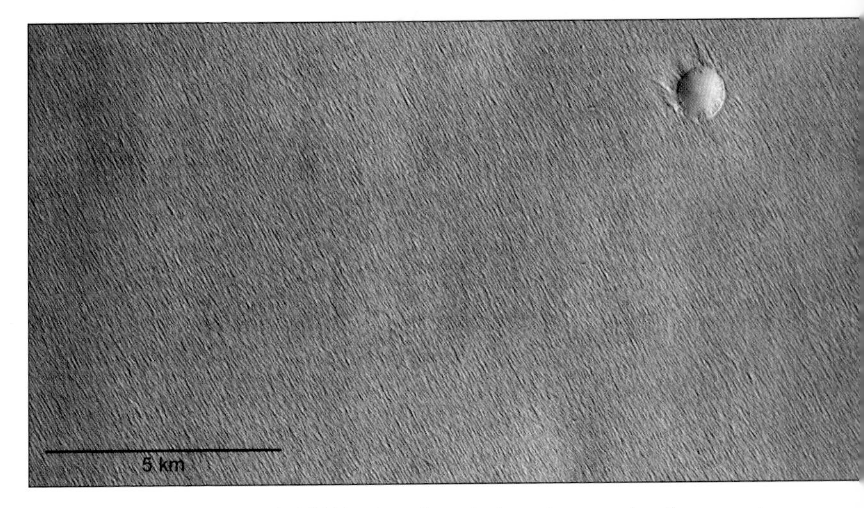

A lonely crater in windswept Amazonis Planitia. The surface of this plain is believed to consist of eroded ridges or dunes, aligned from SE to NW by prevailing winds. One fresh crater is seen, but its ejecta blanket has been obliterated. Absence of other impact craters suggests that the wind-scoured surface may be less than 50 MY old. (149 40N, Mars Odyssey visible-light image 20021125.)

Planitia and Amazonis Planitia have seen active lava flows oozing and steaming their way across the Martian surface during that most recent "moment" of Martian time.

These conclusions fit perfectly with the Martian meteorite evidence. If four to eight random Martian impact sites (locations unknown) have given us rocks with ages such as 170 MY and 325 MY, then it stands to reason that a search for the *youngest* eruptive sites should yield some sites less than 170 MY old. Our ages on the order of 10 to 100 MY thus support the meteorite evidence for young lava eruptions and vice versa.

CONTEMPORARY ERUPTIONS?

More important, these results from both meteorites and crater counts confirm our earlier arguments that volcanism is a continuing phenomenon on Mars. The rates, in terms of cubic kilometers of lava formed per millennium, are perhaps a hundred times lower than in the first third of Martian history, but if lava has flowed within the previous 10 MY, it is likely to flow again within the next 10 MY. Volcanoes on Mars could erupt at any time, although the likelihood in any given century is uncertain.

The confirmation of very young lava flows raises an interesting point about strategies for future robotic rover landers or even human landings. If we want to go to a place with the best bedrock exposures for rock sampling, uncomplicated by later sediments, the Amazonis and Elysium plains may be the place to land. Dating samples from such a site would be helpful in calibrating the crater-count dating system, and observations of fresh lavas might answer many questions about Martian volcanic processes. On the other hand, the volcanism there may be so much like Icelandic and other terrestrial examples that we would see only what we've already seen on Earth, so that such a site could be a disappointment from other perspectives. This will be an issue for future explorers and mission chiefs to decide.

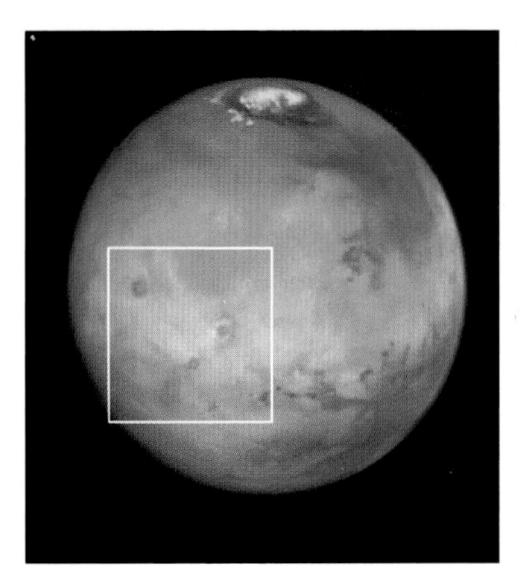

CHAPTER 27

THARSIS

Land of Spectacular Volcanoes

Much of the western hemisphere of Mars—from longitudes about 60° W to about 170° W—is dominated by a vast volcanic highland desert, a bright region mapped in the 1800s. Its name, Tharsis, chosen by Schiaparelli, has a colorful history. Judeo-Christian texts recorded Tharsis as a place that produced silver which was traded in ancient times, and Herodotus recorded that ancient Mediterranean sailors heading for Egypt were driven off

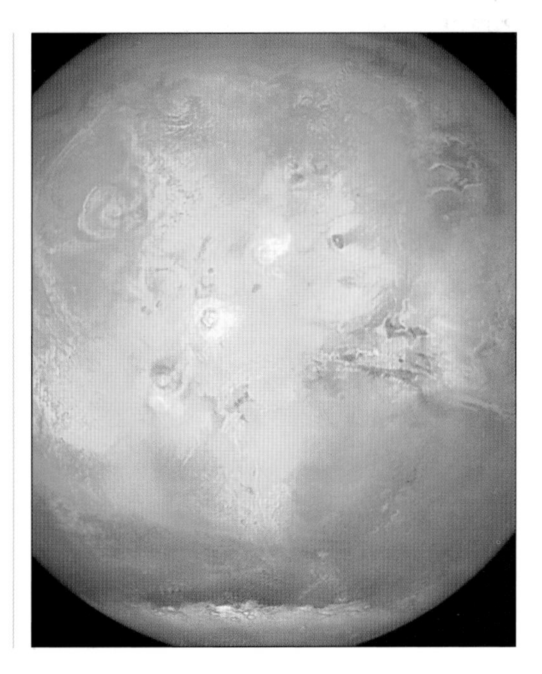

The Tharsis region of Mars is a bright desert marked by white cloud masses at the locations of the four tallest volcanoes, which include (from the top down): Olympus Mons (upper left, faint white patch), Ascraeus Mons (upper center), Pavonis Mons, and Arsia Mons (lower left center.) The visual view alone gives no hint of the huge, dome-shaped profile of this region of Mars. (MGS global mosaic, Malin Space Science Systems.)

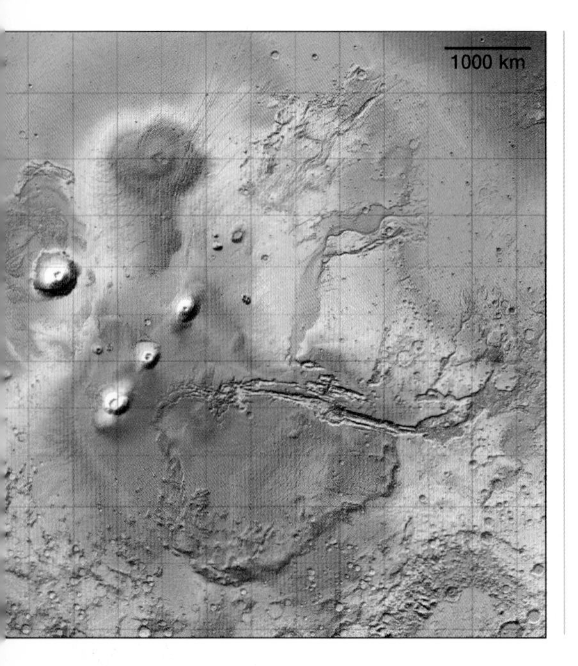

Laser altimeter map shows the enormous high dome of the Tharsis plateau. White is highest (peaks of the four largest volcanoes); brown, red, and yellow are the next highest, marking the fractured top and young lavas of the plateau. The green zone is of moderate elevation, marked by many downslope drainage channels. Blue marks the low floor of the Valles Marineris fracture system and the surrounding low northern plains. (MGS laser altimeter team.)

course and pushed through the Pillars of Hercules, where they came to a place he called Tartessus, a town on the Spanish coast just west of Gibraltar, founded in 1200 B.C. The town was eventually destroyed in 500 B.C. The town name now lives on, on Mars.

Topographic maps show that this region, centered roughly on the equator at about 115° W longitude, is a huge, gently swelling dome of lavas, rising 9 km (29,000 feet) above the mean elevation of Mars. The main part of the dome is about 3,000 km across. It could cover Europe or the western United States. You could call Tharsis the High Plains of Mars, but that hardly does it justice; these are high plains at the height of Mount Everest! The lava-covered plains themselves are called Tharsis Planitia. It is a single enormous volcano-tectonic construct.

The history of tectonics—or the fracturing, faulting, and deformation of terrain, presumably involving mighty earthquakes—is especially interesting. The lava-dome is surrounded by radial fractures that stretch from the equator to as far as 70° N latitude and 70° S latitude, clearly indicating unusual stresses on the crust and focused on Tharsis. The famous canyon complex, Valles Marineris,

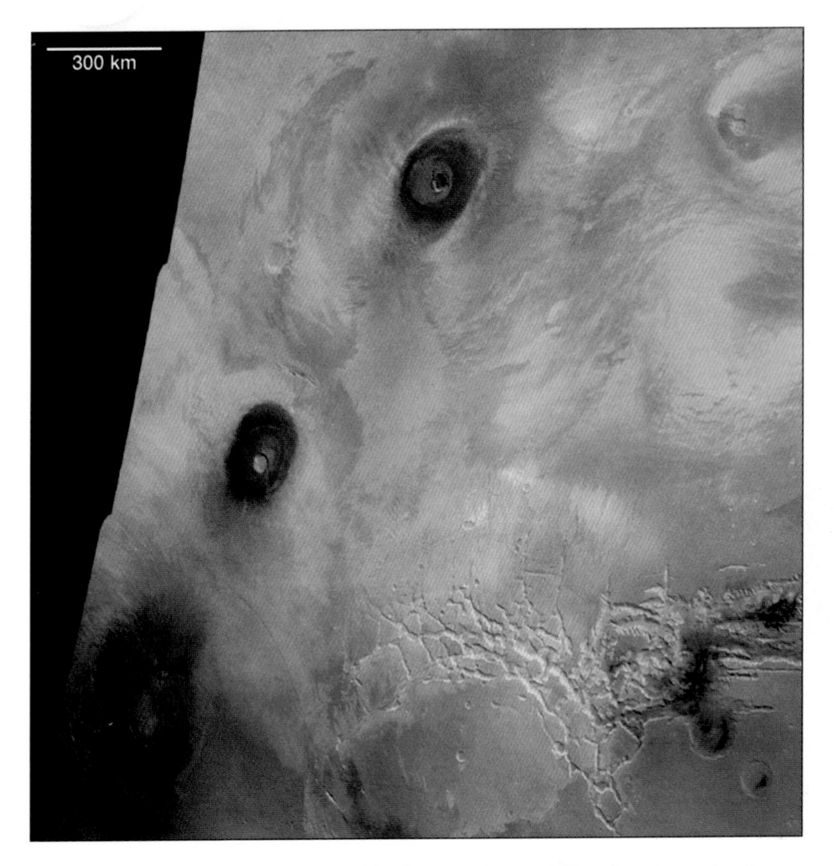

A view of eastern Tharsis, showing three of the giant volcanoes protruding above a low morning haze. At the SE corner of the view is a mazelike fractured area known as Noctis Labyrinthus, associated with the W end of Valles Marineris. Complex fracture systems are also found at the ends of similar size terrestrial rift valleys, such as the Red Sea. (NASA Viking I orbiter mosaic.)

is part of the system, pointing like an accusing finger from the heart of Tharsis. Most of the other fractures are basically simple cracks, but Valles Marineris has widened into a major canyon. This entire fracture system stretches 8,000 km across Mars.

The outer regions of Tharsis are mostly Hesperian-age lavas with bits of highly deformed Noachian surface visible here and there. The central parts are covered with young lavas. Along with Amazonis and Elysium, they are a possible source area for the 170- to 1,300-MY-old basaltic lava meteorites from Mars.

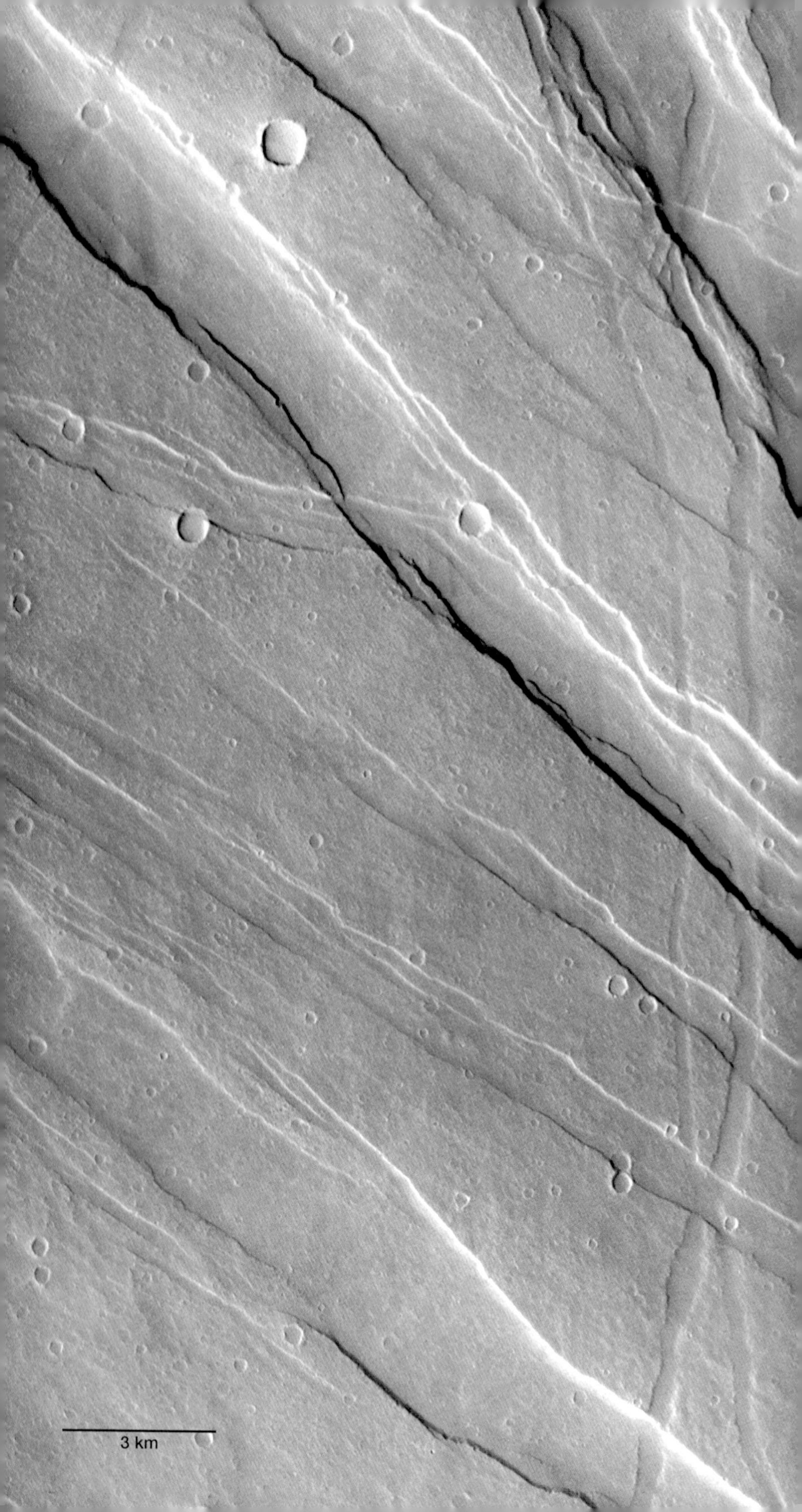
3 km

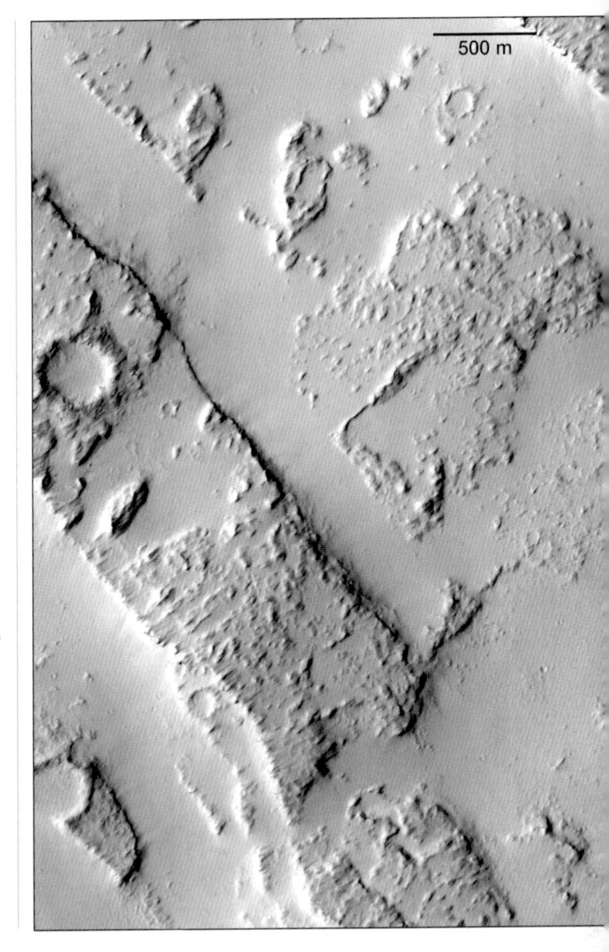

Part of the Tharsis area (facing page) is highly fractured in response to the strong tectonic forces that acted there. This swarm of parallel fractures forms sunken valleys separated by raised ridges. (Mars Odyssey Themis camera, visible image 020522.)

Dust and sand have mantled much of the topography of the Tharsis area. The low spots among the rugged fractures and craters in this image are silted in with fine, smooth material. (118W 8N, M02-01875.)

Strangely, Tharsis and Amazonis are not the dark color of basalt, but appear as bright deserts. As mentioned earlier, thermal spectrometer mappings show that these regions are not bare rock, but are mostly covered with a thin layer of fine dust. Global dust storms eject fine dust into the high atmosphere far above the summits of Tharsis, and this eventually sifts down and settles onto all surfaces. Perhaps the high dome of Tharsis, with its very thin air, lacks strong enough winds to blow the dust away once it arrives. In any case, the dust cover in most areas is thin (centimeters or meters), because we can see the fresh lava textures through it. In other areas, we can see thicker masses of dunes encroaching on recent lava flows.

A COMPLEX EVOLUTION

The young lava plains are only the cap of the great dome, the frosting on the cake. The dome itself has attracted two broad theories of development. In the first theory, it was uplifted by ascending currents from below; in the second theory, it was built by accumulation of lavas piled up on the surface. Geophysicists set to work to distinguish which model was right—in effect, to identify the major forces shaping Mars. For example, in the first theory, fractures around the dome would be the result of an *upward* flexure of the crust; but in the second, the lavas would create a great weight on the surface, creating a *downward* flexure. Given enough time, in the second view, the crust would deform and the lava pile would sink, like a cannonball settling into a barrel of asphalt on a hot day. The data fit this second theory. Researchers have actually mapped downward flexure around the outskirts of the Tharsis dome.

As for chronology, the dome is believed to have begun piling up in Noachian times from the eruption of lavas. This conclusion comes from the fact that the dome seems to have affected and deformed Noachian topography before the end of the Noachian period. A number of Mars geologists, notably Brown University's James W. Head III, James Dohm at the University of Arizona, and Kenneth Tanaka of the U.S. Geological Survey mapping program, have tried to use the geologic maps of different units to piece together the volcanic and tectonic history of the Tharsis dome construction. The results suggest a succession of different activity centers, starting with volcanic lava flows in the Noachian era, the opening of Valles Marineris in the late Noachian or early Hesperian era, and continued development of five individual Amazonian volcanic mountain cones that are dotted across the dome.

THE LAVA-LAMP EFFECT

The goal of such studies is to learn fundamental, underlying planetary processes. Vast symmetric patterns are clear at Tharsis, but what do they mean? The fractures, for example, clearly

The Tharsis dome is covered by lava flows of various ages, many of them very young. Here, NE of Olympus Mons, a dark, craggy, fresh flow (left) has flowed over an older region where several lava flow fronts can be seen (right). (126W 25N, E10-03393.)

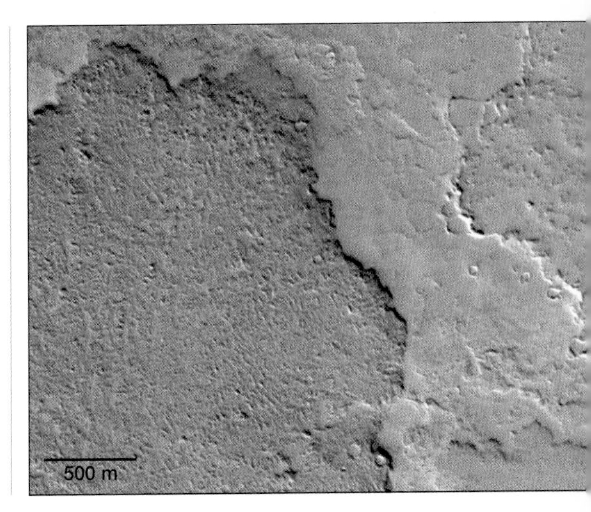

express a radial symmetry, and yet in detail they deflect around volcanoes and many seem to wander off in mysterious curves away from a direct radial line. Most geologists and geophysicists suspect that the underlying story of Tharsis involves a **mantle plume**—an ascending current of hot material in the sluggishly churning and convecting rocky material below the planetary crust. The plume of hot molten rock, or **magma**, rises toward the surface like a blob in an aptly named lava lamp. This rising plume does not in itself push up the surface, but where it hits the underside of the crust, a hot spot is created, pockets of magma are formed below the crust, and volcanic eruptions result.

The ultimate creator of plumes is heat released by radioactivity in the minerals of the rocky body of the planet. The heat accumulates inside, so that the planetary center becomes hot. The analog in the lava lamp is the hot lamp at the bottom. Just as in the lava lamp, circulation patterns called **convection currents** are set up. Hot materials rise. At the same time, material that has reached the surface cools and sinks. Convection patterns can be complex, with many ascending currents; or they can be simple, with just one ascending current that rises on one side of the planet and flows to the other hemisphere, where there is one descending current. One group of geophysical models of Mars calls for a simple pattern, with a long-lived, stable plume under Tharsis, constantly feeding heat to the area and providing masses of magma that find outlets to the surface through various fractures and volcanic vents.

Earlier we discussed how continental drift drags Earth's crust across mantle hot spots, likely causing a string of midsize volcanoes to be created, rather than one enormous mountain. Because the Martian currents are not strong enough to drive continental drift, the crust is relatively fixed and the long-lived mantle plume remained under Tharsis. Lava kept erupting, and what we see today is a 4,000-MY accumulation of flows.

DEFORMING A BALLOON

Around parts of edges of the Tharsis dome can be seen dim outlines of a broad, ragged trough, especially on the east side, which, in turn, defines key parts of many fluvial drainage systems on Mars, including the great waterway. This trough is believed to mark the sagging of the crust of Mars, due to the enormous weight of the piled-up lavas in the Tharsis dome, acting like a pile of bricks in the middle of a stretched rubber trampoline. Like the trampoline, the crust of Mars—or, more precisely, the outer layer of relatively stiff, elastic rock known as the **lithosphere**—deforms due to added weight.

But the trampoline analogy does not adequately cover the situation. Because Mars is a full globe, a balloon is a better analogy. If you press your finger into one side of an inflated balloon, the rest of the balloon does not maintain its shape. It deforms to adjust to the depression on one side. In a similar way, the enormous pile of lava on the Tharsis side of Mars has not only created a sag immediately around Tharsis but has also deformed the entire planet.

Geophysicist Roger Phillips of Washington University in St. Louis, working with a team of ten other geophysicists in 2001, made computer models of the shape of Mars as it responded to the accumulation of lavas in Tharsis. They concluded that it would cause a bulge in the crust of Mars on the side opposite to Tharsis,

Two different lava textures on Earth, named after the Hawaiian terms for these types of flows. The differences are related to lava fluidity, temperature, and composition. These two types account for some of the differences between smooth and craggy flows seen on Mars. Top: *"Pahoehoe" flows are smooth with a texture like molasses. (Mauna Kea volcano summit, Hawaii; photo by author.)* Bottom: *In "aa" flows, the crust solidifies and then is broken into incredibly rough, craggy slabs as the underlying flow continues to move. Such flows would make dangerous landing areas! (Pinacate volcano, Sonora, Mexico; photo by author.)*

and that this bulge actually exists—namely, in the ancient high crust forming the region of Arabia. Of course, this idealized planetary shape has been modified by other geological events. The Hellas impact basin punched into the south part of the Arabia highland, and the trough around Tharsis has been partly filled with sediments by Martian rivers draining into it. As Phillips's team pointed out, many of the Noachian valley network systems around the planet show drainage directions that seem to be influenced by the topographic distortions due to Tharsis, which is part of the argument that the main lava mass of the Tharsis bulge was in place by the end of the Noachian era.

DID THARSIS AFFECT THE CLIMATE OF MARS?

Phillips and his colleagues go on to connect the beginnings of Tharsis with the early atmosphere of Mars. They estimate that 300 million cubic kilometers of lava had to be erupted to form the main bulk of the Tharsis bulge during the Noachian era. In this way, Tharsis brings us back full circle to climate issues that arose when we visited Noachian sites. Lava eruptions from Earth's upper mantle release carbon dioxide and water, and so the Phillips group suggested that the total carbon dioxide emitted by infant Tharsis could have been enough to transform the Martian atmosphere. They estimate that the total gas could amount to 1.5 times Earth's present atmosphere (i.e., 1,500 millibars of pressure), assuming it was not lost into space as it was erupted. Similarly, the total emitted water vapor would have been enough to make a global water layer 120 meters deep (on a flat Mars), or, more realistically, seas that were a few hundred meters deep, covering the low areas of Mars. These numbers are consistent with evidence discussed in earlier chapters, indicating that early Mars may have had atmospheric pressures of several hundred millibars as well as abundant water.

There is a second way in which Tharsis may have affected the climate, a more subtle and perhaps ultimately more profound way, mentioned briefly in chapter 23. Remember that Mars has a 25°

axial tilt, very similar to the 23½° axial tilt of Earth; and as explained in chapter 23, the Martian tilt changes by substantial amounts. The tilt affects the climate, because high axial tilt points the summer polar cap toward the sun, allowing the entire polar ice cap to melt or sublime. This probably raises the total air pressure only slightly, but it puts much more water vapor into the atmosphere and allows more water mobility on the planet.

There are interesting additional wrinkles in the story of Tharsis's evolution. As early as 1979, Colorado dynamicist William Ward and researchers at Cornell University pointed out an intriguing effect. Today's Mars, with the Tharsis dome, is like a billiard ball with a lump of dirt stuck on one side: The extra mass affects the spin. The 1979 team "removed" the Tharsis mass (in the computer model!) and recalculated the axial tilt changes. The team concluded that the axial tilt variations may have been greater before Tharsis formed. Estimates of the effect call for an axial tilt range from about 10° to 50°, so the warming effects on polar ice and high-latitude ground-ice deposits, during the time of maximum tilt, could have been greater in early Noachian time before Tharsis formed. Remember that the timescale for axial tilt changes is a few MY, so the pre-Tharsis Noachian climate may have had wild variations every few million years. At zero axial tilt, both poles would be cold all the time; early water vapor would have remained frozen there, as in our arctic regions. At high axial tilt, a few MY later, all this would have burned off. The combination of massive CO_2 and H_2O gas release into the atmosphere during the initial Tharsis buildup, before 3,500 MY or even 4,000 MY ago, plus the total burn-off of polar ices every 10 MY or so during cycles of high axial tilt—not to mention a cooler early sun combined with a greater heat flow from inside primordial Mars—surely produced a different Mars from the one we know today.

Tharsis may thus be a key to the radical climate change on Mars. As Tharsis grew, it temporarily thickened the atmosphere; at the same time, it reduced the range of axial tilt.

According to speculations by Carl Sagan and others, the volcanic episodes and/or axial tilt cycles could even cause abrupt flips from a cold, dry Mars to a warm, wet Mars. The logic is this: Suppose an eruption pumps enough CO_2 into the atmosphere to exceed a certain critical threshold of greenhouse warming, at which point CO_2 and H_2O ice at the poles melts or sublimes (this could be especially efficient if it happens during a time of high axial tilt when the summer poles are tipped toward the sun). Then more greenhouse gases go into the atmosphere, which causes still more greenhouse warming. A runaway greenhouse effect thus develops and creates a warm, wet Mars, until the volcanism subsides. Then, as the greenhouse gases are lost (for example, by molecules drifting off into space due to Mars's low gravity), the greenhouse effect subsides and Mars grows colder. That causes the CO_2 and H_2O to freeze, and a runaway cooling ensues, in which Mars reverts to its cold, dry state. The quick oscillations, perhaps on timescales of a few MY at a time, could explain some evidence for sudden shutdowns of terrain softening and for sporadic, last-gasp flows of water in the Hesperian river channels.

All of this at the moment is only hypothesis. Improved measurements from Mars, such as the mass of the Tharsis lavas and various dynamical properties, together with improved theoretical understanding of the axial tilt oscillations, may help us to refine our knowledge of the possible climatic side effects of the Tharsis dome.

MIGHTY VOLCANOES

Actually, we've glossed over the most striking aspect of Tharsis. Dotted across its central rise are the mightiest volcanic cones in the solar system. Four distinct conical volcanoes rise to about 22 km in altitude or more (about 70,000 feet in altitude), and a fifth volcano to the north is lower but larger in diameter. We'll describe them more in the next chapter, in which we visit the tallest and most famous: Olympus Mons.

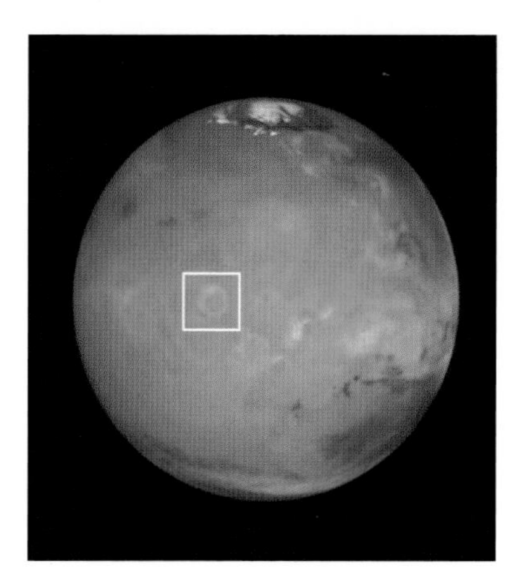

CHAPTER 28

OLYMPUS MONS

Largest Volcano in the Solar System

A visit to Mars is incomplete without a tour of the extraordinary mountain known as Olympus Mons. Olympus Mons is not only the highest mountain on Mars, but the largest volcanic cone in the whole solar system. Like Mount Fuji, Mount Shasta, and other famous volcanic peaks of our own planet, it is basically a huge pile of lava flows that erupted at one spot, piling upon one another to build a mountain. But Olympus Mons dwarfs all mountains we Earthlings have seen, volcanic or not. Its discovery is a 100-year odyssey of Martian exploration.

DISCOVERY OF AN ENIGMA

Big as Olympus Mons is, it is not big enough to have been seen or understood by telescopic observers in the era before the age of spacecraft exploration. Still, something odd was noted there. Tenacious astronomers, peering through their telescopes of glass and brass in the 1800s, kept records of changes they saw on the red planet, and the region of Olympus Mons emerged as an early puzzle. Sometimes observers saw a bright spot there, like a strange white beacon in the broad orange Martian desert of Tharsis. That's why in 1879 Schiaparelli named it Nix Olympica, or Snows of Olympus. The name turned out to be wonderfully prescient, because without any knowledge of the true nature of

Mars, Schiaparelli had named Mars's grandest peak after the grandest peak of classical history—snowcapped for much of the year—where the mythic Greek gods had made their everlasting home. Schiaparelli was not necessarily claiming that Nix Olympica was a mountain. He probably chose the Greek peak more for its mythic luminosity. Homer, for example, wrote of Mount Olympus as being bathed in a "white radiance" that glittered like a frosty "plain of snow."

For decades, Nix Olympica continued to be viewed as a variable, bright patch. It's shown, for example, as a small, very bright spot in the middle of the Tharsis desert on the map of Mars published in 1930 by the Greek-French master observer, E. M. Antoniadi. Other bright spots were seen in the Tharsis area, and many observers reported that they could behave strangely. One such incident was reported in 1951 by Japanese amateur observer Tsuneo Saheki, who observed a bright "flare" roughly 1,200 km (700 miles) east of what we now know as another Tharsis volcano, Pavonis Mons. He saw it as "a very small and extremely brilliant spot" that within half an hour brightened to rival the polar ice cap and then faded in another half hour. Astronomers debated the meaning of such observations. Could such a "flare" be the flash of a volcanic eruption? Or a glint off ice? In those days, advanced life-forms on Mars still seemed possible. Could the "flares" be something artificial—a signal from a Martian civilization? Frustrated astronomers of the 1950s could only speculate about the enigmatic red planet.

DECIPHERING THE BRIGHT SPOT

The bright spot called Nix Olympica turned out to be caused by the clouds that frequently form atop Olympus Mons. Transient clouds can form over our own terrestrial mountains in this way and are called orographic clouds. Over mighty Olympus Mons, they can swell to prominence in just a few hours, then dissipate in the thin Martian air—all in response to the ebb and flow of wind and thermal conditions around this towering but now frozen inferno.

Clouds were proposed as the cause of Nix Olympica's bright spot as early as 1907 by E. C. Slipher of Arizona's Lowell Observatory. In that year Slipher made 13,000 photographs of Mars, hour after hour, with an 18-inch-wide telescope in Chile, where the planet was favorably positioned for observation. The photos showed a wealth of detail, including temporary whitish veils that occasionally masked some of the familiar dark markings. Most curious of these bright patches was a recurring pattern of four or five distinct bright spots in the Tharsis desert, most visible in photos made in violet light. These spots formed a shape that was vaguely like the letter W. One of the patches corresponded to Schiaparelli's Nix Olympica. The growth and dissipation of the white spots led Slipher to suggest they were clouds. But there was no direct proof.

The enigmatic W became famous among Mars observers. Slipher called it "the most remarkable meteorological phenomenon ever observed on Mars." It was photographed again at later oppositions, most notably in October 1926 and most spectacularly in June 1954, when astronomers at Lowell and Mount Wilson Observatories documented it. The mysterious clouds appeared faintly at first around two o'clock in the Martian afternoon. They gradually developed and brightened until, toward sunset, they became so prominent and bright as to rival the Martian polar ice cap—as reported a few years earlier by Saheki in Japan. On the next morning, the spots were entirely gone, only to reappear in the early afternoon and gradually repeat their behavior in the same place during the afternoon.

Caltech astronomer Robert S. Richardson, who independently rediscovered this formation in 1954 at Mount Wilson, joked that he could not decide whether the pattern was really a W, which could stand for the god of war, or whether it should be viewed the other way around as an M, which could simply stand for Mars.

THE FIRST CLOSE-UPS

Everyone hoped that the first space probes to Mars, in the 1960s, would help solve the mystery of the white "flares" and the W-spots. However, none of the twenty-two photos snapped by Mariner 4 showed the region of Nix Olympica during the first flyby in 1965. The close-up pictures by Mariners 6 and 7, in 1969, again missed the Nix Olympica region, but a few distant shots on approach to Mars showed the whole planet much more clearly than from Earth, and Mariner 7 caught a good view of the Tharsis side of the planet. Mysteriously, Nix Olympica showed up as a murky, bright, ring-like feature. This confirmed the brightening but didn't reveal a cause.

The discovery of the mountain itself and the cause of the clouds—and the conversion to the topographic name Olympus Mons—came about only by further spacecraft exploration, a century after Schiaparelli. Mariner 9 arrived at Mars in late 1971, went into a preplanned orbit around the planet, and made the first close-up global mapping of the Martian topographic features. As a member of the imaging team on that mission, I recall the excitement of the day when the true nature of Olympus Mons was revealed. By chance, a huge, planetwide dust storm was raging when "we" approached Mars. (Spacecraft scientists and engineers always talk as if they themselves are out there in space, not just our robotic probe.) The dust pall covered up all the traditional features, and in our approach photos the planet showed a disappointingly featureless disk. Closer examination of the accumulating photos showed four mysterious dark spots on the otherwise bland surface—in the region of Tharsis. What were those four dark spots? We kept looking at each new set of images, trying to make sense of fixed spots in the midst of a dust storm.

In those days of Vietnam and Watergate, we could not instantly forward an image from one computer to another, but had to rely on TV images being turned into photographic prints. Unable to link by e-mail or JPEG images, the scientists on the team had gath-

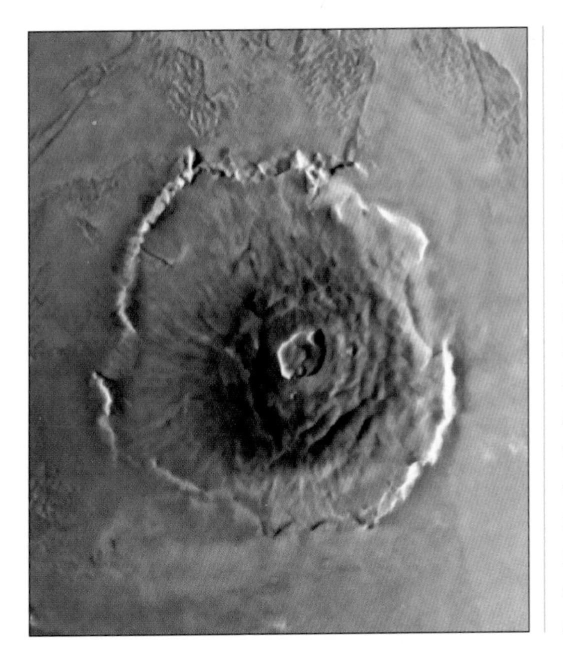

Looking straight down on the largest volcano in the solar system. This view shows the summit caldera (center), a faint radial pattern of lava flows extending downslope, and the cliff that surrounds the base of the mountain. Reddish brown colors on the summit area are slightly more pronounced than they are farther downslope because the summit area protrudes from haze in the lower atmosphere. (Viking I mosaic.)

ered at Caltech's Jet Propulsion Lab, where they waited for each day's photo prints. One day, Carl Sagan, another team member, came running into the project office, waving a Polaroid photo, the best image yet of one of those puzzling dark spots. In its center was a crater! More specifically, it was not an impact crater but had the morphology of a volcano's summit collapse caldera. In other words, in the space of a minute, we had the answer to the century-old mystery of Nix Olympica: it was a huge volcano, so tall that its top protruded above the dust clouds. The photos soon showed that all four dark spots were the tops of giant volcanoes with summit calderas.

Over the next few weeks, the dust slowly settled back to the surface of Mars, and the lower flanks of Olympus Mons were exposed in all their majesty. We soon saw that this mountain was high enough to cause the occasional formation of clouds over the summit, confirming that the notorious transient bright spots were short-lived massive clouds. Similarly, ground fogs sometimes collected around the cliff that marked the base of the mountain, creating the misty circular ring seen by Mariner 7.

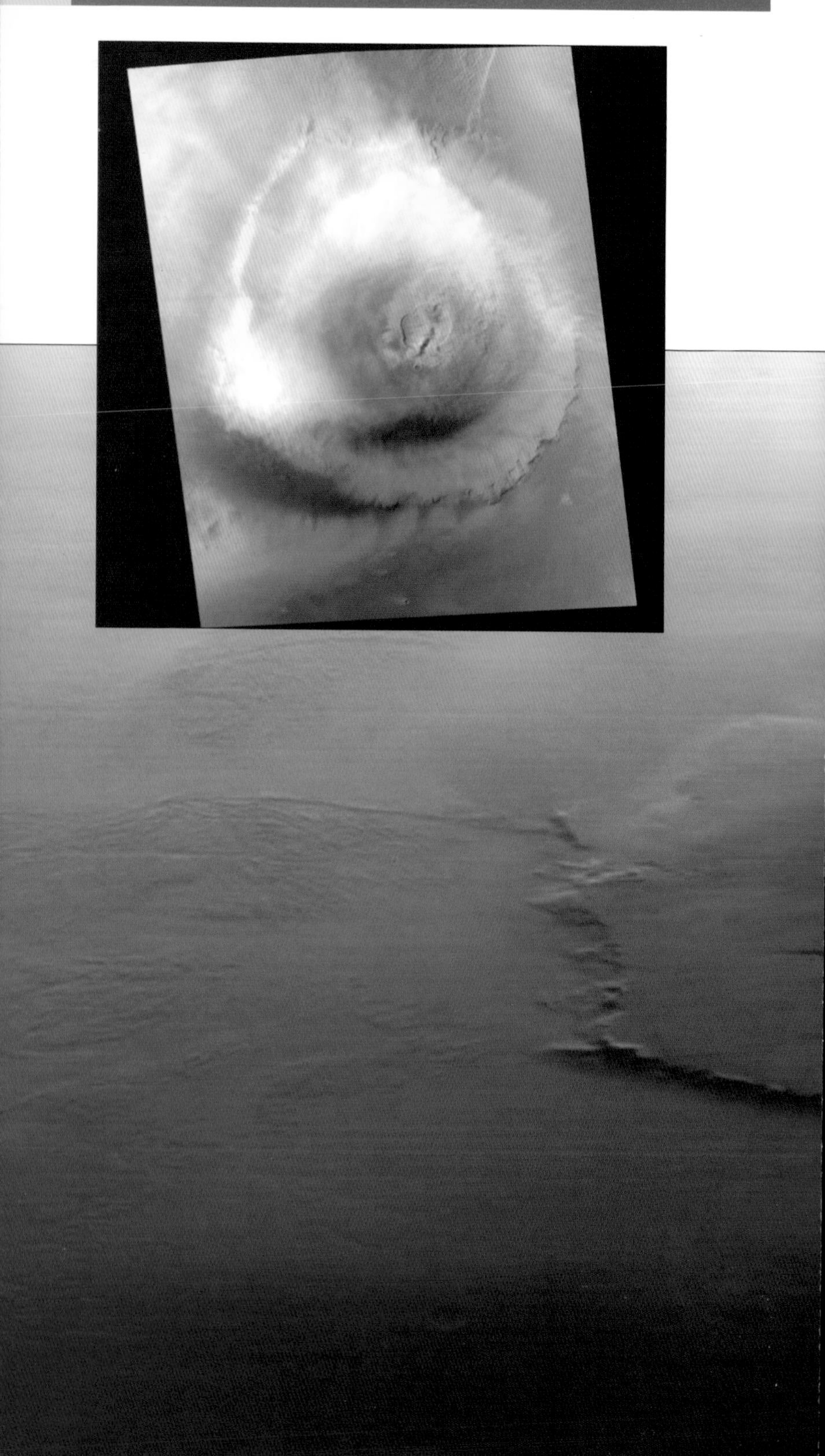

Late-afternoon view of Olympus Mons from orbit on a hazy day shows late afternoon bluish-white clouds forming over the summit and extending to the northwest (inset). Such cloud masses formed the bright spot sometimes seen from Earth, which gave the original name "snows of Olympus" to this feature.

Olympus Mons at dawn, looking east from orbit by the Mars Global Surveyor orbiter. The sun is rising on the plains at the top of the image, and illuminating the east flank of the volcano. The west flank and foreground are deep in dusky twilight shadow (below).

The entire slope of Olympus Mons is made up of overlapping lava flows. In this image of the southwest slope, lavas have flowed from the top (upslope) toward the bottom. Each flow moves downhill until it cools and solidifies in a tongue-shaped lobe, like wax spilled from a candle. Several of these "flow front" lobes can be seen in the upper two-thirds of the image. Similar surfaces exist on the slopes of large volcanoes on Earth, such as Mauna Loa in Hawaii. The sparse scattering of impact craters suggests an age of no more than a few hundred MY. (136W 13N, Mars Odyssey image 20021017.)

Impressive cliffs at the southeast base of Olympus Mons include raised blocks of faulted terrain, engulfed by lavas on the slopes. The process that formed the cliffs is unknown. Box shows the location of the next image. (131W 15N, M08-01934.)

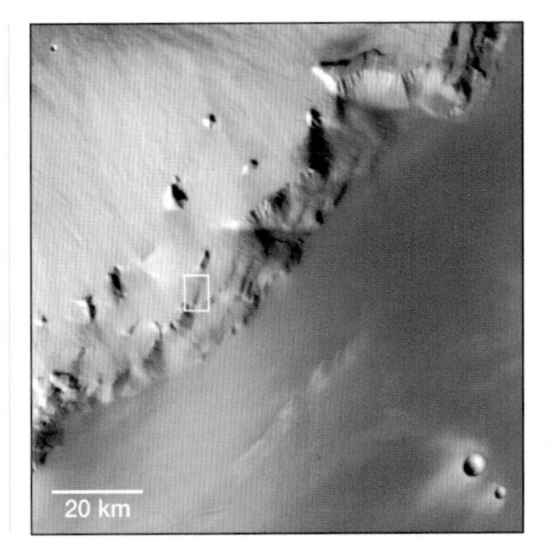

"Motionless torrents, silent cataracts." This quote from Samuel Taylor Coleridge (about glacial crags, and inscribed by English composer Ralph Vaughn Williams on his Antarctic Symphony), beautifully fits this view of frozen lavas cascading over the cliff at the base of Olympus Mons. This image shows numerous examples of lava levees, which are parallel ridges formed when the lava flow solidifies at its edges, while the middle, molten part of the lava then flows on downhill, leaving a depressed valley in the center. A good example can be seen in right center, flowing from above the cliff (top), down the slope, and then swerving toward the left. (131W 15N, M08-01933.)

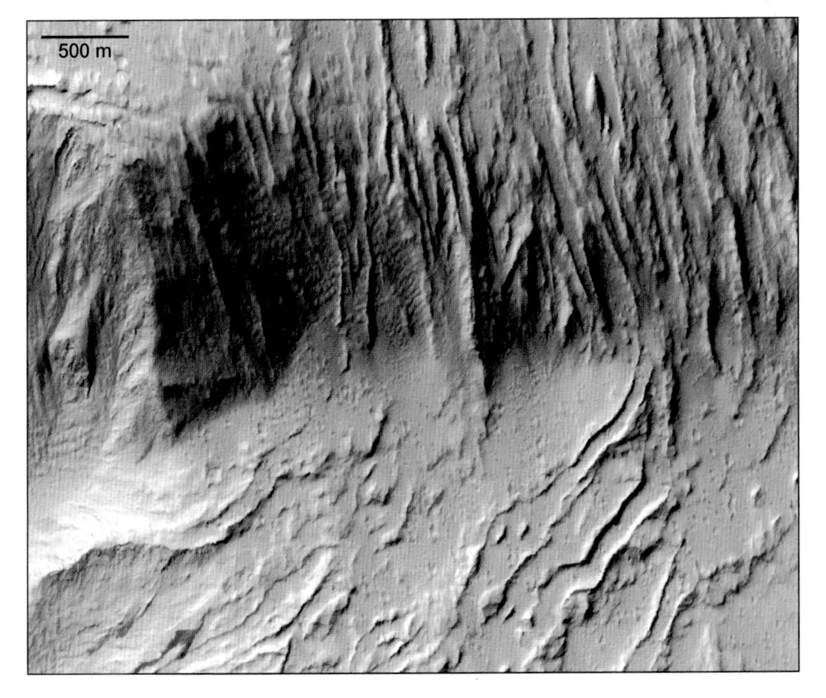

THE NATURE OF THE BEAST

A swollen lava heap about 600 km (400 miles) in diameter, Olympus Mons rises about 21 km (13 miles or 69,000 feet) above the mean Martian reference altitude and roughly 18 km (58,000 feet) above the immediately surrounding plain. These altitudes were measured with astonishing precision by the MGS laser altimeter—the altimeter team actually reported the height as 21,287.4 meters, implying a precision of 10 centimeters (a few inches). The last one or two decimal places hardly mean much in real life; they would be changed by the movement of a boulder here or there!

Here, then, is a single mountain mass three times as high as Mount Everest, with a width that would straddle the whole Himalayan range. Its base is defined by a cliff rising a kilometer (around 3,000 to 4,000 feet) and extending all the way around the mountain. The volcano would approximately cover the state of Missouri.

East of Olympus Mons, in a nearly north-to-south row, lie three more volcanic peaks that are similar in size, if not quite as impressive. From south to north they are Arsia Mons, Pavonis Mons, and Ascraeus Mons. According to the same laser altimeter measures, their heights are 17.8 km, 14.1 km, and 18.2 km, respectively. (The precision-reported height of 14,057 meters for the south rim of Pavonis Mons comes in just 69 meters lower than Elysium Mons, making Elysium the fourth-highest mountain on Mars and Pavonis the fifth highest.) Farther north is a fifth prominent volcano, rising only about 2 km (6,500 feet) above the mean surface but broader than the others and surrounded by an enormous system of curved fractures. Its name, Alba Patera, is based on the same root as *albino* and refers to "the white region," another reference to clouds in the area. This volcano is about 540 km (340 miles) across, wide enough to cover the state of Nebraska, but its semi-concentric fracture system could cover all of Texas. Together, these five peaks and a number of smaller satellite volcanoes literally cap the grandest known volcanic province.

Paradoxically, Olympus Mons is so big that it would be impossible for an astronaut to see it from the surface of Mars. There would be no single vantage point from which you could perceive its overall shape. If you were up close, you could see part of the cliff around the base, or you could hike along twisted lava flows on its gently sloping flanks, hardly sensing you were on a Missouri-size mountain. If you backed off to "see it from a distance," it would recede in the haze and disappear over the horizon. It would be like trying to see the entire Rocky Mountain chain or the Great Lakes from one spot on the ground.

You can do better from space. From orbit above Mars, we've looked down on the mighty structure and seen it whole—a circular, shield-shaped cone rising at a shallow angle from the Tharsis desert. Often it is capped by a mass of thin clouds that form near the summit in the same way that clouds build over mountains on Earth, as winds push air upslope, causing it to cool so that ice crystals or water droplets condense. On the summit, an orbiting observer can see a prominent volcanic caldera, big enough to enclose a city the size of New York or Los Angeles. This depression is about 70 km (42 miles) across and around 3,000 meters (10,000 feet) deep.

The closest terrestrial analog to Olympus Mons is the volcano Mauna Loa in Hawaii. It rises to 4 km (14,000 feet) above sea level, but only its summit sticks above the waves of the Pacific. It is really a giant cone of lava that rises 8 km (28,000 feet) above its base on the seafloor. Like Olympus Mons, it has gentle slopes formed by overlapping flows of a fluid basalt lavas. Also like Olympus Mons, it has a complex of overlapping calderas on its summit, but these are only a few km across. Mauna Loa as a whole is only about a third as big as Olympus Mons. The cause of the cliff around the base of Olympus Mons, shown on page 305, is uncertain. The other volcanoes lack such a sharp cliff.

My Martian Chronicles

PART 11: THE VIEW FROM ORBIT

Nearly 30 years after the fledgling efforts of Mariner 9, I found myself a member of the imaging team on a new Mars orbiter named Mars Global Surveyor.

This spacecraft had a very high-resolution, or telephoto, camera designed by Mike Malin of Malin Space Systems in San Diego. Mike's a fascinating guy who followed an unusual path after graduating with a Ph.D. in planetary science. Instead of the usual route of climbing the academic ladder, he got one of the MacArthur "genius" grants and then formed his own company to build spacecraft cameras. In telephoto mode, his camera system could photograph features far smaller than ever seen before on Mars. The beauty of the design, however, is that the system also had a wide-angle mode, which could show wider views of various large-scale features.

After a few months on the team, I discovered that the meteorologists in our group used the system at maximum wide angle to get photos of the cloud patterns on the whole planet, and the geologists zoomed in to use the maximum telephoto, narrow-angle capability, looking straight down to shoot images of rocks and dunes and lava flows. Few, however, were using the intermediate capabilities. We were ignoring the chance to shoot moderate-angle images out to one side, like an ordinary 35-millimeter camera view out an airplane window. Some basic but obscure law of human nature dictates that people will zoom their lenses either all the way in or all the way out; few use the intermediate focal lengths. One dream I had as I joined this team was to get the kind of astronaut eye view you might get out the window of a space shuttle orbiting Mars: an oblique view looking out toward the horizon, showing Olympus Mons or similar features in a more human format. I talked to Mike and to his assistant Ken Edgett, who headed the actual programming of the camera. We debated:

"But if you take that kind of picture, you aren't getting any better resolution than Viking got back in the seventies. What's the point?"

"We don't have many images of Mars geology seen in this way. We're just getting straight-down high-res images."

"But what you want is just a pretty picture." Among planetary scientists, the term *pretty picture* had somehow become a dismissive name for something that might appeal to the masses or to the public relations office but had no real technical value.

"But it would help us see Mars as we see things on Earth. When you see things in a more familiar way, it spurs

A wide-angle astronaut's eye view of the south half of Olympus Mons, looking across the volcano toward the hazy Martian horizon. This view was obtained by Ken Edgett and the author by programming the MGS camera in an unusual side-viewing mode.

(continued on page 312)

(continued from page 310)

new ideas about them, associations with things you already know."

"It uses up a lot of bytes of memory in the system."

"It's worth it. These pictures will be reproduced more than most because people will understand them . . . There's lots more orbits to get the usual kinds of pictures."

In the end, Mike yielded to his philosophy that team members should be able to use the camera in whatever creative way they preferred. Ken and I targeted a few of these "tourist snapshot" views of Mars. There was only a brief opportunity to obtain them in the months when we were in our initial high orbit. Various geometric and other restrictions involving spacecraft speed prevented this scheme after we got into the final "mapping orbit" at low altitude. One of our best successes came when we targeted Olympus Mons. We used much of the available computer memory on one orbit to get a shot of Olympus Mons viewed looking east obliquely under low sun. Alas, a glitch in the transmission lost the north half of the volcano, but the image that came back was exquisite and was reproduced in *Astronomy* magazine, *Sky & Telescope* magazine, and other venues. I had fun teasing Mike when he showed up at a planetary science meeting and used this image as the lead slide in his talk.

"See, I told you it would be impressive!"

"I agreed to do it, didn't I?" To his credit, he did.

DATING THE LAVA FLOWS: IS OLYMPUS MONS AN ACTIVE VOLCANO?

Mars Global Surveyor views showed exquisite detail on the slopes of Olympus Mons and revealed that some of the individual lava flows have almost no impact craters! This meant they must be very young in geological terms. As we counted craters on these features, we affirmed that, although the mean age of the volcano's flanks may be a few hundred MY, the youngest flows could have formed within the last 10 MY. This is extremely contemporary, less than half of 1 percent of the age of the planet. It's very unlikely that volcanism shut off just before we humans got our instruments to Mars, so the likely conclusion is that Olympus Mons is not extinct, but just dormant. Volcanism continues in modern times on Mars! Eruptions may still be many thousands of years apart, so Olympus Mons is unlikely to be active in our lifetime (though, in private moments, one wonders if some of the flare clouds just might be billows of steam from a volcanic eruption!). In any case, the fires of Olympus Mons shall rise again.

TREKKING OLYMPUS MONS

Imagine a party of future astronauts reaching the summit of Olympus Mons. What would they see and what would be their exploration goals? If they stood on the rim of the summit caldera, the view would be awesome. Looking across the city-size floor, the explorers would make out not a singular round depression but a complex of several intersecting calderas of various depths, a battlefield of geologic forces, suggesting an elaborate history of subsidence. Each depression is a volcanic sinkhole, produced when lava drained from a subterranean chamber beneath the summit's surface and the surface collapsed. How terribly the "earth" must have trembled during such an event! How fantastic were the resultant clouds of steam and volcanic dust roiling up into the high Martian atmosphere, obliterating the tack-sharp stars from the sky and catching rays of the sun well after it set below the horizon.

Hiking downslope, the pioneers of Olympus Mons would forget they were on a mountain. Explorers could traverse 10 to 20 km of lava flows and dust deposits before they came to a steep embankment overlooking the next terraced slope below. Hiking along a single flow, they would come upon serpentine channels—lava riverbeds—that stretch a few hundred meters across, tens of meters deep, and tens of kilometers in length. Among these flows they might find lava tubes—long, narrow caves evacuated by lava—that could be sealed and used for habitats. In other places, vast prairies of bare rock stretch downslope to the top of the basal cliff, which stands a thousand meters above the orangish surrounding plains, the Tharsis desert. Even there, the flows do not end but swerve in nervous paths around obstacles and cascade down the cliff face onto the surrounding plains, where many of them are hidden in vistas covered by windblown dust and sand dunes.

CHAPTER 29

VALLES MARINERIS

The Grandest Canyon of All

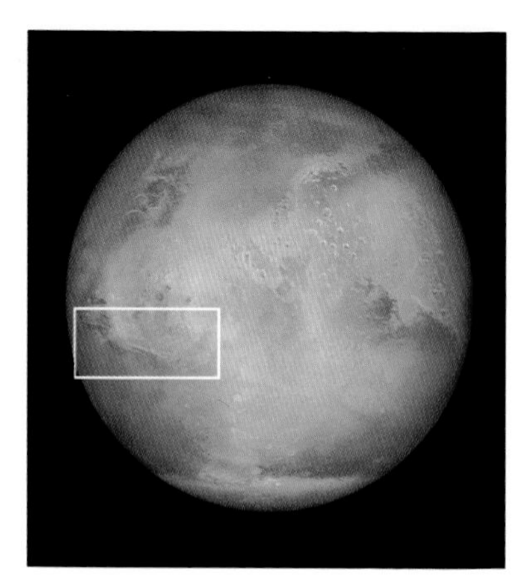

Amid the hundreds of so-called canals that Percival Lowell, Giovanni Schiaparelli, and their cohorts mapped on Mars in the late 1800s was a short, stubby one extending into the Tharsis desert, ending in a bright, oval region that contains a dark patch with the resounding name Lake of the Sun. The obscure canal itself was given the name Coprates by Lowell in 1894, after a river in Persia, now Iran.

When Mariner 9 began the first close-up mapping of Mars in 1971, no one had any reason to think that the Coprates canal was any more significant or real than the others—wind streaks at best and illusions at worst. Yet, soon after Mariner 9 arrived in orbit, its cameras showed an odd thing in the region of Coprates. As the impenetrable dust pall of the 1971 dust storm settled, a dim whitish band began to appear in the position of Coprates canal, stretching a thousand kilometers (600 miles) east-west near the equator. Here was a mystery! Why should a feature mapped as a dark line show up as a bright streak? As the dust continued to settle and clear, the answer became apparent: this bright streak was airborne dust, filling a fantastic canyon system. Even as the rim of the canyon became clear, the denser air near the canyon floor remained filled with airborne dust for many weeks. Parallel bright streaks appeared because, although there is one main canyon, the whole complex includes several parallel smaller canyons and fissures.

Mariner 9 scientists wanted a more imposing name than might be suggested by the classic nomenclature. They decided to name it after their spacecraft, and it was christened Valles Marineris (Valleys of Mariner).

ANALOGS ON EARTH

Although Valles Marineris is usually called a canyon, this term is misleading. The "canyon" is so big that Arizona's Grand Canyon would be a mere crack in the sidewalk by comparison. According to the altitudes measured by the MGS laser instrument, parts of the Valles Marineris floor lie as much as 5.7 km below the mean surface of Mars, making it one of the lowest regions on the planet. The whole system is not primarily a water-eroded canyon, like our Grand Canyon, but is ultimately caused by fractures, or, as geologists call them, **faults**. The cause of the fracturing relates to the evolution of the huge Tharsis volcanic system,

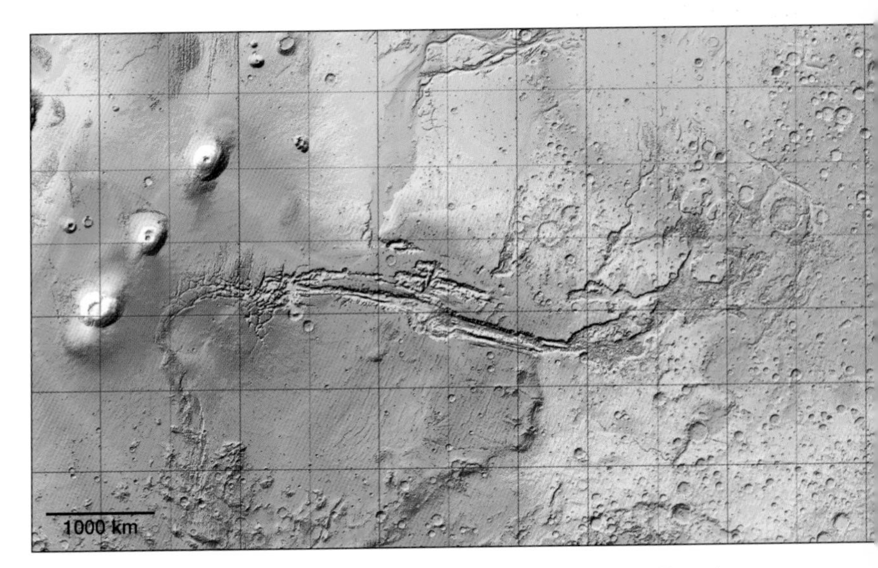

Laser altimeter map shows the enormous extent and complexity of the Valles Marineris canyon system. Blue marks the lowest elevations. Numerous parallel fractures can be seen, the larger ones widened by collapse. A general downhill trend runs from the high Tharsis elevations (brown, left) to the east (right), where the canyon and adjacent depressions drain into river channel systems. (MGS laser altimeter team.)

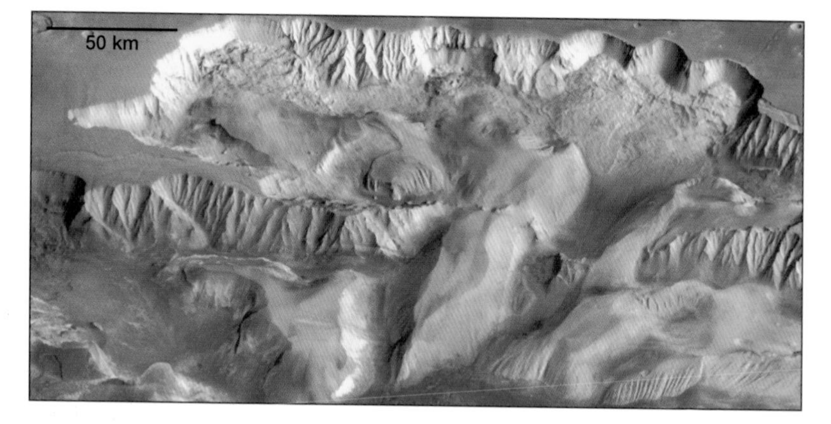

Valles Marineris started as a system of narrow fractures, but it widened as unstable portions of the walls collapsed into the floor of the canyon. In this image, scalloped cutouts in the north wall of the canyon show where whole sections of the wall collapsed and formed a huge landslide onto the canyon floor (good examples are in upper right). The hummocky landslide deposits can be seen at the foot of the wall. This is the "Candor Chasma" section of the canyon (northernmost canyon segment, upper center of previous radar image). (Viking mosaic, National Space Science Data Center, NASA.)

as remarked earlier. Valles Marineris fractures radiate from the center of the Tharsis volcanism. Perhaps the fractures resulted from the accumulated weight of the massive lavas, stressing the crust of Mars, or perhaps there was some incipient plate-tectonic fracturing, associated with the mantle plume under Tharsis.

The closest terrestrial analogs are great rift valleys opened by the mighty fractures associated with plate tectonics on Earth. One example lies in North America, where Baja California is being split off the mainland of Mexico, creating the Gulf of California. The notorious San Andreas Fault near Los Angeles is part of the same fracture system that is separating the California coast from the mainland and widening the Gulf of California. Indeed, the Gulf of California is about the same size and shape as Valles Marineris. Also the same size and shape is the Red Sea, of biblical fame; it, too, was created by plate-tectonic faulting as the Arabian peninsula separated from Africa. The fractures along these faults allow magma to reach Earth's surface, and both the Gulf of California and the Red Sea are dotted with volcanoes, supporting the connection between the Valles Marineris and the volcanoes of Tharsis.

THE HISTORY OF THE CANYON

Valles Marineris has had a marvelously complex history, encompassing many Martian phenomena. Fracturing, collapse, landslides, and water flow all seem to have been involved. At the tops of the cliffs, blocky walls rise vertically like castle ramparts. In the first months of the MGS mission, camera team members were excited to target these cliff walls with the camera and the spectrometer. They discovered layer after layer of flat-lying beds in these walls, like strata in the Grand Canyon. What kinds of layers were they? Lava beds? Sandstones from ancient windblown dust layers? Or perhaps massive limestone layers from the floors of ancient seas? The spectrometer ruled out sandstones or limestones, and indicated the typical spectrum of Martian igneous rocks. The layers are probably individual basaltic lava flows seen in cross section. Although there was much commentary about this at the time, it is not surprising in retrospect. After all, Valles Marineris is a canyon system cut into the biggest volcanic dome in the solar system. University of Arizona team member Alfred McEwen and his colleagues from the MGS team pointed out after this discovery that given the total depth of presumed lava beds seen in cross section in the Valles Marineris walls, the total volume of lava must have been enormous, confirming intense volcanic activity in early Martian history.

Some detailed puzzles remain. For example, U.S. Geological Survey researcher Baerbel Lucchitta used Viking images to point out in the 1980s that the Valles Marineris cliffs have some outcrops of a dark layer that looks as black as a coal seam in certain pictures. In some places, the dark material seems to be weathering out of the canyon wall and dribbling down the cliff face. Of course, as we learned at the White Rock, it is dangerous to assume that a digitally processed image tells you anything about extreme black or white

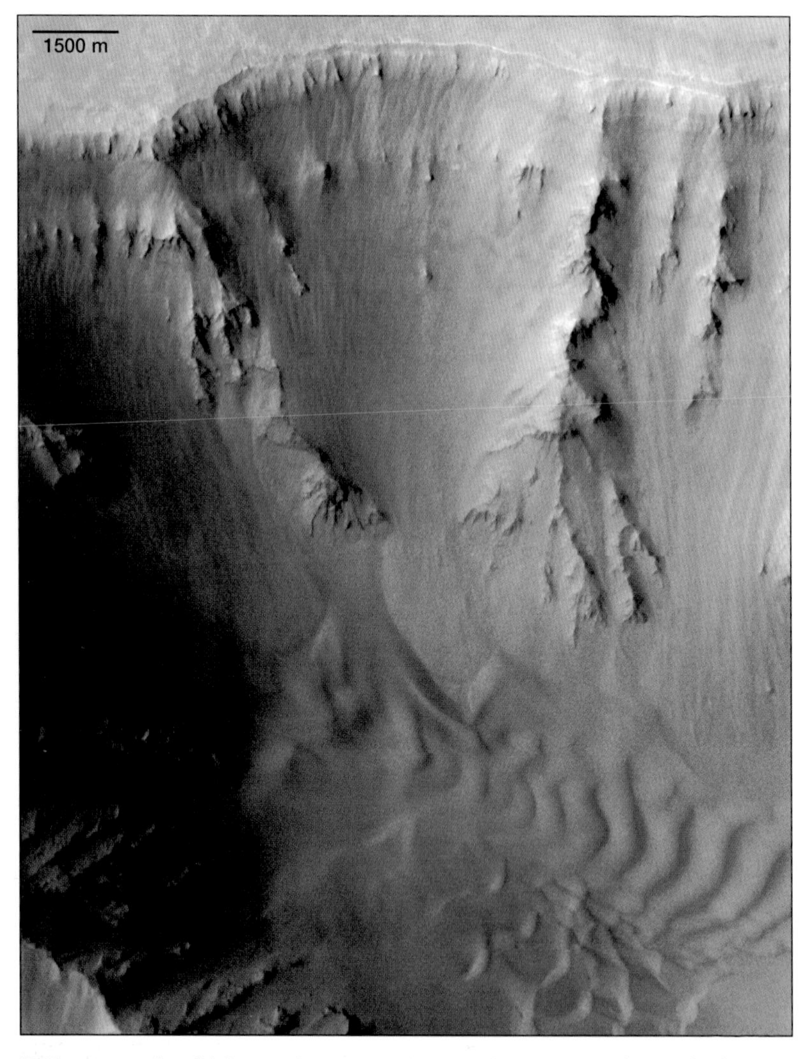

Towering walls of Valles Marineris. The cliffs rise as much as 4 km (13,000 feet) above the dune-covered canyon floor (bottom). The cliff face reveals flat-lying layers extending 2 to 3 km below the rim (top). (58W 15S, Mars Odyssey Themis camera image 020621.)

tones; the contrast may be stretched so much that a darkish gray layer looks black. Still, the layer seems different from most of the rock wall units. What is the explanation of Lucchitta's dark layer? She is inclined to think that it may be ashy or fragmental lava material. Such a material can be darker than normal basalts, and the particulate nature could explain why it seems to erode easily

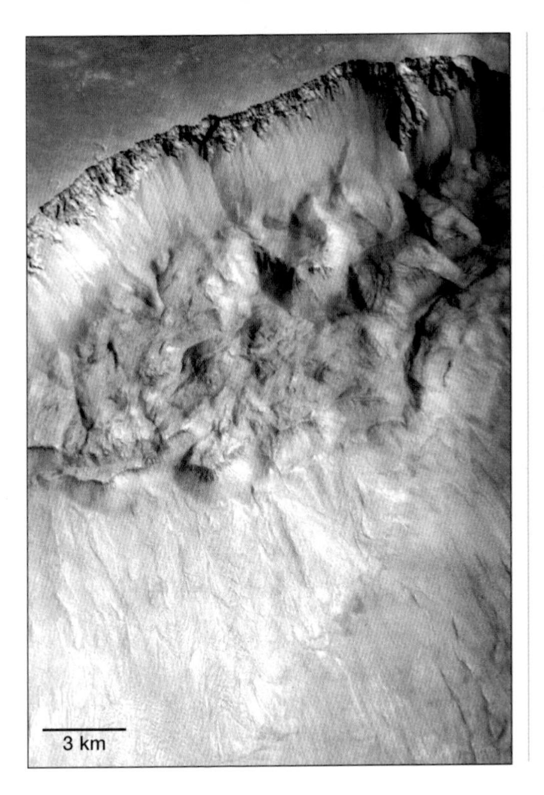

Close-up of a collapsed section of the south wall of the "Ganges Chasma" region of Valles Marineris shows a stratified rim (top), hummocky landslide deposits (middle third), and deposits on the canyon floor (bottom), which were striated by outrushing material during the landslide. (45W 9S, Mars Odyssey Themis camera image 020329.)

out of the cliff wall. The material is so dark that it tends to be underexposed in most images of the area, and it could still contain some surprises.

THE ROLE OF WATER IN VALLES MARINERIS

Once the first fractures began to open the canyon, water must have been an agent in widening or deepening it. The best evidence lies at the east end, where the canyon opens into an extensive area of water erosion near Ares Vallis; here, it looks as if water had emptied out of the canyon itself.

The cliff walls themselves have been widened by enormous landslides. Perhaps water erosion undercut the walls or earthquake activity shook them down. Huge, amphitheater-shaped alcoves in the walls mark hollows where masses of the cliff have dropped away, and at their feet can be seen the fallen wall material, collapsed into jumbled, hummocky terrain typical of large-scale landslide deposits.

Elsewhere along the floor and lower walls of Valles Marineris, layered deposits are draped over the terrain. These are different from the flat layers cut by the walls, and they seem to be layers of material deposited in the canyon after it began to form. Researchers argue whether they represent wind-borne dust layers

Overlapping striated deposits from landslides (facing page) caused by collapse of the south wall of Valles Marineris. Striations are believed to mark the direction of flow of the material as it surged onto the canyon floor. A whole city could be buried by one such mass of debris. (78W 8S, Mars Odyssey Themis camera image 20021113.)

or ancient sediments deposited in water. The canyon floor of Vallis Marineris, particularly at the eastern, outflow end, is deep enough to have higher than normal air pressure, which would allow water to be stable in liquid form even on today's Mars. This is consistent with the idea that water flowed through these parts of the canyon system.

Valles Marineris is so complex, in terms of its tectonic evolution, layered walls, stratified floor sediments, and water outflow, that dozens of scientific papers have been published in an attempt to interpret its history. More will come in the future; but for the best answers, we should look forward to the day when humans are able to climb around the vast interior of the system and begin to sort out the answers firsthand.

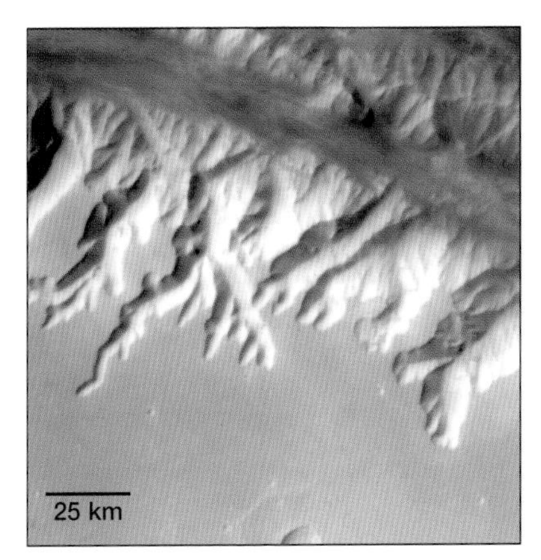

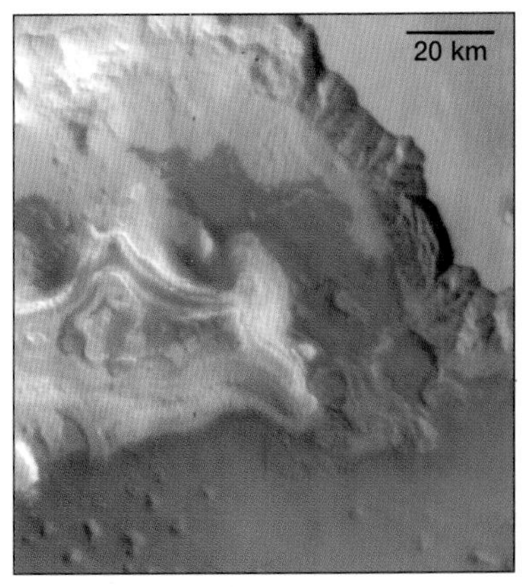

Evidence of probable water flow into Valles Marineris comes from the dendritic channel (above left) leading off the nearby plains into the canyon system. This may be a sign of groundwater sapping, which was discussed earlier in the text. (81W 9S, M02-03602.)

Above right: Massive layered deposits on the floor of Valles Marineris appear to be sediments currently being exposed by erosion. Arguments continue over whether the sediments were formed by deposition due mainly to wind or water. (50W 7S, M03-02942.)

CHAPTER 30

CERBERUS FOSSAE

An Incipient Valles Marineris?

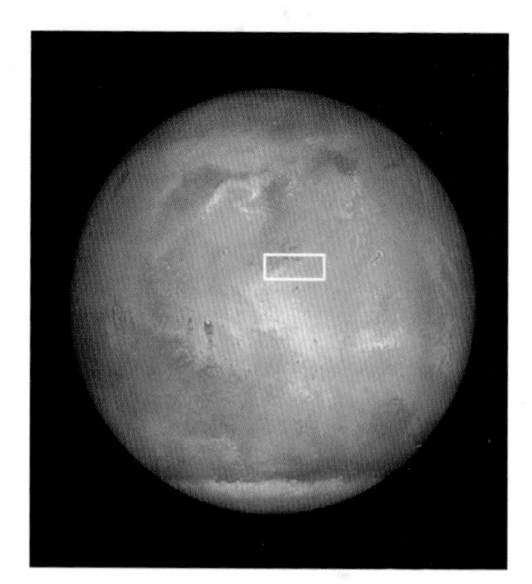

On the southeast side of the Elysium volcanic province are strange parallel fractures, oriented northwest-southeast and radiating from Elysium Planitia lava plains, south of the volcano Elysium Mons. The fractures lie in an area called Cerberus, and they carry the topographic name Cerberus Fossae. Cerberus is an odd name introduced by Schiaparelli in 1882; it descends from the name of a three-headed dog who guarded the entrance to hell in Greek mythology. Wagging his great dragon-like tail, he seemed to be a friendly enough pooch, but he had a nasty habit of refusing to let anyone leave.

The name has another peculiarity. The fracture system was originally called Cerberus Rupes, but "Rupes" turned out to be an error in naming. It refers to a cliff face because early orbital photos made these fractures look like cliffs. But MGS photos show that the main ones are deep, narrow, straight fissures, hundreds of meters wide, looking something like slot canyons. The system is now called Cerberus Fossae, with the term "fossa" referring to a fissure or narrow groove. In many places, the Cerberus fractures come

Astronaut's eye view of a northern outlier of the Cerberus fracture system, looking east. One of the fractures stretches as a dark line from a foreground crater toward the bright, hazy horizon. Oblique views of this sort are less useful than vertical views for mapping and scientific measurements, but are more familiar in terms of human experience. (195W 14N, SP1-25705.)

in sets of two or more closely spaced fissure-canyons, sometimes with one ending and another picking up at an offset nearby.

The canyons are mostly very dark-colored, much darker than the surrounding plains. The cause of the color seems to be that the plains are covered by the usual light-colored reddish dust of Mars, while the canyons cut through layers of fresh, dark-gray basalt, which is continuing to weather out of the canyon walls and cover the interior with darker gravel and debris.

AGE AND FORMATION OF THE CRACKS

One of the astonishing aspects of the Cerberus fractures is that although they are located amid extremely young lavas, the lavas do not flow into the fractures. Rather, the fractures cut through the lavas. The canyon rims are sharply edged, as if recently cut by a knife, and the walls show clear layers, probably earlier lava flows. This means the fractures are still opening up, with cliff walls collapsing into the cracks, creating the sharp edges. At the Planetary Science Institute, graduate student Dan Berman and I counted craters in the area and estimated that some of the lavas cut by the fractures are no more than a few hundred

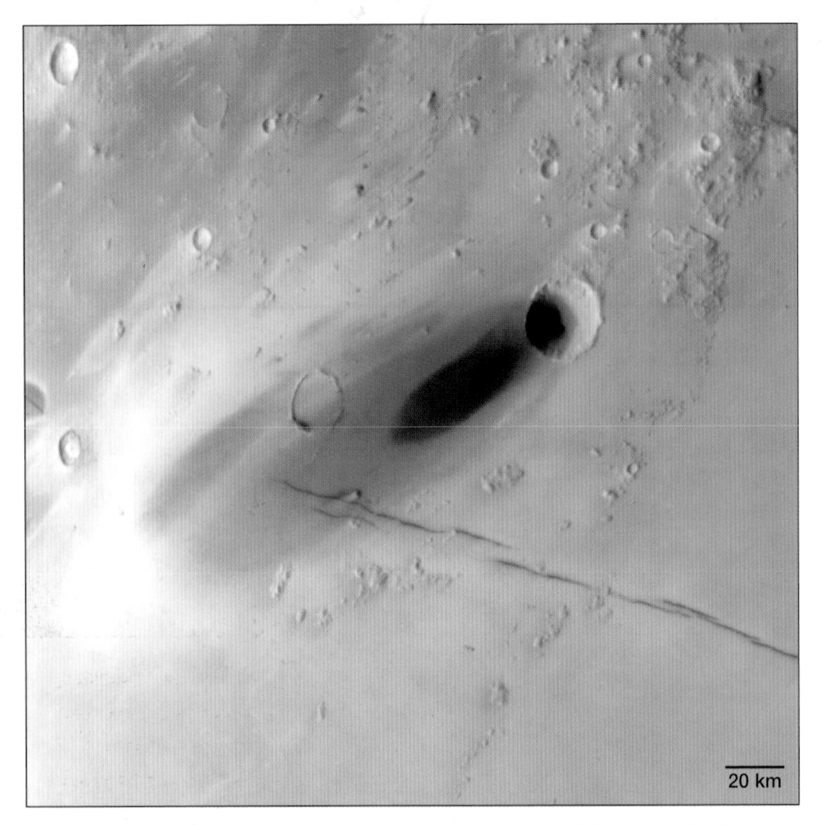

Regional view of the main Cerberus Fossae fracture (dark lines) shows that it consists of short, parallel fractures. The canyons are too narrow to be resolved at this scale. This view includes a spectacular dark wind tail from the crater in the upper right center, with a dark core and a broader diffuse dark tail around it. (192W 8N, M04-00823.)

MY old. Thus, the fracture system itself has been forming in the last few percent of Martian history.

The mode of formation is indicated by the MGS high-resolution photos. In some places, you can see fractures in the wall rim, as if whole blocks (large enough to contain a rim-side visitor center and hundreds of tourists, were they in an American national park!) are poised to collapse into the canyon. More intriguingly, other images show places where a fissure ends, but the line of the canyon fissure is extended by a row of collapsed pits. In another area, an old, isolated hill is broken by an early row of collapsed pits parallel to a younger nearby fissure. These observations suggest the following scenario. First, the Elysium plains are like a multilayer cake, each layer composed of a lava flow, perhaps overlain by thin layers of soils composed of dust and gravel. Second, the layers are being stressed regionally, causing long fractures to open at depth, probably accompanied by major earthquakes. As the crack widens, it may propagate all the way to the surface, or, if it doesn't reach the surface, it may simply remove support from the overlying layers, causing them to collapse into the fractures in rows of few-hundred-meter-wide pits. This explains the collapsed pits in some areas. Third, once the fractures open and make a narrow canyon on the surface, the underlying cracks continue to widen. This withdraws support from the canyon walls, causing new sections of the rim to collapse into the floors. In this way, the canyons are still widening and eating their way out into the surrounding areas. Some are a kilometer or more in width and hundreds of meters in depth.

Collapsed pits extending beyond the end of one of the Cerberus Fossae fractures (bottom left) suggest that the canyon formed as subsurface fractures opened, causing the surface to collapse. Coalescing collapsed pits formed the canyon itself. (198W 16N, MGS FHA-01651.)

GLOBAL TECTONIC FORCES AT WORK

The radial orientation to the 700-km distant Elysium volcanics raises the question of whether or not these fractures are a junior version of Valles Marineris, radiating from Tharsis. In other words, can we say Cerberus Fossae are to Elysium as Valles Marineris are to Tharsis? The Cerberus fracture system is about

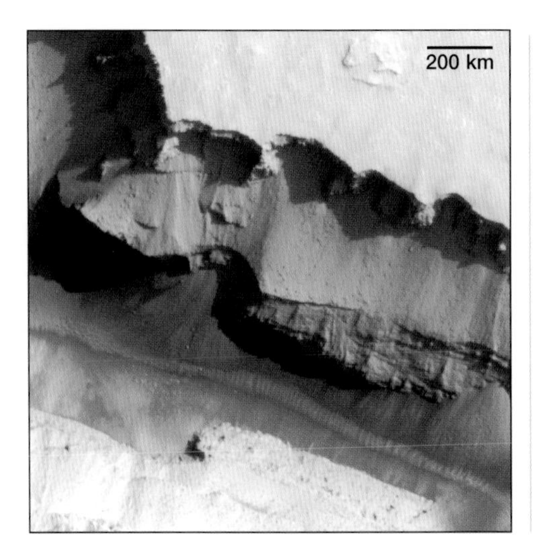

The spectacular gorge of the Cerberus Fossae canyon. This telephoto view shows the wall of the narrow canyon exposing layered deposits (lava flows?) with bright dusty plains outside the canyon and dark (basaltic lava?) debris inside. Narrow band of medium-bright dunes marks the canyon floor. (202W 10N, M07-03419.)

60 percent as long as the main Valles Marineris canyon system, and the main fractures are tens of kilometers apart. If the Cerberus fractures widened, merged, and lengthened, we would have a huge canyon system 100 km wide and 1,200 km long, rivaling Valles Marineris. In other words, Cerberus Fossae may give us a picture of what Valles Marineris looked like when it began.

As early as 1986, MIT geophysicists J. Lynn Hall and Sean Solomon and Brown University geologist James W. Head III, writing in the respected but overcrowded *Journal of Geophysical Research* (at that time it was publishing more than 13,000 pages per year on various geological and space subjects; now the number is more than 30,000!), studied whether the weight of accumulating Elysium lavas had created Elysium's own tectonic fracture pattern. They concluded that the weight of the Elysium volcanics did create some observed concentric fractures and down-warping of the relatively rigid crustal rock layer, about 50 km thick, but that the Cerberus Fossae fractures were not due to this load alone. There may have been different stages of tectonic forces, including not only uplift from a mantle plume under Elysium but also stresses from the formation of the massive Tharsis dome to the east. The Cerberus fractures trace back eastward to the center of the 4,800-km distant Tharsis eruptive complex, as well as aligning

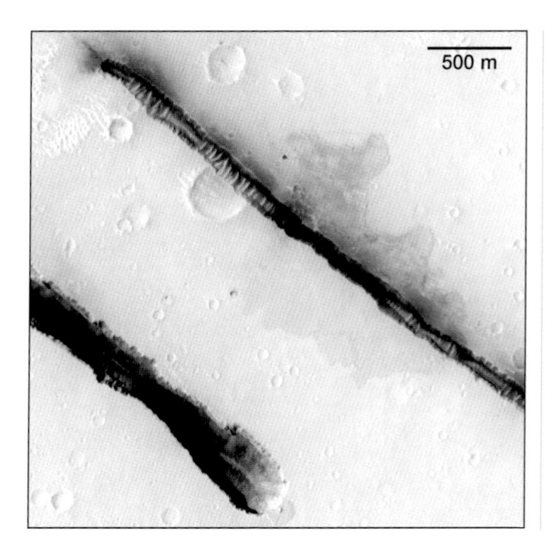

Dark, molasses-like material north of the upper fracture (and fainter similar material to the south) may mark flows of lava or even mud that gained access to the surface through the fracture. (204W 10N, M02-01973.)

with Elysium. At that distance, they lie on the far outskirts of the Tharsis radial system, but because they are on a line between Elysium and Tharsis, they may be a product of the combined stresses emanating from the two eruptive centers.

LAVA, WATER, AND MUD FROM THE FRACTURES?

Early workers who mapped the area prior to the MGS mission thought the Cerberus cracks might be the source of either the young lavas of Elysium or even a source of water that coursed down a nearby youthful channel system, Marte Vallis, which we'll visit in the next chapter. They reasoned that as cracks opened, lavas or water from volcanically melted ice would find access to the surface. If that were true, then we'd expect to see small spatter cones of lava along the fissure and/or areas where eruptions surged out of the cracks, draping lava over the wall layers and rim and creating flows running out from the canyons.

The MGS photos showed enough detail to resolve these ideas. Intriguingly, the fractures weren't lined by spatter cones, nor were there many, if any, images of flows either coming out of or draping back into the chasms. Instead of the lava issuing out of the crack, the crack, along most of its length, cuts the lavas, as stressed earlier.

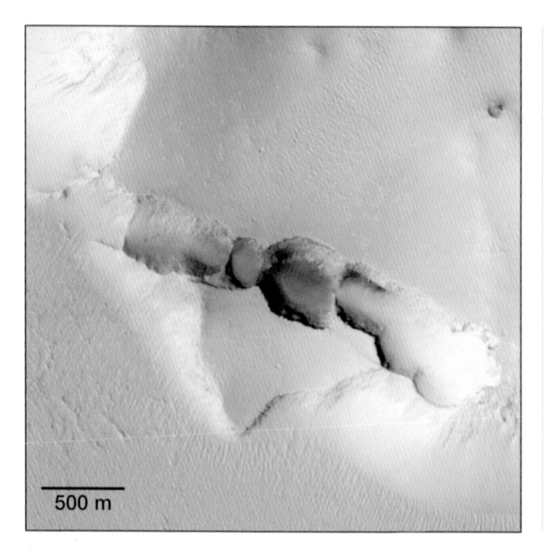

Another portion of the Cerberus Fossae system shows the fractures expressed not as a continuous canyon, but as a row of collapsed pits. The canyon probably grew by enlargement of such pits as their walls collapsed. (195W 9N, M07-03839.)

Nonetheless, there is some evidence of emanations from the cracks. In 2001, Devon Burr and Alfred McEwen at the University of Arizona and Susan Sakimoto at NASA pointed out an MGS image (p. 327) that shows a flow or puddle of dark material extending from one of the Cerberus Fossae fissures a few hundred meters in both directions. The material is likely to be a small lava flow, though considering the level of detail of the image, it could even be a mud flow. Either way, this key image confirms that the Cerberus Fossae fractures have occasionally provided access for fluids (lavas? mud?) that were released on the surface. Nonetheless, it's interesting that we don't see many such features along the fracture. Nor do we see indications that water flowed along the canyon floor. Similarly, it's surprising that we don't see young lavas that flowed from nearby sources, reached the fissures, and cascaded over the walls into the canyon. So Cerberus Fossae seems to be not simply a passive conduit for young lava or water eruptions, but rather an active fracture system that is still developing today and modifying the surrounding terrain.

The geophysical and tectonic study of Mars is still a youthful field, and more careful study of the craggy Cerberus canyons may clarify the story of the Elysium fractures, the sources of lava and water, and the degree to which Cerberus Fossae may be showing us an infant version of Valles Marineris.

CHAPTER 31

MARTE VALLIS

Recent Floodwaters?

Just south and east of the Cerberus Fossae fractures, on the Elysium Planitia plains, is an extraordinary feature: a large outflow channel system that has eroded through recent lava flows. It is referred to as the Marte Vallis system. The reason it is so surprising is that outflow channels generally date from the Noachian or Hesperian periods, before about 2,000 MY ago, when there was much more water loose on Mars. But this system cuts through very young, Amazonian lavas.

The channels are shallow and hard to trace in Viking images, and until the MGS mission it was hard to say much about their age or nature. As a result, they were not traditionally listed among major channel systems. Earlier books on Martian aqueous features—such as *The Geology of Mars* (1976) by Viking mission scientist Thomas Mutch and others, *The Channels of Mars* (1982) by hydrologist Victor Baker, and *Water on Mars* (1996) by U.S. Geological Survey researcher Michael Carr—give this system scant or no attention, favoring more prominent outflow channels and valley network systems such as Ares, Nanedi, and Nirgal Valles. A few early technical studies, however, did indicate something interesting in the area, and images and data from MGS instruments finally catapulted the Marte Vallis system into prominence.

As mentioned in chapter 26, Viking-era maps showed a large depression in southern Elysium Planitia, and some investigators believed the flat plain here marked a large, ancient lake bed, associating the faint river channel with the ancient lake. MGS radar altimetry reduced the size of the basin, although there is a small, shallow low region along the channel where water might once have ponded. As early as 1990, Caltech researcher Jeff Plescia used Viking images to document lavas in the region as late Amazonian features and noted that they are unusually young.

THE NATURE OF THE CHANNELS

The MGS altimetry is so precise that it reveals the outlines of the channel even where it is only a few meters deep, allowing us to trace its extent and source regions. There are two broad parts to the system. At the west end, the source channels lie near the west end of the Cerberus Fossae fracture system, and this part of the system has recently been given the name Athabasca Vallis. As

Laser altimeter map of the Marte Vallis region shows a low, extremely smooth plain (blue, center) formed by lava flows. Absence of craters shows the very young age of the surface. Floodwaters cut a channel system (too shallow to show here) from the upper left, across the center of the image, and then NE out of the upper right. (Laser altimeter team; map processing courtesy Nicholas Hoffman.)

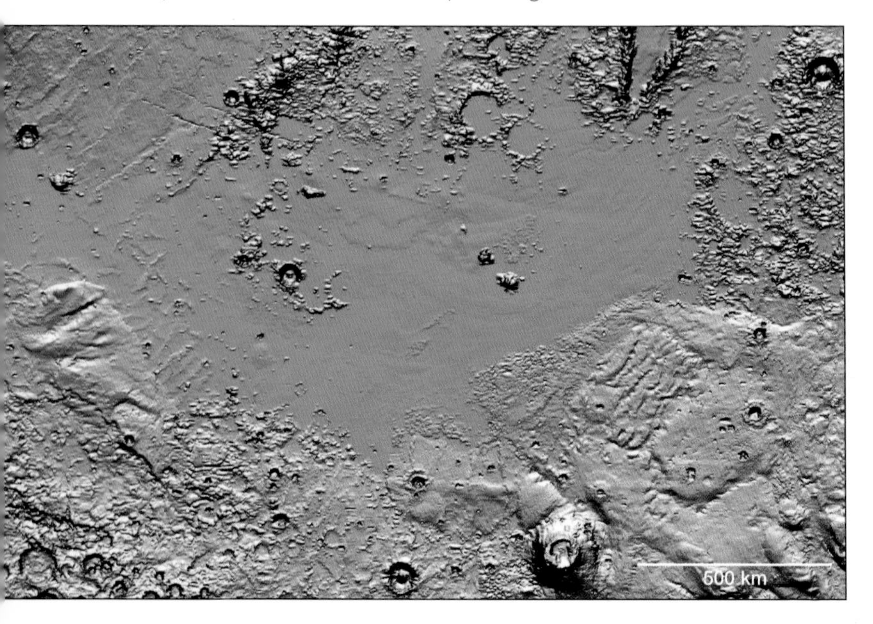

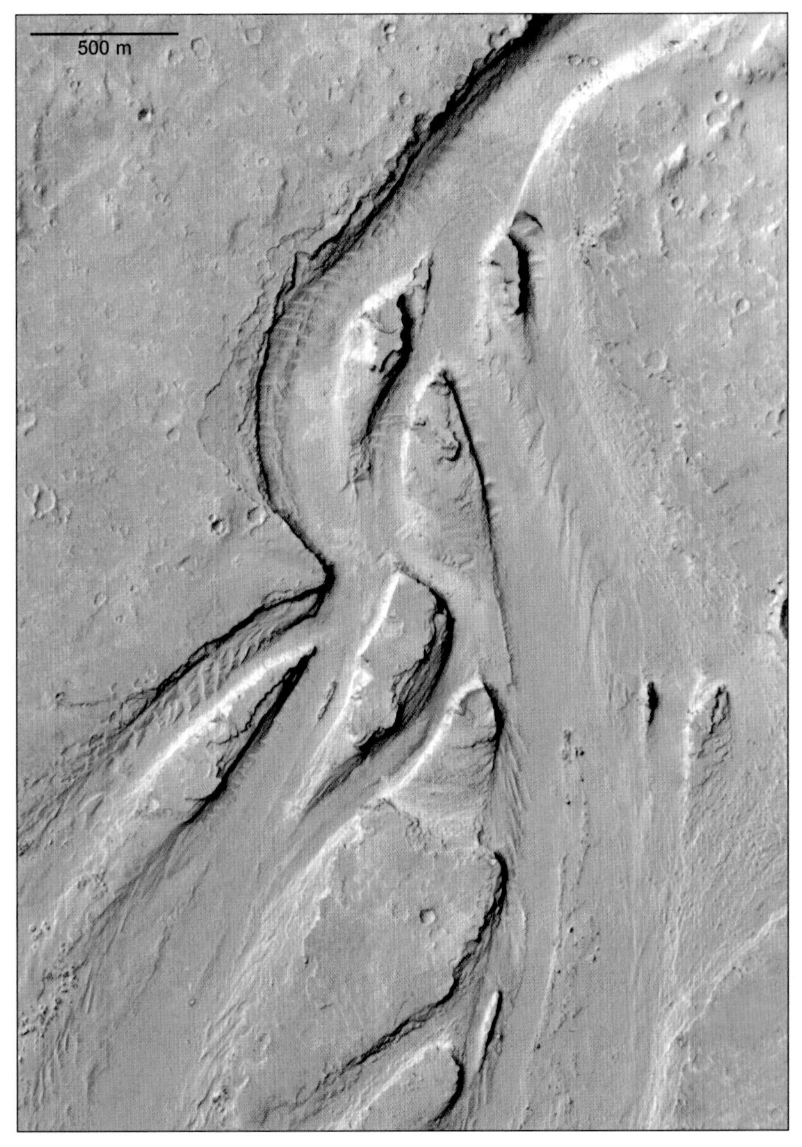

South-flowing water in the Athabasca part of the Marte Vallis system eroded a maze of streamlined islands in the middle of the channel. Absence of craters in the channel floor indicates a young age for the last water flow. (206W 8N, M21-01914.)

noted in the last chapter, Devon Burr and her colleagues suggest that some of the water actually came out of the Cerberus Fossae fracture system. Farther south and east, these channels grade into Marte Vallis, the name still used for the downslope end of the sys-

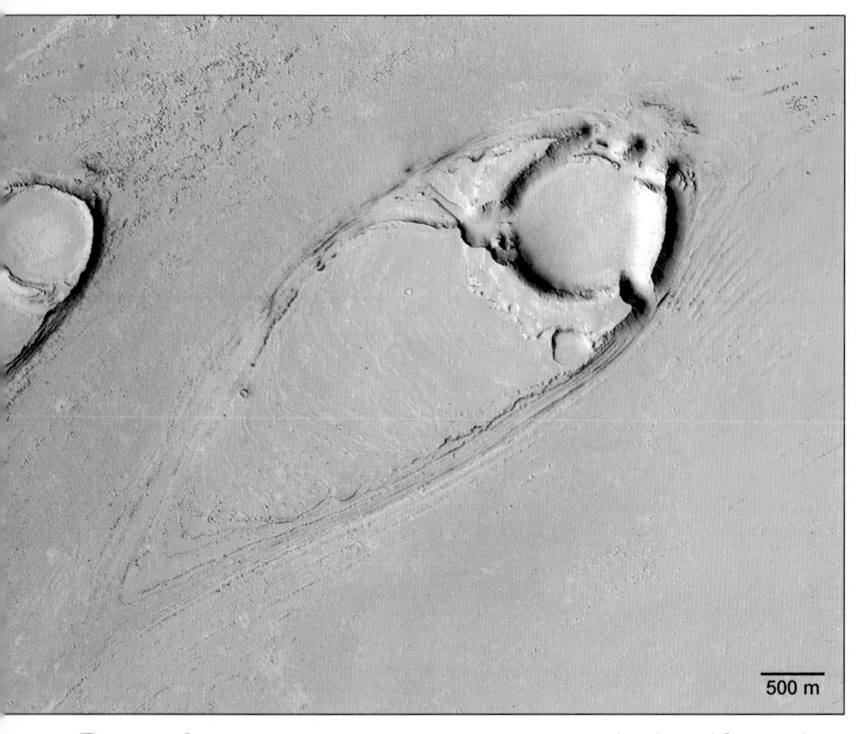

Portrait of a streamlined island. An ancient impact crater in the path of the Marte Vallis floods divided the waters, creating a teardrop-shaped island in the midst of the river. Even this high-resolution view shows no craters on the channel floor, implying a very young age for the last water flow. (204W 10N, M07-00614.)

tem, which ultimately curves northeast and empties onto the very young lava plains of Amazonis Planitia, where it appears to fade out. Perhaps the water was lost by evaporation and sinking into the soil in those regions, or perhaps fresh lava flows have buried the end of the shallow channel system. MGS images, shown here, reveal striking erosion features, such as streamlined islands cut around craters by the flowing water.

Devon Burr, working at the University of Arizona with Alfred McEwen and Laszlo Keszthelyi, along with Jen Grier at the Planetary Science Institute and Susan Sakimoto of NASA, have looked at the possible water-flow rates in the channels. They estimated flows in Athabasca Vallis, possibly emanating from the Cerberus Fossae fracture, at 1 to 2 million cubic meters per second, compared to a typical flow rate of the Mississippi at about 0.04 million cubic meters per second. In the Marte Vallis system, they estimated as much as 5 million cubic meters per second. The flow rates are thus a few percent of the rates estimated for Ares Vallis but still around a hundred times that of the Mississippi.

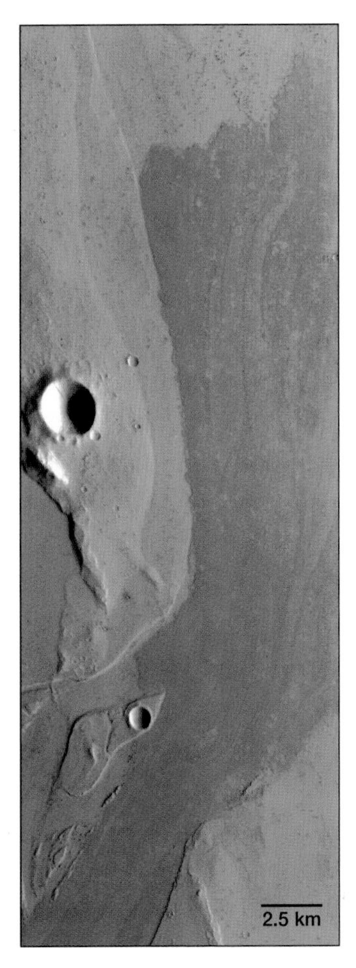

A lava flow in a riverbed. A young, uncratered lava flow entered the Marte Vallis riverbed after the last water flow, moving north down the channel from the bottom of the image. At the top, the lava solidified and stopped, forming a flow front with convex lobes, confined within the riverbed. (206W 8N, M21-01914.)

The water floods, lava flows, and fractures have been locked in a close interplay in this area. For example, in one striking MGS image, we can see that a fresh, dark, nearly uncratered lava flow entered a shallow riverbed, flowed downstream for at least 40 km (25 miles), and then solidified in the middle of the channel.

DATING THE FLOOD

Telephoto MGS images, with a tenfold improvement in detail over the best Viking images of the area, showed exceptionally few impact craters that postdated the lava flows. The river water that cut into those flows must be very young. Crater counts by Dan Berman and me, published in 2002, indicated that many lava flows in the region are probably less than 200 MY old and that parts of the river floor in the channel systems may have been shaped by water flow within the last 20 MY.

Devon Burr and her coworkers published independent crater counts and age analyses in 2002, attempting a more detailed study of the different surfaces. They derived ages of 2 to 10 MY for the last flow in the Athabasca end of the system, 35 to 140 MY for the last flows in Marte Vallis, and 10 to 40 MY for some unnamed channels just north of the other two. They also estimated an age of 35 to 140 MY for a lava flow that covers part of Marte Vallis. These results are comparable to ours.

Taken together, these findings are a rude shock to our understanding of large outflow channels on Mars, most of which are thought to date from the Hesperian era. This system, with its indications of large-scale water outbursts, apparently was active in the last 1 percent or so of Martian time.

A final but important observation concerns the time intervals between episodes of lava formation and river channel formation: they were probably less than a few million years old. We don't see major differences in crater densities between the lavas and the riverbeds.

TRACING THE WATER

The big mystery with Marte Vallis and Athabasca Vallis, as with Ares Vallis, is the source of the water. Geologist Jeff Plescia, in 1993, estimated the total volume of erupted lava in the area; he assumed a plausible water content of 1 percent for the lavas and then calculated the total mass of water vapor that would have been released during the lava surfacing of the Cerberus plains. He came up with a whopping thousandfold increase in the planet's atmospheric mass of water, although probably not enough total gas to raise the total atmospheric pressure noticeably (the water vapor is only a small part of the total carbon dioxide atmosphere). Plescia suggested that as the extra water vapor erupted into the atmosphere, it could have condensed and precipitated as rain or snow, which could have led, in turn, to local runoff and channel formation. On the other hand, the water that formed the channels seems not to have been primarily runoff, because we don't see tree-like valley networks of shallow runoff tributaries leading into the main channels. (Some might have been covered by later lava flows, but we'd expect to see some traces of valley networks.) Also, if these are recent and ordinary lava flows, like those elsewhere in Elysium, Amazonis, and Tharsis, we have to wonder why such young channels are not more common.

On the other hand, if we appeal to Cerberus Fossae as the source of the water, as do Devon Burr and her colleagues, then we still have to ask where the water came from before that. If it was ancient water, stored in the area since early times as ground ice, then why didn't it melt a long time ago during earlier eruptions of lava in the area? If the area was charged or recharged with ice only recently, then what was the mechanism of water transport and recharge?

This line of reasoning leads to radical suggestions that the ice has not been just a static reservoir since Noachian or Hesperian times, when it first formed, but that it continually re-forms after underground water flows back into the regions after each lava flow, freezing in the cavities among the flows. Underground lava tubes and porous cinder layers may allow water to flow over large distances of hundreds of kilometers, to recharge the pore spaces and cavities, creating ice deposits ready to be melted by the next volcanic episode. My best guess is that this is the key process, perhaps aided by variation in water abundance and frost condensation during recent cycles of changing axial tilt. Although this is my favorite answer among the alternatives, I'd estimate its chance of being right is only 30 percent. The source of water that could cut a recent channel on Mars remains, by and large, a puzzle.

We've already discussed one remaining clue that provides independent proof of underground mobility of water, or at least water vapor and moisture, on modern Mars. It is the fact that Martian meteorites contain salts and datable weathering products of water that got into the rocks, as recently as 670 MY ago in the case of the nakhlite meteorite named Lafayette. Somehow, under the "dry," frozen deserts of Mars, water still moves.

CHAPTER 32

CYDONIA

and the Face on Mars

When we visited the ancient hematite deposits and fossil craters of Terra Meridiani in chapter 12 and the sediments of crater Crommelin in chapter 13, we found that massive layers of material have formed in some areas of Mars, and in some places they have been stripped off again. This applies over wide regions of Mars, where eroded remnants of old layered deposits now stand as isolated mesas. The border of the southern highlands and the northern lowland plains is full of this kind of terrain, as if the uplands have been eroded back and dissected into plateaus and valleys. A notorious example is a stretch of plains and isolated, broken hills called Cydonia, after an ancient seaport on the island of Crete.

Cydonia lies at about latitude 32° N to 42° N, where the cratered uplands of Terra Meridiani and Arabia break up into the lower plains of Acidalia and Chryse. *Break up* is an apt term. The area is rather like megascale chaotic terrain, where the old uplands and sediment deposits have been carved into blocks, perhaps by combined fracturing and water erosion. As we saw while visiting Vastitas Borealis (chapter 11), some of the low northern plains may have been ancient seafloors or lake beds, covered later by sediments and lavas accumulating hundreds of meters deep. Later, those layers began to be stripped away, leaving remnant hills. What we see now are remnant masses of layered sediments scattered throughout the area, often showing flat, resistant summit layers, or **caprock**. Often we can see that the caprock layer is an old cratered surface, proving that the old summit surfaces thus mark preserved fragments of once extensive upland cratered plains a few hundred meters above the present surface. These plains must have been exposed for hundreds of millions of years in order to accumulate the observed crater density before erosion cut into them and reduced them to freestanding mesas. Some of the caprock layers may be old lava flows that formed on weaker surfaces, which would explain their resistance to erosion. Among some of the other mesas, the caprock layer has eroded through, producing mesa summits with broken crags, pits, and boulders.

The geography is not unlike that of the Four Corners area of the American Southwest, where the Colorado Plateau has eroded back, leaving isolated mesas, often topped by resistant caprock layers, such as the famous mesas of Monument Valley, immortalized in any number of early Western films.

THE FACE FRENZY

The Cydonia area has been the subject of a time-wasting, deliberately orchestrated, and pseudoscientific frenzy about one of the mesas. Above all, the incident illustrates the foibles of twentieth-century life in the presence of powerful but mediocre news media. It all started in the 1970s, when Viking cameras photographed one of the more eroded Cydonian mesas at a time of day when shadows on the rough summit produced a crude face-like pattern, with two small shadows making eyes, and another longer valley shadow making a mouth. Wags on the Viking team, following planetary scientists' penchant for whimsical nicknames, named this hill the Face on Mars.

The problem was that a ragtag band of tabloid editors, promoters, and amateur naturalists took this nickname seriously . . . or pretended to.

Soon, like politicians wrapping themselves in the American flag, a few pseudoscience writers cloaked themselves in an aura of

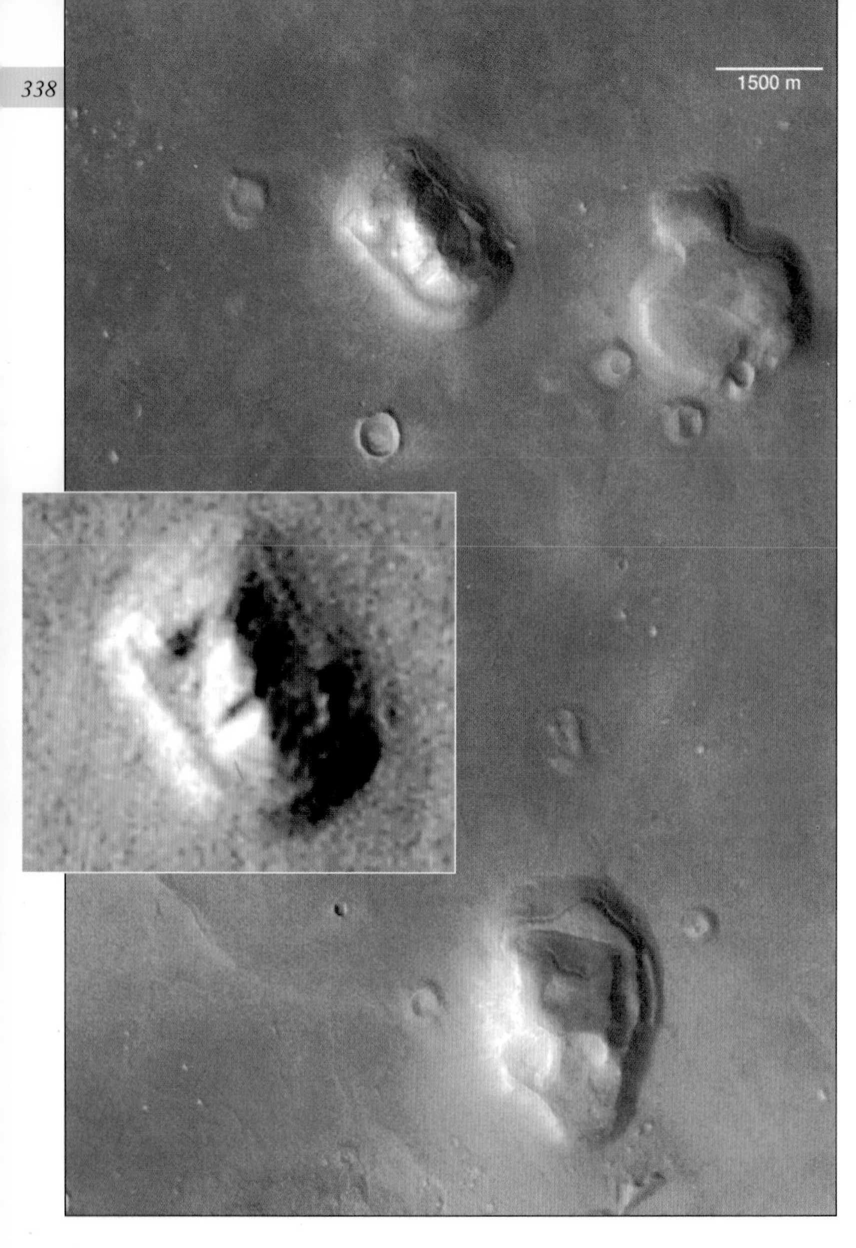

Inset: *The notorious image that created a tabloid mythology about a face on Mars. At the low resolution of the Viking Orbiter images, the fuzzy view of shadow patterns resembled a face. (10W 41N; Viking, NASA.)*

Main image: *Which one is the "face" on Mars? This view shows three of the many eroded mesas in the Cydonia area. The northernmost one (top center), under certain lighting, held shadows crudely resembling a face. The mesas show thick layering, especially on their NE sides where the slanting light and shadows emphasize strata. The whole area may have been covered by sediment layers, of which these mesas are the only surviving remnants. (10W 41N, Mars Odyssey mission Themis camera image 020413.)*

sober scientific curiosity and wrote about the mysterious monument on Mars that was carved with the likeness of an enigmatic face. It's important to understand how the pseudoscience media work. If a sensation-monger decides to stand up and say NASA is hiding "a face on Mars," it makes a great story for the less discerning media. A symbiotic relationship develops. The promoter gets free attention, and less reputable media get an easy story. Was the face left on Mars in the dawn of time as a signal? Was it, like the canals of Mars, a remnant of a supercivilization? Tabloid journalists and "sleaze" TV producers soon generated feature stories and documentaries about the strange stone face. According to them, "careful analysis" showed not only the alien face, but a mysterious hairdo as well. For these folks, the story had self-propagating feedback. The fact that reputable scientists complained about the flap over the Face on Mars simply proved that a conspiracy was afoot to ignore it! Why were there no official press releases about the face? Was NASA trying to cover up this fabulous, troubling discovery? Why were "they"—the all-powerful "they" of American myth—trying to delude us? By the 1980s, a whole cottage industry had sprung up around the face and the way establishment scientists were "suppressing" secret discoveries about it. Junk Web sites abounded.

Soon Mars scientists, myself included, were receiving unsolicited manuscripts, purporting to prove that the region of the face was crowded with pyramids and mysterious geometric configurations. Well-meaning amateur authors would take an old Viking picture, draw a line from the center of one mountain peak to the southeast corner of another, then to an upland promontory, and then to a point halfway between two craters and—voilà!—it was a perfect rectangle, a sure sign of intelligent design. The broken mesa-land of Cydonia soon became a veritable city full of ancient monuments. The nadir came in the early '90s, when NASA's Mars Observer mission (carrying the first version of what later became the MGS camera) failed upon approach to Mars due to a fuel-line rupture. Web-based rumormongers proclaimed that NASA had

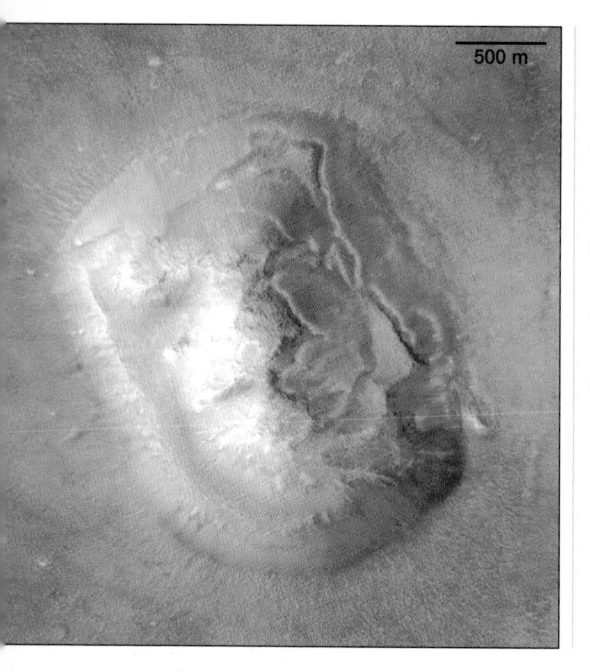

Is it an alien sculpture? Close-up of the "face" shows only a highly eroded mesa. Eroded rubble on the summit (center) lies atop a resistant layer that forms a cliff around the outer edge (prominent on the left and bottom of the mesa). (10W 41N, E03-0824.)

deliberately sabotaged its own spacecraft to prevent the public from learning "the truth" about the ancient Martian monuments. It was a trying time for Mars researchers. You'd come home from a sensational scientific meeting about new spacecraft findings, Martian meteorites, and possible discovery of fossil microbes, then go to a cocktail party where, once they found out that you studied Mars, all that people wanted to hear about was the Face on Mars.

Even before MGS got into orbit, we targeted its cameras to photograph the face and the Cydonia region along with thousands of other features on Mars. We looked forward to seeing what our cameras would show in terms of the features that created the face-like pattern. Eventually, we got images at several angles and resolutions. Not surprisingly, the face turned out to be just another of many eroding hills. As for pyramids, it was obvious to most scientists from the outset that they were not symmetric, Egypt-like pyramids but various-shaped, eroded peaks whose somewhat flat sides were probably associated with landslides and wind erosion. They were quite common all over Mars. The supposed special geometric patterns, like rectangles, were ludicrous from the outset because in a random field of complex features, once you allow yourself not to pick centers but various corners and other points, you can usually find symmetric rectangles, triangles, or other shapes.

What people forgot in the face frenzy is that erosion very commonly produces geologic formations that under certain lighting or at certain viewing angles look like faces, babies, skulls, or other such features. In fact, numerous formations on the Moon are already known to amateur observers by various nicknames. Certain telescopic observers in the 1800s even worked themselves up over a supposed lunar city with rectangular bastions—which turned out to be merely a mass of faulted hills. You need only look

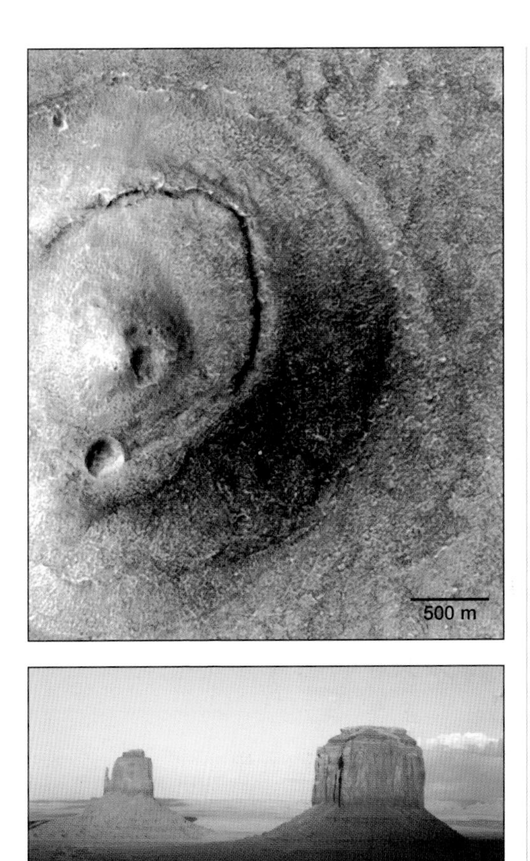

A conical butte not far from the Face on Mars shows a similar two-layer natural structure—a lower slope capped by a pronounced resistant layer or fracture (dark ring), surmounted in turn by an eroded hill. (9W 41N, E03-01950.)

Below left: *The famous mesas of Monument Valley, Utah, are a product of erosion not unlike that of the Cydonia region on Mars. Hundreds of meters of sediment covered the area (note distant background cliffs). Erosion carved this mass into separate mesas, but resistant caprock layers tend to protect individual buttes. (Photo by author.)*

at the design for the 2000 American quarter from New Hampshire to see a famous American natural feature, the Old Man of the Mountain. This modest cliff's profile bears a likeness to the craggy face of an older gentleman. Yet no one claims that space aliens were running around New England a million years ago!

YOUR TAX DOLLARS AT WORK

The whole affair of the Face on Mars was an exercise in public gullibility. A blessing and a curse of living in a free country is that people can say what they want—in the tabloids, on billboards, on junk TV, or on the Web. Many in the media laugh off this situation. It's fun. It's just summer silliness. Everyone knows it's a joke.

In my more curmudgeonly moods, I lose patience with this attitude. A good example of the real damage occurred during the MGS mission. A few months after our spacecraft arrived in orbit, the media, led by tabloid reporters and pseudoscience TV producers, were begging for new photos of the Face on Mars. The problem was that the camera normally looked more or less straight down, and the orbit of the spacecraft took it over any given area only so often, so that we would have to wait some months before any new face photos were obtained. The face was on our high-priority target list, but that wasn't good enough.

The facemongers accused us of stalling. The pressure from the media was so strong that NASA headquarters began to feel the heat. Since public tax dollars had sent the camera to Mars in the first place, critics said, the public should have its picture of the Face on Mars. NASA was suffering a bout of fear that if the spacecraft unexpectedly conked out, the agency would face another round of innuendo that they had deliberately suppressed world-shattering images of the alien buildings on Mars.

So the word came down from NASA headquarters that, in effect, we were to drop what we were doing and reprogram the spacecraft to look off to one side in order to get an image of the face. The decision forced the science teams to set aside their scientific priorities. Malin Space Science Systems, which built and operated the camera, had to reprogram its activities. Funds that would have been used for data processing and public release of images had to be reprogrammed to figure out how to point the camera at the face now instead of waiting a few months. As a result, a picture was obtained a few months early, confirming that the so-called

Like the idea of flying saucers, the idea of the so-called Face on Mars was far more about sociology and human behavior than about credible evidence for intelligent aliens. It's a story about misconception, silliness, promoters, and the insatiable drive of the less reputable media to exploit any quirky story to sell papers or attract viewers.

Face on Mars was nothing more than a pile of dirt. The shadows that made eyes and a mouth in the face were created by various gullies in the eroded caprock.

It's important to understand how the media foment a flap like this from behind a shield of "objective balance." Throughout this episode, virtually no working scientists believed in the face. Among several hundred members of the Division of Planetary Sciences of the American Astronomical Society, I doubt you could find more than one who thought there was any real scientific issue there. But many journalism students are taught a distorted version of objective balance, which can end up being substituted for more thorough reporting. This version of objective balance amounts to the notion that you should find at least one person on each side of an issue. So while the reality may be that 999 out of 1,000 scientists scoff at the idea of a nonnatural face on Mars, what you end up seeing on TV is the talking head of the thousandth scientist, saying that there is a bizarre mystery, while one of the 999 dissenters says there is not. One for, one against. Balance. Not a word about which side represents dominant opinion and why.

Of course, a better protection for someone who wants to know the truth is the scientific method of review and testing, worked out over the past 500 years and relatively unknown in our popular culture of sensationalism. One aspect of the scientific system is that any claim must be reviewed and criticized by peers before it is published in reputable media. It is evaluated not by whether it makes a good story but according to whether any valid *evidence* is presented in its favor. Further, it is scrutinized to see if it is a claim that can actually be tested. It would be terrific if science reporters asked less about opinions and more about evidence, and if their editors, not to mention their bottom-line-obsessed corporate bosses, would give them time to dig deep enough into the story to separate nonsense from news. Good reporting should explain the situation, not just appeal to an authority from each side.

CHAPTER 33

PROMETHEI TERRA

Mysteries of Melting Mountains

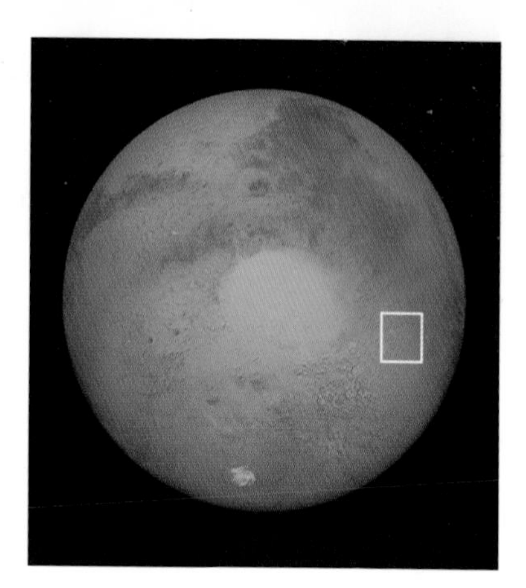

In chapter 8, we visited the softened terrain of ancient Noachian Mars, formed possibly by flow or creep of buried ground ice. What we left out of that discussion are striking features evidencing much more recent deformation, apparently by recent ice flow. Just as Marte Vallis shocked us with evidence of recent major water flow, these features shock us with ongoing terrain softening.

Good examples are the **debris aprons** scattered all over the higher latitudes of Mars. Spectacular cases can be found east of Hellas in Promethei Terra, a region named after Prometheus, one of the Titans of Greek mythology, who stole fire from Zeus in the heavenly realm and gave it to humans. Has Promethean fire, in the form of volcanic heat, melted ice in these regions?

Many mountain groups and isolated hills in Promethei Terra are surrounded by an apron of debris, as if the hill had melted and material had flowed down the hillside, out onto the plain. Researchers such as Mike Carr, G. G. Schaber, Steve Squyres, and Baerbel Lucchitta argued as early as the late 1970s that these so-called debris aprons indicate some kind of downslope flow or erosional movement that softened all terrain, and most researchers assume that the flow involves ice.

Debris aprons are often spectacular, as if a whole mountain had partly melted and started to ooze out onto the surrounding plain

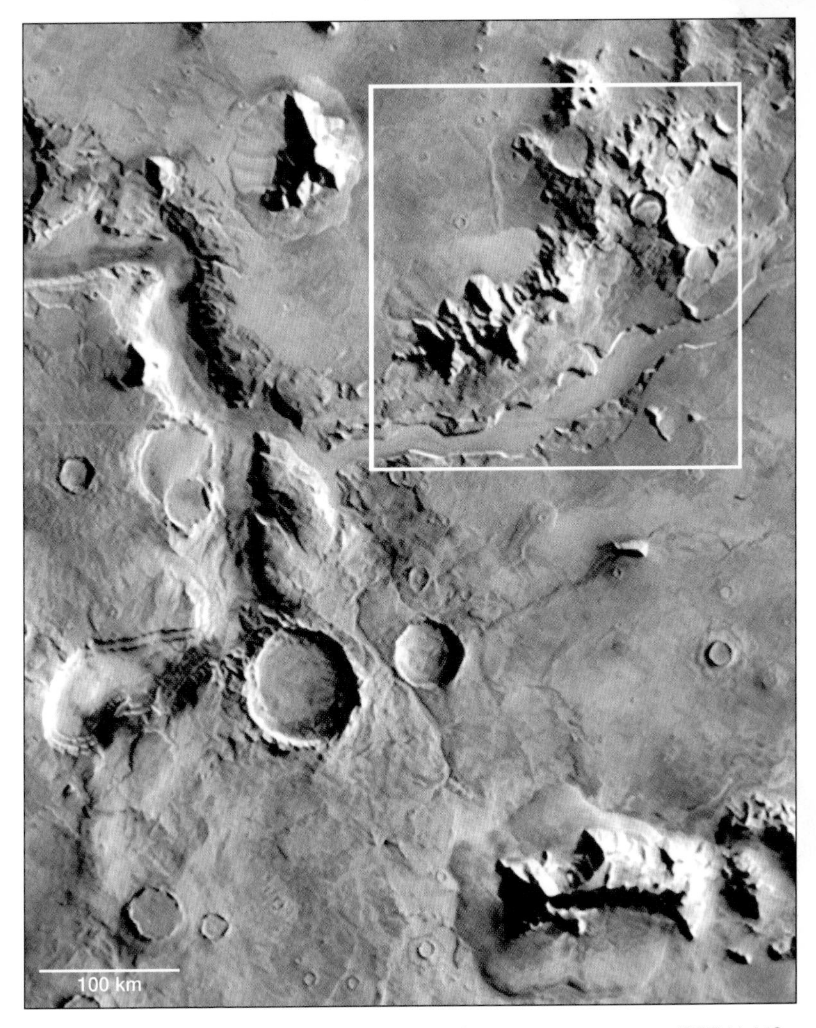

A mysterious region of melted mountains, straddling Reull Vallis (upper center). Major peaks (bottom right, upper left, and box) are surrounded by shallow aprons of debris spreading onto the adjacent plains. Box shows location of next image. (257W 44S; Viking mosaic, courtesy David Crown.)

like molasses. Yet the peaks and ridge lines are muted, not sharp. How can this be? Highly detailed images from MGS show telltale lineations on the apron surfaces that seem to confirm glacier-like movement and reveal the course of the flow—downhill, then fanning out at the margins. Such flow may be carrying material away from the lower mountain slopes, initiating new landslides that keep the crest lines sharp. Planetary Science Institute geologist David Crown, working with Texas colleague Timothy Pierce, has pointed out that some aprons seem to emanate from large scallops in the sides of these mountains—telltale scars where part of the steep slope broke away in an avalanche, leaving sharp ridge lines behind. Crown and Pierce think that such avalanches may initiate debris aprons by carrying icy debris from the mountain downslope into an ice-rich deposit at the base, which then flows glacially and deforms into a debris apron.

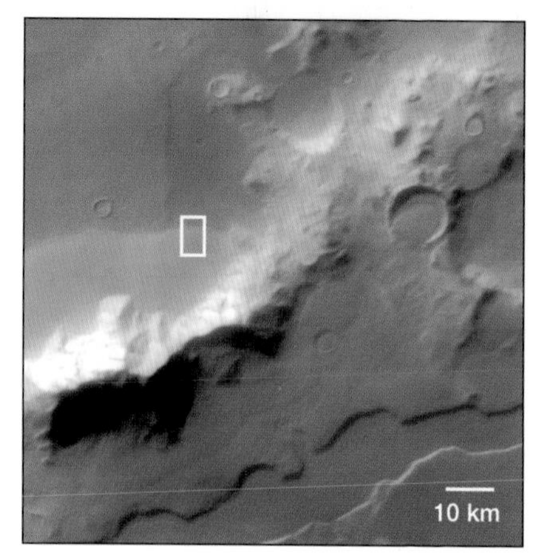

Ridge shown in previous image has a prominent debris apron on its N side. Box shows location of image on facing page. (256W 41S, E01-01295.)

On Earth, the closest analogs are features called **rock glaciers**, mentioned in chapter 8. Among terrestrial mountains, rock glaciers flow as masses of soil and rock fragments impregnated with ice. They can begin to resemble Martian debris aprons, but not exactly. Pierce and Crown, writing in 2002, measured slopes of Martian debris aprons and found them on average shallower (1 degree to 7 degrees) than slopes of terrestrial rock glaciers (5 degrees to 27 degrees). Perhaps the Martian conditions produce gentler slopes, or perhaps the aprons are not really analogous to rock glaciers.

The debris apron shown in the frame right flows N and displays prominent lineations marking the direction of its sluggish flow. It extends onto a smooth plain (top) with scattered old craters showing the flattened rims typical of terrain softening (top center). (256W 41S, E01-01294.)

The most astonishing thing about Martian debris aprons is that some of the flow activity seems to be relatively young. The surfaces of the debris aprons, seen at high resolution in the MGS images, have

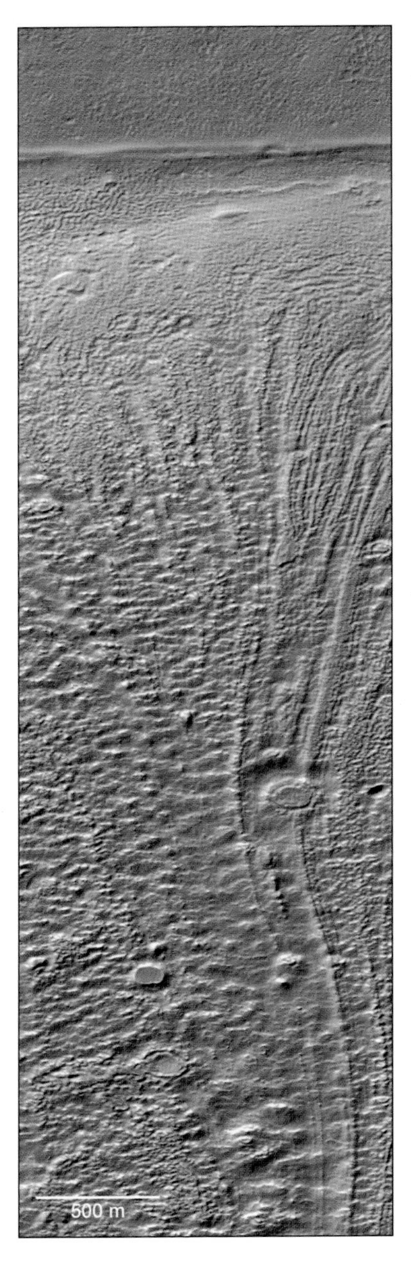

only a few fresh-looking craters with diameters in the range of 20 meters to 100 meters, and depths in the range of 3 meters to 30 meters. This means the upper tens of meters of the debris are constantly deforming and obliterating the small craters. Crater counts by Dan Berman and me suggest the ages of 10-meter-scale structures to be as young as a few MY. Ages for 100-meter-scale structures may be more like 10 to 100 MY. According to earlier counts by David Crown's group, larger craters are still older, with ages of several hundred MY. The bottom line is that the upper tens of meters of material in the debris aprons have been flowing within the last 100 MY.

In support of this, the high-resolution photos show lineated patterns in some apron surfaces, looking very much like flow lines in young, terrestrial glaciers. The scale of these lineated patterns is around 10 meters to 100 meters, consistent with the idea that the flow layers involve depths of this scale. Theoretical calculations of ice behavior on Mars by Elizabeth Turtle at our institute, working with Asmin Pathare of UCLA, imply that the timescale for ice flow on steep Martian hillsides is on the order of 1,000 to 100,000 years—but probably longer on the shallow slopes of the debris aprons. This suggests that the debris aprons are not just left-over relics of Noachian terrain softening, but something much more recent. If they involved Noachian ice deposits on isolated mountains, they should have flowed 4,000 MY ago and completed their deformation in less than a few MY, whereupon they would have accumulated very heavy cratering. Instead, the deficit of small-scale craters seems to imply that the surface layers are continually re-forming and that at least some of the ice is recent.

TONGUES OF FLOWING ICE

The larger region of Promethei Terra has even more dramatic evidence of recent ice flow. There is an unnamed 100-km (62-mile) crater at latitude 39° S and longitude 247° W, only 400 km (250 miles) east of the debris aprons illustrated above, where MGS images showed some of the most stunning glacier-like features on the planet. On the south wall of this crater, various valleys run downhill toward the floor, and each valley is filled with strange, striated material that resembles terrestrial glaciers or rock glaciers. The chevron-like pattern of the striations is typical of an ice-flow mass because the center of a glacier flows fastest, while the ice at the edges of the valley is held back by the walls. The virtual absence of craters larger than the 10-meter resolution limit implies that these features are deforming by flow that removes craters faster than they can form—which limits the age to less than about 10 MY. The age could be considerably less. These features look even more like glaciation than anything among the debris aprons.

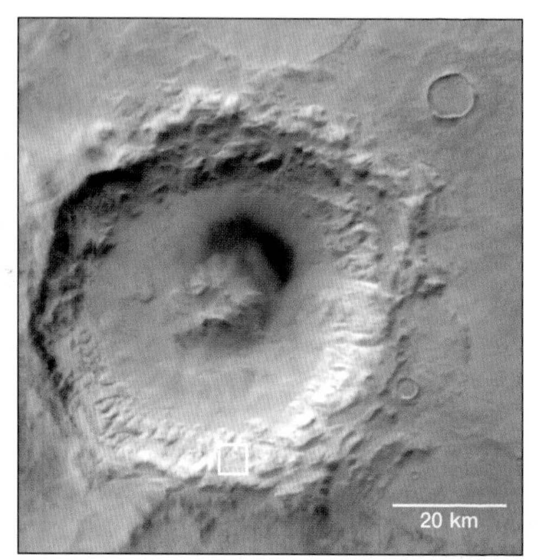

On the north wall of the crater is a still more striking feature, which produced my single favorite image from MGS. Dan Berman and I found this feature while researching images at the PSI—we each think we found it first—although Ken Edgett was probably first to see it as he processed images at Malin Space Science Systems in San Diego. The surprising feature is an extraordinary tongue of material, draping down the crater wall. By luck, the MGS perfectly frames the feature. The tongue emerges indistinctly from uphill sources among ridges on a crater wall, spoozes down the wall, and ends in sharp U-shaped ridges at its downhill tip. Beyond this edge are softer U-shaped hills, curving around the edge of the tongue. It's impossible to look at this feature without thinking that the material has flowed down the wall. It has an area of about 4 square km on which there are few, if any, craters larger than 20 meters—implying an upper age limit of about 10 MY or less.

Unnamed crater (facing page) NE of previous views has many extraordinary features indicating glacier-like ice flow. Box shows location of next image. (247W 39S, mosaic of MGS images M00-01618 and M18-00898.)

Crater in this image shows strong suggestion of rock glacier flow down a valley in the S wall. Uphill (bottom) material seems to have flowed down the main valley toward the crater floor (top). (247W 40S, M08-07937.)

1 km

Unique MGS image reveals a tongue-like feature extending from the N rim (top) of unnamed crater downslope toward the crater floor (bottom). Lack of impact craters suggests an age less than a few MY. The feature may be a very recent glacier. (247W 38S, M18-00897.)

What is the strange tongue-like flow feature? The fact that such a feature is so uncommon on Mars suggests that it may be short-lived. My own wild speculation is that water may have erupted high on the crater rim and ponded in a hillside depression, freezing into a lens of ice. (If a recent water eruption sounds strange, wait till you see the next chapter!) Such a lens of ice may have been held back by only a thin ridge of crater-wall rubble. If the pressure of the water or ice, or subsequent erosion, breached this ridge, then the ice may have suddenly found itself perched on a slope and flowed downhill within only a few thousand years. As this rock glacier of ice and dust flowed, it was self-destroying, because the movement of the ice kept churning the ice near the surface, where it sublimed into the thin Martian air. This would leave a "fossil glacier" of ice-poor dust, which might soon blow away in the Martian dust storms. This short lifetime for the feature would explain why we see so few features of this type.

This scenario is wild speculation, as I say. There is no proof about the origin of any of these features. None of them has the brilliant white color of fresh ice, but that does not refute the theory that they involve ice flow. They are likely to be ice, soil, and rocks, and to be covered by a mantle of red dust dumped by millennia of global dust storms. Their concentration in one region of Mars suggests unusual water activity in that area. Between the crater and the debris aprons we have discussed lie the headwaters of Reull Vallis, one of several winding riverbeds that empty into the Hellas basin.

WHENCE THE ICE?

The melting mountain slopes and glacier-like features of Promethei Terra bring us back to the question that keeps nagging us in this book (and the younger the ice features, the worse the nagging gets): Where did the recent water and ice come from?

If ice-rich debris aprons are flowing in modern times because of ice content, how did the ice get up into the mountain source areas and then into the apron material? How did glacier-like masses form recently on crater walls?

The short answer is that no one knows. The nature of recent fluvial processes is perhaps the biggest mystery of modern Mars.

Hypotheses are wide-ranging. For example, Mike Malin and his associate Ken Edgett, writing about debris aprons in 2001, backed away from all interpretations mentioned so far. First, they prefer the simpler term *apron,* arguing that we don't even know they are composed of debris. Furthermore, they deny that the molasses-like forms or the striated surfaces are caused by flow, and suggest that the aprons, instead, are eroded exposures of layered substrate, i.e., hard stratified layers that underlie the mountains around which the aprons occur. This view, however, does not seem to explain why the aprons universally extend radially out to limited distances from the parent mountain, nor does it explain the radial striations that resemble other flow phenomena, such as those found in glaciers.

Another radical possibility is that transient climate variations within Amazonian times have been extreme enough to allow the formation of ice deposits on the slopes of mountains and crater rims. The cycles of axial tilt have been invoked as a possible cause. For example, in the scenario of French theorists F. M. Costard, F. Forget, and N. Mangold, water-frost deposits may form at midlatitudes every few MY due to the climate cycles caused when high axial tilts burn ice off the polar caps and add water vapor to the atmosphere. In extreme versions of this idea, there might have been rain or snow in mid-latitudes at certain times. The rain or snowmelt could soak into the debris, freeze to form ice, and then start glacial movement.

Remember, practically every hypothesis that has been raised to explain the young ice flow in debris aprons or glaciers, or young water flow on hillsides, or even a young outflow channel such as Marte Vallis, is a challenge that flies in the face of the conventional view of Mars as a dry, well-frozen place.

CHAPTER 34

HILLSIDE GULLIES

The Wet Planet Mars

On June 19 and 20, 2000, rumors began to fly in the world of Mars buffs. The Internet buzzed with whispers and theories. Something was going on. The rumors said that a big discovery had been made by MGS imaging team head Mike Malin and his associate Ken Edgett, the Mars geologist who was the first to examine each MGS picture as it came in.

For several days, the rumors escalated with each new e-mail. The White House had been briefed. Malin and Edgett had found water on Mars. A frozen sea. A lake bed. Evidence of life. No, a pond of water, maybe in an old crater in Noachis. Or was it a big deposit of carbonates? One major professional space-related Web-based news agency reportedly said it was a lake of water on the floor of Valles Marineris. An announcement was imminent. A press conference next week.

The situation reminded me of rumors about UFOs. Years earlier, I had been asked to be involved in an Air Force–sponsored study of the UFO phenomenon, the famous Condon Report of 1969. I learned an important lesson from a retired British intelligence officer who spoke at one of our meetings about rumors of odd happenings. What he learned from his work of analyzing reports from North Atlantic shipping lanes during World War II, trying to track German submarine activity, was that if you get reports that something unusual has happened, something probably has happened, but not necessarily what the rumors say. That rule applied here.

Most of the Web-based rumors were grotesquely off base, but by June 21 CNN had the story essentially right, and the formal announcement came from Malin and Edgett the next day at a NASA-sponsored press conference. They announced the astonishing discovery of hundreds of fresh-looking water-erosion gullies on Martian hillsides at high latitudes. This discovery had been known by members of the imaging team for some months. Malin and Edgett had presented it to us at a team meeting in

One of the images used by Malin and Edgett to announce the discovery of recent erosion gullies on Mars shows these features at a moderate latitude of 30° S, in the walls of Nirgal Vallis. These gullies show classic features, originating in blocky layers near the top of a cliff, running fairly directly down the cliff, and ending in triangular delta fans, which in this case appear to be deposited on top of dunes on the channel floor. (39W 30S, M03-02290.)

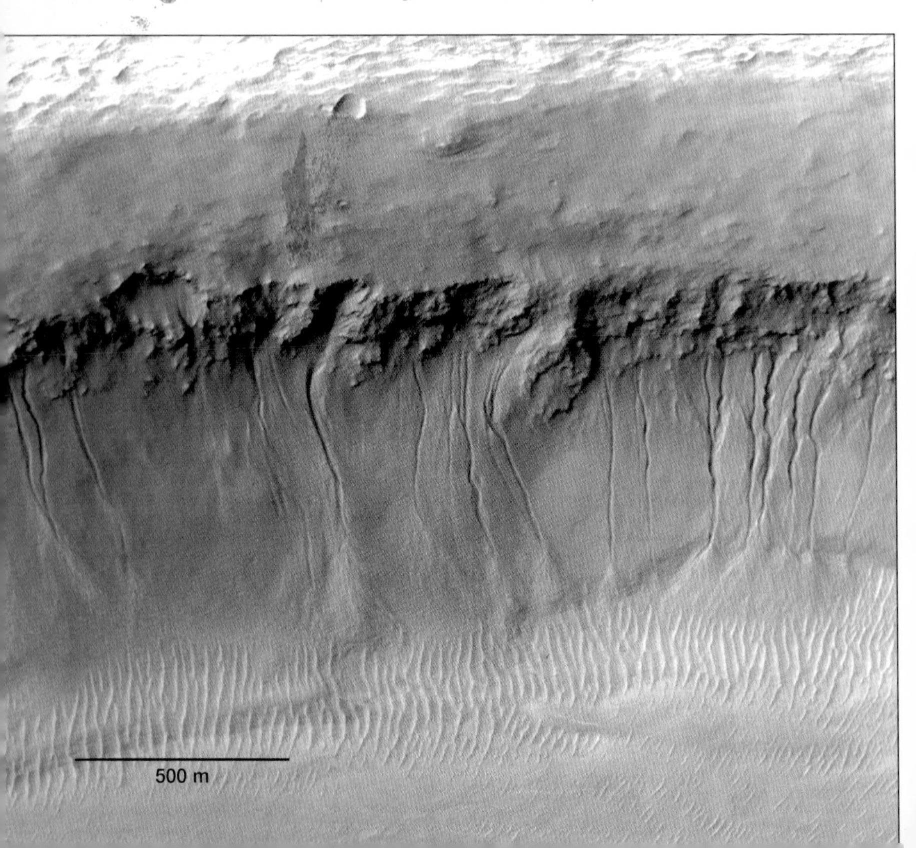

January 2000, as perhaps the most exciting discovery made by MGS, but (over objections from a few of us) they delayed a public announcement for a few more months until they could assemble all their best images and write a paper for publication. Their paper was scheduled for the journal *Science* at the end of June. Normally, announcements would be held up until paper publication, but the press conference was moved forward a week because the rumors were reaching absurd proportions.

What Malin and Edgett found was revolutionary in the context of the prevailing idea that Mars has been frozen and dry for millions or billions of years. At high latitudes, especially in the southern highlands from around 30° or 40° nearly to the edges of the permanent polar ice caps, many hillsides and crater inner walls displayed small, fresh-looking gullies that seemed to be carved by water that flowed down the slopes. These were much smaller and younger-looking than the other water-flow features, such as valley networks and outflow channels. The typical gullied hillside was a smooth slope of talus, or debris lying nearly at the angle of repose and topped by blocky outcrops forming a cliff or crater rim. Such cliff forms are very common on Earth as well, at the edges of eroding plateaus.

Malin and Edgett noted that gullies typically had three parts. At the top end, they usually originated at a disturbance or oval depression, which Malin and Edgett called an alcove. The alcove was most often located at the top of the talus slope, next to the base of layered rock outcrops. It might be a couple hundred meters in width or length. The middle section of a gully was a narrow, fairly straight channel, tens of meters wide and a few hundred meters long, running directly downhill. At the bottom of the slope, some gullies showed a delta-shaped fan of debris, emptying out of the channel and deposited on top of the level plain below the cliff. Other gullies narrowed and faded out before reaching the bottom of their hillside. To summarize, the gullies looked like places where water had somehow appeared, creating a mudflow of debris that eroded a channel down the slope and deposited the muddy slurry of debris at the base.

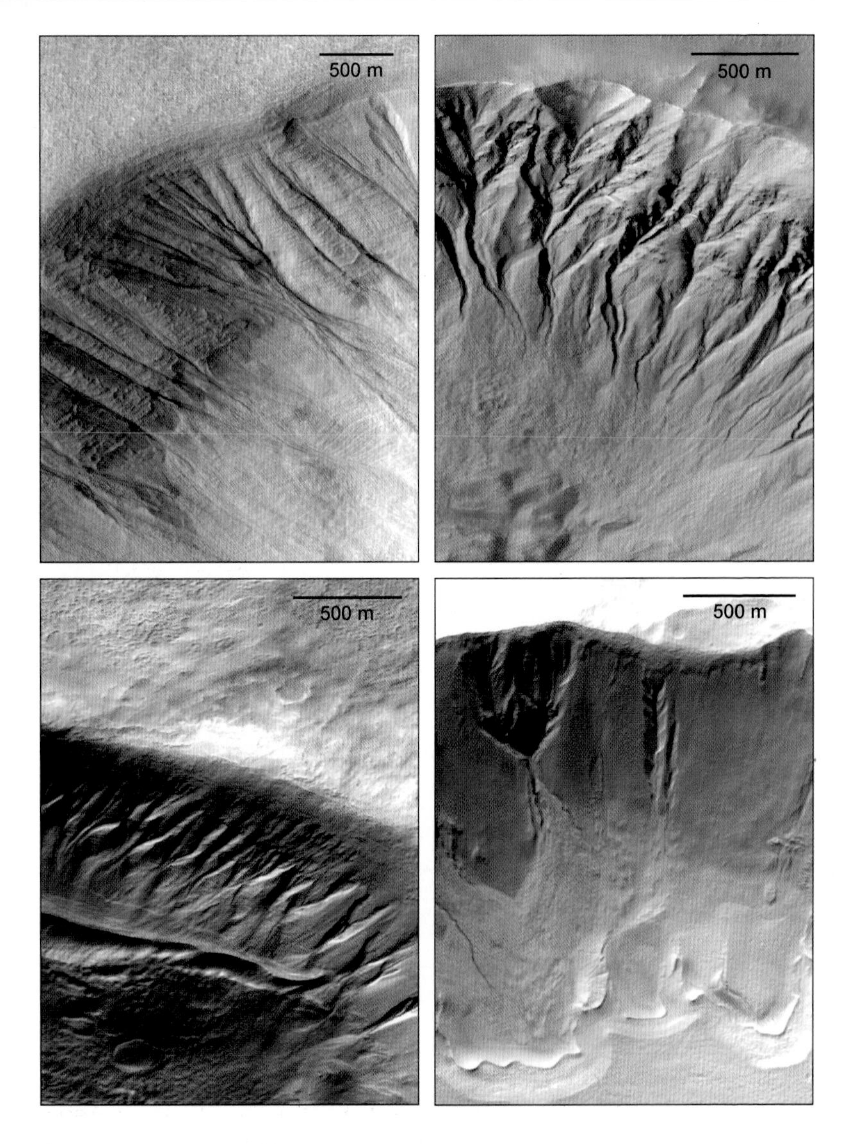

A potpourri of mid-latitude Martian gullies. These gullies show various degrees of erosion into the Martian hillsides. In most cases, the water source and uppermost erosion appear to be in blocky rock layers exposed near the top of the cliff slope. (Upslope is toward top of each frame.) Crescent-shaped ridges at the base of some gully systems resemble those at the foot of the glacial tongue shown in the last image of Chapter 33, suggesting that some gully formation occurred during the melting of ice on these slopes. (Top images: 267W 33S, M09-02875; 168W 37S, M07-01873. Bottom: 158W 37S, M18-00576; 166W 39S, M18-00303.)

This hillside has been almost entirely cut into canyons by gully systems, and the original smooth talus, or debris cover, has been carved by erosion. Crescent-like ridges line the bottom of the slope, suggesting glacial deposits. (162W 43S, M19-00651.)

FRESH TRACKS?

The discovery of a new form of water erosion was exciting enough, but even more dramatic was the fact that the gullies seemed very young. The narrow gullies themselves were too small to be well dated by the usual method of counting superimposed craters. For that technique, you need a larger area in order to get a good statistical sample of younger craters. But the broader hillside areas where gullies

SCIENCE AND PUBLICITY

The possibility of modern water release prompted NASA to feature Malin and Edgett in a full-blown Washington, D.C., press conference, in conjunction with the publication of their discovery in *Science*. The press conference put the MGS discovery on the front pages of national papers.

While I felt good about the increased public interest in Mars, I could not help but have some serious misgivings about the emerging practice of orchestrated press events for scientific discoveries. The e-mail rumor-mongering and suspense that preceded the news conference was fostered partly by outdated, self-serving rules followed by the leading international general science journals, *Nature* (U.K.) and *Science* (U.S.). Both journals insist that once your paper is accepted for publication, you should avoid any public discussion of it until it appears. This practice is called embargoing. It might have made sense in the nineteenth century, when discoveries traveled by snail-mail in wax-sealed envelopes carried by the horse-mounted royal post. In the twenty-first century, embargoing of scientific news is a silly pretense because, prior to the press conference, virtually every published paper has been presented at some level in a scientific meeting, discussed with colleagues in lunchrooms and bars, debated with colleagues 10,000 miles away via e-mail, or circulated among friends as a preprint. Yet the journals pretend that they live in an era when scientists get their news only by unwrapping a long-awaited journal that's finally arrived by stagecoach. It's a delusion, this idea that the journals are breaking fresh news. Journals are not newspapers, but precious repositories for final results.

appeared were typically sparsely cratered and seemed fresh. More important, the mudlike deltas were often deposited on top of sand dunes, meaning the gullies were younger than the dunes. Although it is theoretically possible for dunes to be very ancient, they are normally regarded as transient, changeable features, evolving over thousands of years as the winds move quantities of sand. For these reasons, most investigators see the gullies as very young in terms of total Martian history. Ages of a few MY or less cannot be ruled out.

This discovery made yet another radical reduction in the age of the most recent Martian water activity. Is it possible that water has seeped out of hillsides within the timescale of human history? Was the

The motives for the embargo system are not hard to find: it benefits the funding agencies, the journals, bureaucratic institutions, and authors—but not the public or the other scientists. NASA (or any funding agency) is happy because the buildup to an announcement calls attention to their programs. The scientist is happy because he or she receives relished press attention. The scientists' employers back home (university, institute, agency) are happy because it makes them look like they have important staff, and their own press departments get to play up to the local press. And of course the journal is happy, because it gets to pretend it "broke the story." The trouble is that the press conference bears as much relation to the real events as wrestling does to a true sport.

In an earlier era, it was actually unseemly for a scientist to pursue such events. Scientists did their science, and science journalists did their job of reporting it; if a scientist appeared in the press at all, it was likely for a testimony to interesting work. Today, university PR departments routinely crank out hyperbolic accounts of new research. The PR flacks writing the press releases are often poorly versed in science and ignore earlier work while making their own faculty sound like candidates for the Nobel Prize.

My feeling is, let's get rid of the foolish, archaic embargo policy among scientific journals and let scientific discoveries make their own way to public consciousness through high-quality Web sites and responsible press coverage.

water coming from underground sources, like springs on terrestrial mountain hillsides? Does Mars have underground water-filled aquifers and active springs? Here was a headline for future astronauts! Could underground liquid water be common on modern Mars? And if so, could ancient Martian life-forms have survived to the present?

HOW DO THE GULLIES FORM?

The gullies immediately produced a number of questions and controversies. When Malin and Edgett announced their find, they and others were puzzled that these features were not at the equa-

500 m

Comparison of advanced gully erosion on Mars (left) and in Iceland (below). In both cases, the gullies have cut into a smooth talus slope, eroding canyons that cut it into triangular wedges. (Martian example at 163W 41S, M15-01616. Icelandic photo by author.)

tor but at high latitudes. The reasoning was that the equator is the warmest part of Mars, so if ice is going to melt anywhere and run down hillsides, it should be at the equator. On the other hand, several writers pointed immediately to the various kinds of data, including rampart craters, showing that equatorial ground ice is hundreds of meters deep, whereas ice at higher latitudes is much nearer the surface. As mentioned earlier, the Odyssey mission in 2002 confirmed the presence of ice within the top 2 meters at latitudes around 65°.

Within months after the discovery of the gullies, researchers pointed out an interesting fact. Nearly perfect duplicates of Martian gullies exist at high latitudes on Earth. French researchers led by F. M. Costard, at a planetary center in Orsay, near Paris, found "Martian gullies" in Greenland; NASA researcher Pascal

Lee found them in northern Canada; and I wrote about examples in Iceland. These analogs supported the theory that the Martian gullies were caused by water, but the exact mechanism remained uncertain. Was the water from springs? Snowmelt? Or what?

Then a monkey wrench was thrown into the works when Australian researcher Nick Hoffman promoted his carbon dioxide theory of Martian geology, mentioned in chapter 22. He and several American researchers supported Hoffman's idea that the fluid coming out of Martian hillsides was not water at all, but liquid carbon dioxide. This theory was soon criticized as people realized that liquid CO_2 exists only at pressures above 5 bars (compare atmospheric surface pressure on Mars, which equals 0.006 bar), so that the hypothetical Martian liquid CO_2 would have to erupt out of a hillside at nearly a thousand times the ambient air pressure. It wouldn't flow tidily down the hill in a nice channel; rather, it would explode. CO_2 advocates countered that certain high-gas pressure flows could develop, similar to flows in some terrestrial volcanoes. But the perfect water-based analogs on Earth seemed to end that argument.

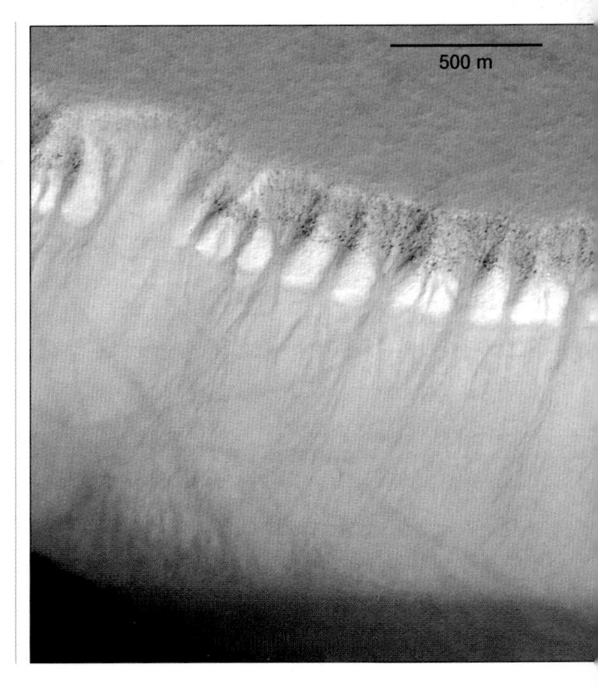

Gullied hillsides may expose distinctly different layers. In this view, the uppermost rubbly layer at the rim of the cliff is peppered with boulders. The layer below that is very bright, smooth material. The upper layer might be eroded lava; the lower layer, a stratum of wind-blown fine dust. The slope below these layers is not only cut by faint gullies, but also crisscrossed by dark streaks marking paths of dust devils. (358W 71S, M12-02093.)

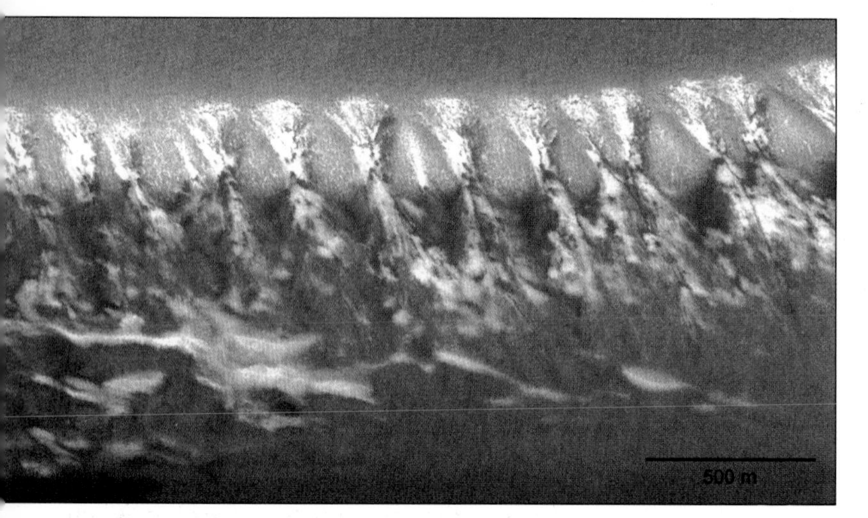

Frost-covered south polar gullies. On this hillside, the alcove at the top of each gully is coated by bright frost. Whether the frost is frozen H_2O or CO_2 is not certain. The water eroding the gullies may come from underground sources, but some theories evoke melting of frozen water from snow or ice cover. (359W 71S, M09-06352.)

MADE IN THE SHADE?

In their original discovery paper, Malin and Edgett pointed out a very strange property of the Martian gullies: they are much more likely to form on slopes that face toward the pole than on slopes facing toward the equator. This was another surprise, because you would think equator-facing slopes would get more solar heating than pole-facing slopes. After all, the south side of an office building in Minneapolis gets more sun than the north side, right?

Several researchers, including Malin, Edgett, and me, all writing in 2001, developed ideas along the following lines. Underground ice layers could melt occasionally as a result of enhanced geothermal heat flow, and the water could flow regionally through underground aquifers, perhaps a few hundred meters below the surface. Wherever the underground aquifer layer intersects a cliff side or a crater wall, water could escape onto the surface, therefore explaining water sources on hillsides. What happens next? In most parts of Mars, water trickling out of a hillside spring would sublime rapidly into the atmosphere when daytime sunlight warmed the soil layers, so water would normally not be released fast enough to flow down a hill and erode a gully. This explains the absence of hillside gullies in most areas. At higher latitudes, the equator-facing slopes would face the sun and get warmed in the day, so that any water arriving by aquifer would tend to sublime away each day. At high latitudes, however, the daytime sun is low in the sky and pole-facing slopes could be shaded all day, like the north side of a northern building. Thus, in our scheme, the soil layers on the pole-facing slopes of a crater would tend to be below freezing, and water approaching those slopes in an aquifer would freeze and seal the aquifer with an ice plug. When the hydrostatic pressure built behind the ice plug, or perhaps when

A hillside outburst. A sudden release of water on a Martian hillside, perhaps issuing from an underground aquifer, would probably be accompanied by a cloud of steam. This is due to the rapid evaporation or boiling caused by extremely low ambient air pressure on the planet. Sufficiently rapid delivery of water could lead to downslope runoff, carving gullies. Some theorists believe that water release episodes may have coincided with periods of thicker atmosphere, and possibly bluer skies. (Painting by author.)

the midnight sun of the winter months heated the soil and thinned the ice plug, the plug would burst and a large, pent-up volume of water would escape too fast to sublime. It would run down the talus slope of loose gravel, erode a channel, and deposit its muddy slurry at the bottom.

A BETTER THEORY?

In 2002, the French team headed by François Costard, who had found matching gullies in Greenland, published an alternate, perhaps better, model of gully formation. Calculating that the Martian poles can nod as much as 45° toward the sun every few million years, the French team came up with a remarkable fact. During these times, Martian high-latitude craters and cliffs get long periods of summertime midnight sun, with the result that slopes with maximum total solar heat input are not the equator-facing slopes (as supposed in the preceding theory), but rather the pole-facing slopes. During midsummer, the pole-facing slope would actually be the one place where Martian daily average soil temperatures would rise above freezing! According to the French calculations, the greatest warmth in the top few meters of soil occurs exactly where Malin and Edgett found the gullies.

The French team assumed that under some axial tilt conditions, frost or snow would be deposited on high-latitude slopes, charging the soil with ice mixed in the surface soil layers. Then, during the part of the axial tilt cycle described above, the ice would melt and saturate the soil layers with water, lubricating the soil layer and causing failure of cohesion. Patches of soil would break loose and create a debris flow that would erode a gully—a process observed on Earth.

According to this idea, Martian surfaces at high latitudes could be recharged with ice every few million years, causing new cycles of recent water activity. There is a possible drawback to this theory: on Earth, debris-flow gullies of this type often start at random spots in the middle of slopes, wherever the water-lubricated soil breaks loose, whereas on Mars they usually start at the top of the slope, just at the base of the blocky cliff-top rock layers, as if the water might be coming from seeps associated with a porous aquifer layer in the rock strata.

The important difference between these two theories of gully formation is that the seep theory invokes underground water, but the axial tilt theory invokes water added directly from the atmosphere, like frost or snow deposits on Earth. The first has more implications for possible Martian biology, because it evokes underground water-rich layers in which microbes might have survived. The second theory gives a more plausible mechanism for explaining why gullies could form so widely in very recent times. Since the maximum axial tilt happens every few million years, it also predicts that the last round of gully formation would have been very recent, within the last few MY—consistent with the observations.

These are two rather different pictures of Mars—a planet with occasional underground water sources versus a planet with aboveground climate cycles that recharge ice in high-latitude surface layers. Conceivably, both theories may be true, and gullies could form as water is delivered to hillsides by either mechanism. Or perhaps there is a combination of effects, in which gully formation requires aquifer delivery of water only during times of maximum axial tilt. Future exploration will focus on the exact mechanisms for forming the mysterious gullies and will try to clarify what the gullies are telling us about water on Mars. That's the beauty of science: each discovery leads us on, like a compass pointing toward the next area of exploration.

My Martian Chronicles

PART 12: MARTIAN GULLIES IN ICELAND

About a month after Malin and Edgett announced the discovery of Martian gullies, an interesting international conference was held in Iceland. The topic was geologic processes in the polar and near-polar regions of Mars.

Iceland was the perfect venue—a country of volcanoes and ice—and Mars experts knew that Iceland has many analogs to Martian features. From the air, giant Icelandic flows of fluid basaltic lava duplicate the very young lava flows in Elysium Planitia. Icelandic provinces where lavas erupt through glacial ice sheets have large, Mars-like outflow channels carved by floods of water that were created as lavas melted ice. Iceland straddles the mid-Atlantic ridge, where the European tectonic plate is moving away from the North American plate, and therefore the country is split by countless fractures and fissures that resemble fracture systems on Mars. There are places in Iceland where you can stand with one foot on rock that is part of the European landmass and the other foot in North America.

The venue was perfect in a more human way. Iceland is a splendid country, tiny and yet full of vast empty vistas under enormous skies. An hour's drive from Reykjavík, the capital, one of the fractured valleys contains a natural amphitheater bounded by lava cliffs, in which the world's first parliament of citizen farmers was organized a thousand years ago. Culture abounds. It's a country whose farmers are also leading novelists, poets, and painters. In Reykjavík, attendees were taken to the premiere of an operatic ballet on Norse sagas, by the Icelandic composer Jon Leifs, with the Iceland Symphony Orchestra and an international dance company.

The conference organizer, an outgoing and busy geologist with the perfect Viking name of Thorsteinn Thorsteinsson, kindly invited me to show some of my Mars paintings at the meeting, and during a break we took a drive up to his ancestral family farm, where I painted another small landscape view. On the way, I gazed out at the passing volcanic landscape, and as soon as we approached the cliffs of the Esja plateau, half an hour north of Reykjavík, I was surprised at every turn in the road. "Stop the car!" I exclaimed every ten minutes. "There's a Martian gully!" I would then jump out and snap pictures.

As it turned out, Iceland is full of Martian gullies. I reported these features in a short paper in a 2001 international Mars conference volume, and as mentioned earlier in the chapter, other researchers reported virtually identical features in Greenland and Canada. A

*Water-eroded gullies formed on hillsides in Iceland match the Martian examples in size and form. (*Above: *S face of Esja plateau, photo by author.* Right: *N of Hvalfjordur, aerial photo by Thorsteinn Thorsteinsson.)*

year later, I flew back to Iceland to meet Thorsteinn and an Icelandic hydrologist, Freysteinn Sigurdsson, to attempt a more careful study. We spent several days driving out from Reykjavík to various gullied hillsides. Thorsteinn and I clambered up the slopes to look at the gullies up close. It was like exploring Mars. Many gullies on steep hillsides were a few meters across and hundreds of meters long. You could see where muddy water had surged down the basaltic slopes, pushing football-size rocks aside, forming a foot-high levee on each side of the gully and leaving a muddy, boulder-peppered, fan-shaped deposit at the bottom.

Surprisingly, the gullies had not been well studied as a specific class of geologic features, and it was hard to draw final conclusions about how they formed or that might shed light on how the Martian features may have formed. For example, we could document waterfalls and springs that seemed to gush out of the basalt cliffside layers, where underground water had flowed in underground aquifers. Records suggested that this water was created as winter snowpacks melted, sank into the porous basalt, and then flowed laterally underground. At the same time, some gullies seemed to start not from aquifer water, but from snowmelt on the hillsides. We photographed classic debris-flow gullies, where a 10-meter-wide patch of soil halfway up the slope got so wet from melting snow that cohesion failed and the whole patch slipped away, turned to muddy slush, and flowed downhill. This matched the French Mars-gully model, but it raised the problem of why we don't see more Martian examples of gullies starting halfway down the slope.

I would like to be able to write that we have the final, grand solution to the origin of the Martian gullies, but I think it is more productive to acknowledge that we don't have all the answers and that we may require future missions and even astronaut landings to figure out what is going on and what it means about Mars.

CHAPTER 35

INCA CITY

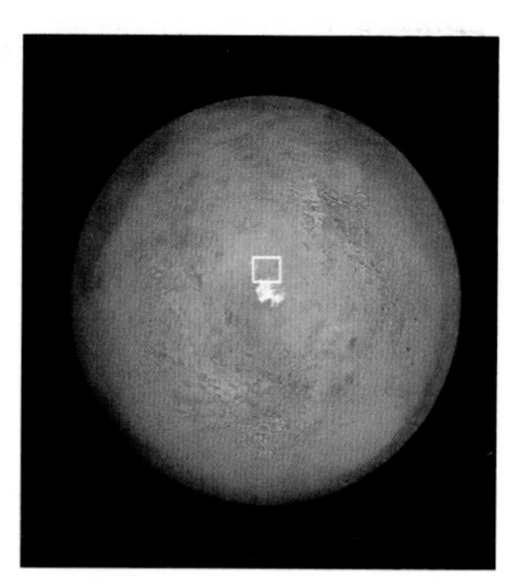

During the Mariner 9 mission, in 1972, when Martian features were being mapped close-up for the first time, imaging team members found a very strange feature at the high southern latitude of 82° S and longitude 66° W. It seemed to be a rectilinear grid of straight walls and squarish depressions, typically 3 km on a side. It was a natural formation, but its origin was difficult to judge. Just as Viking team members later named the Face on Mars with whimsical affection, Mariner 9 team members referred to this formation informally as Inca City. It vaguely resembled an ancient, abandoned urban ruin, complete with walls and plazas. If you wanted to pretend you had an alien artifact on a distant world, Inca City would have been a better candidate than the Face on Mars, but mercifully no one made a cottage industry of ballyhooing Inca City in the tabloids.

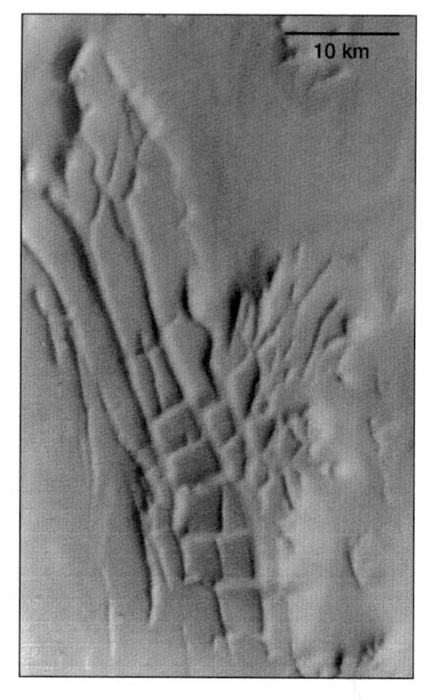

What processes could possibly have formed this pattern? This was never treated as a burning question in Mars science, because the feature was viewed as an oddball natural curiosity, not as a key to planetary evolution. Most researchers were more interested in more fundamental issues, such as the source of water in the channels, the cause of terrain softening, the question of climate change, or the duration of volcanism. Still, Inca City begged for an explanation.

Early speculation fell into two camps. One said that the winds created some sort of giant dune complex. Wind does weird things, and the intersecting ridges might be the product of two different prevailing wind patterns. However, the feature was too big to be convincingly explained in this manner. The other possibility was some sort of fracture or fault pattern. Tectonic forces can often produce a set of parallel fractures, and it is not uncommon to have two sets of forces that produce intersecting fractures. Once the fractures exist, magma may ascend into them, filling them with molten rock, which then cools. At high latitudes such as 82° S, where Inca City lies, we know that there are polar sedimentary lay-

Mariner 9 frame on which Inca City was discovered in 1972 (facing page). Scientists were puzzled with its rectilinear ridge pattern. (69W 82S, Mariner 9 frame 8044333.)

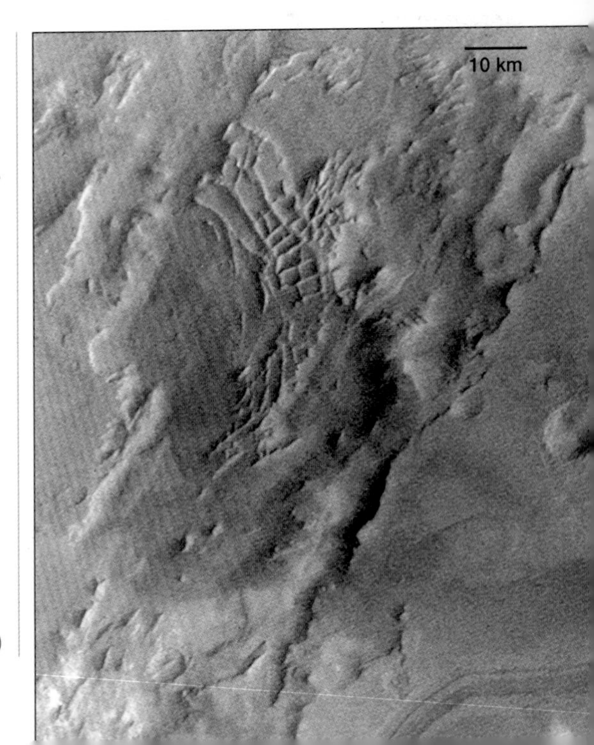

Regional view from Viking 2 orbiter shows layered polar deposits to the south (bottom) and suggests that Inca City has been exposed as overlying smooth layers (right) are peeled away by erosion. (69W 82S, Viking 2, NASA.)

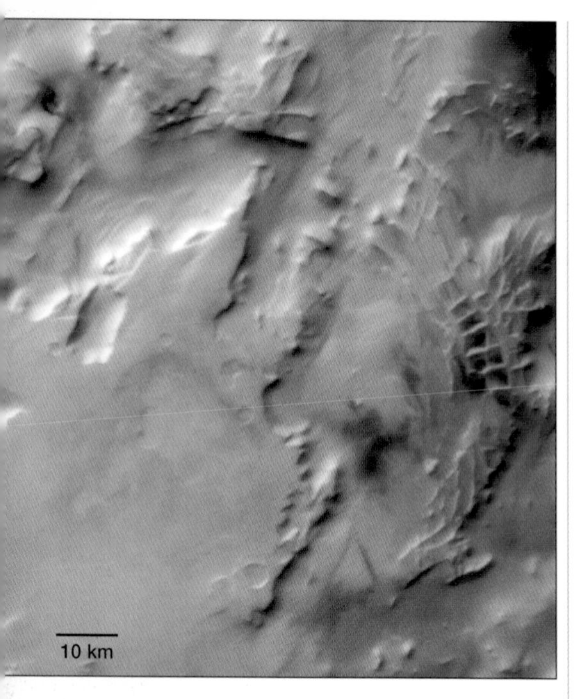

MGS wide-angle regional images revealed that Inca City is part of an 86-km circular structure, probably the root structure of an ancient eroded impact crater buried under smooth sediments (lower left) now being exposed by erosion. The discovery of the circular feature showed how photomapping can reveal new features in an area that has been photographed before. (69W 82S, E09-00186.)

Detailed view of Inca City walls (facing page) shows frosty, flat-topped ridges with some gully formation (lower right corner). The small dark spots may be rock outcrops protruding through the frost or places where the frost cover has sublimed, exposing dark soil underneath. Similar dark spots have been seen in other frosty areas. (69W 82S, MGS image release 7908.)

ers (we'll visit them later), and they are likely to be weaker, more crumbly rock than freshly solidified igneous rock. Thus, as erosion and wind strip away the host-rock sediments, they could leave the igneous rock standing in ridges that marked the original fractures. Resistant rock formations standing in wall-like slabs are well known on Earth; they are called **dikes** because they resemble ragged walls built to hold back water. In short, the second theory was that Inca City was a set of intersecting dikes exposed by erosion. But in that case, why such a localized, striking pattern?

MALIN AND EDGETT TO THE RESCUE

As they ran the MGS cameras, Mike Malin and Ken Edgett kept a practiced eye on the Inca City region as the images came in. In 2002, they solved the mystery of Inca City with a news

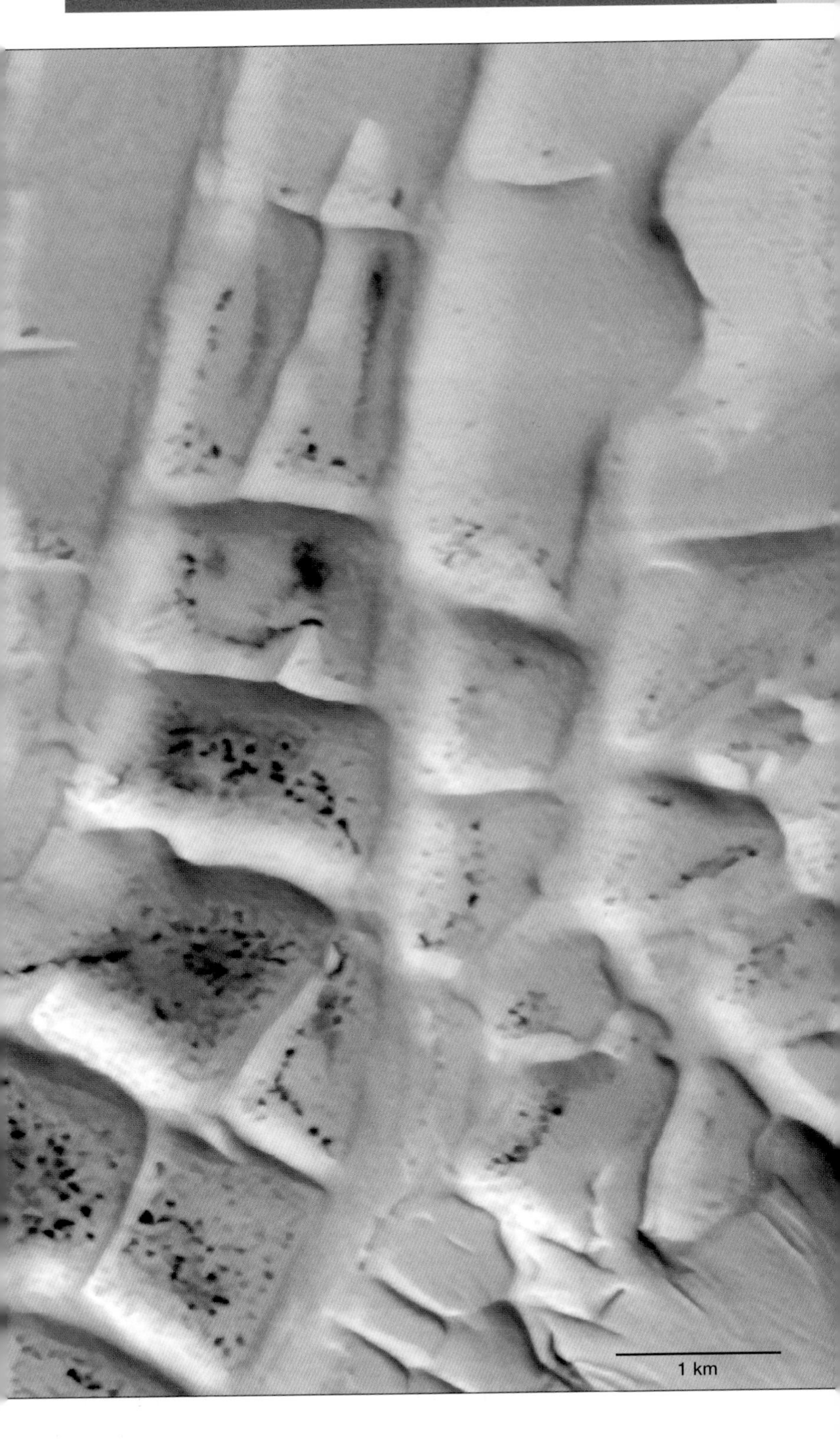
1 km

release on their Web site. A regional shot, under just the right lighting, showed that Inca City was an isolated part of a larger circular pattern, 86 km (53 miles) in diameter. From the images, it appears that most of the circle is obscured by a broad, flat sediment layer, typical of the margins of the south polar cap. Malin and Edgett previously emphasized that modern Mars reveals a lot of exhumation of ancient structures, and they suggested that the circular structure is an ancient impact crater whose roots are just being revealed. Possibly, the underground circular and radial fractures associated with the impact were filled with ascending magma, forming resistant dikes. Probably the crater was then eroded down to its originally buried roots, then covered over, then partly revealed by exhumation. The news release text, however, prudently concludes that while the new image reveals the circular context, it doesn't "reveal the exact origin of these striking and unusual Martian landforms."

My own work helps at this point. As noted in chapter 5, when I was a graduate student in the '60s, I worked on the concentric and radial structures of giant impact basins on the Moon. It is clear that the largest impacts create not only dramatic concentric structures of multiple interior and exterior rings, as well as the main crater rim, but also radial fractures. Often these intersect; and there are places on the Moon, including a broad region near the center of its visible face, where the radial and concentric fractures and rims produce patterns of rectilinear ridges similar in scale and appearance to Inca City. Although the main circular pattern mapped by Malin and Edgett is 86 km across, we don't know if that marks the original rim size of an eroded crater or only the inner circular root structures of a much larger feature. Whatever its original size, Inca City is part of a large ancient Martian impact structure, and its story testifies once again to Mars's very active and complex history of burial and exposure.

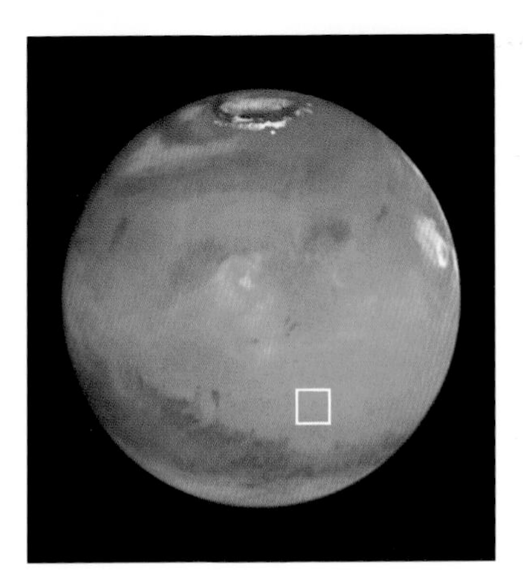

CHAPTER 36

A CLUSTER OF CRATERS

Just east of a river channel system named Ma'adim Vallis lies another odd formation: a tight cluster of several dozen small craters, shoulder to shoulder. The bigger ones are 500 meters to 700 meters across, and there are many smaller ones in between. The cluster spreads over a diameter of about 12 km. It's odd. It can't be a random grouping. The craters in the cluster formed virtually at the same moment, as can be inferred because they share rim structures rather than one crater's rim clearly being superimposed on an earlier crater. Not only are the craters tightly grouped and similar in appearance, but other

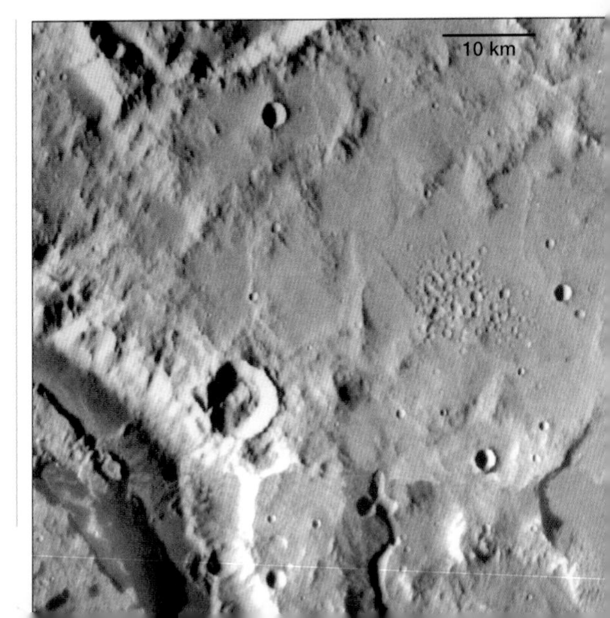

An odd cluster of craters lies NE of the Ma'adim Vallis outflow channel (lower left corner). This cluster includes dozens of craters in a diameter range of 200 to 700 meters. (183W 20S, Viking orbiter mosaic.)

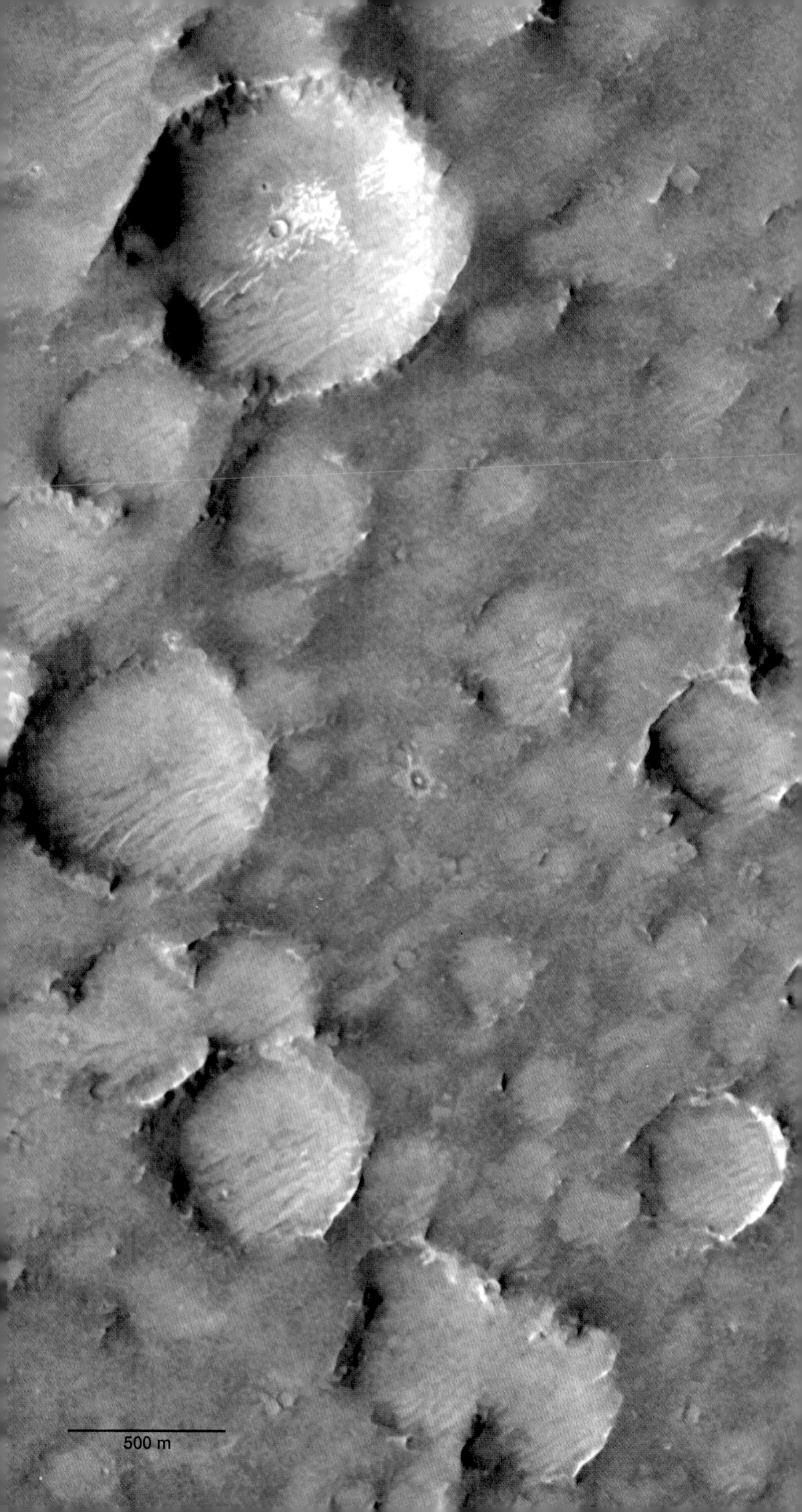
500 m

such clusters can be found elsewhere on Mars. Some unknown process is producing them.

You can get a rough idea of what is involved if you think about how craters form. Depending on the speed of the projectile, the crater may be about three to ten times the size of the body that hits the ground. So we are looking for a situation in which a swarm of 100-meter-scale, office-building-size fragments comes crashing down out of the Martian sky.

It's well known that on the Moon, Mars, and other planetary bodies, crater-forming impact explosions throw out massive blocks of rock or loosely consolidated fragments, which fall back and make so-called **secondary impact craters**. Usually, these are in strings of debris, often roughly radial to the parent or primary crater. For example, the rays, or radial streamers of bright dust blasted out of lunar craters, are dotted with small secondary craters. Therefore,

A close-up of the Ma'adim Vallis crater cluster (facing page) shows numerous 500-m-scale craters. The ones that overlap each other must have formed at the same time, since there is no clear superposition of one on the other. (182W 20S, M12-01425.)

Another crater cluster similar to the Ma'adim Vallis example. (270W 32N, M11-02890.)

the first theory that one might deduce is that these are simply secondary craters from ejecta thrown out of some nearby large primary crater. The problem is that they are not strung out in an alignment that might point back toward a source region, nor does a search of nearby territory suggest a likely candidate for a parent crater.

Once I noticed the cluster on the banks of Ma'adim Vallis, I began to see clusters in many different locations, and I learned that other researchers had seen them, too, though no one had made any systematic study of them. They seem to be randomly dotted around Mars.

THE ICY COMET THEORY

A second theory occurred to me. If such clusters are really isolated in random positions on Mars, they might be the result of breakup of some unusual kind of weak meteoroid in the atmosphere of Mars, analogous to fireballs exploding in the atmosphere of Earth and dropping meteorite fragments across the countryside. (**Meteoroid** is the term used for a small solid body in space or one entering an atmosphere, but before it hits the ground. If it survives all the way to the ground, it is called a **meteorite**.) Because the Moon lacks such random clusters, I thought the unique process on Mars might be the action of the Martian atmosphere on incoming meteoroids, causing them to break up.

The problem is that Mars's atmosphere is thin, and meteorites are strong, so most meteorites zip straight down to the Martian surface, making single craters. However, the stress of slamming into even a thin atmosphere at cosmic speed can be very high, like the stress experienced by a diver belly-flopping from a high dive into what otherwise seems like a nice, fluid, accommodating swimming pool. At first I thought this might be an important discovery: the scattered clusters might be telltale markers, revealing some unknown class of very weak meteoroids in space, possibly related to comets. Comets are interplanetary icebergs that consist of ice and dust. When sunlight warms them, ice sublimes and they spew out gas and dust, forming a tail. Some comets break apart spontaneously in space due to the stresses of the gas jetting off the surface, proving that at least some of them are very weak. My idea was that a comet, perhaps fractured by previous collisions in space, might slam into the Martian atmosphere at a typical cometary speed of 20 km per second (45,000 miles per hour), heat up like a space-shuttle heat shield, and explode in a blast of expanding steam and a shower of 100-meter-scale fragments scattered over several kilometers. Interestingly enough, my colleague Stu Weidenschilling at the Planetary Science Institute had written a paper only a few years earlier, concluding that comets were weak aggregates of primordial 100-meter-scale icy building blocks, formed in the original solar system nebula that spawned the planets. So everything fit. The Martian crater clusters could be proof that very weak ice-aggregate comets really do exist.

The idea immediately hit a roadblock. I could find no experts on meteorites and impacts who believed that an icy comet could explode in that way. At 20 km per second, a meteoroid traverses the main bulk of the atmosphere in only a few seconds, and that's why a shooting star in Earth's atmosphere lasts only a second or two. Even if a meteoroid exploded and the fragments spread at bullet-like speeds of 200 meters per second, the fragments would separate by only a few hundred meters before they hit the ground. Jay Melosh, a Tucson colleague at the University of Arizona who has written the standard text on impact craters, discussed these ideas with me and felt that a meteoroid explosion couldn't create such large clusters as seen on Mars. Two years later, I met some Russian experts in this field and worked with them to make computer models of such events; they calculated that the likely result of a meteoroid explosion in the Martian atmosphere would be smaller-scale clusters of 10-to-20-meter-scale craters scattered over a few hundred meters—which we actually found in the MGS pictures (see page 379). So the larger craters, scattered over 10 kilometers, seem to need another process.

My Martian Chronicles

PART 13: CRATER CLUSTERS, COMPUTERS, AND CAFÉS

Between 2000 and 2002, I had a unique opportunity to study what happens when meteoroids crash through the atmosphere of Mars.

It started in the fall of 2000, when I attended a meeting in Moscow, sponsored by a three-decades-old partnership between Brown University in Rhode Island and Moscow's Vernadsky Institute, a premier geochemistry institute of Russia. Planetary geologist Jim Head of Brown and leaders of the Vernadsky Institute, such as Sasha Basilevsky, deserve tremendous credit for originating these meetings during the Cold War, overcoming many political obstacles raised by government agencies on both sides. Such opportunities for American and Russian scholars to establish working relations was one of the factors that spurred the liberalization of the Soviet Union under Gorbachev, contributing to the final collapse of Soviet oppression.

The hallway and auditorium walls at Vernadsky looked about the same both during and after Communism—the same kinds of posters and mineral displays as at an American university geology department. Rocks are no respecters of ideology. As I sat in the audience in the 2000 meeting, a Russian researcher, Ivan Nemtchinov, appeared in the next seat and was soon tugging on my sleeve and inviting me to visit his lab. A few blocks away, among birch trees and apartment buildings, the lab turned out to be a once secret institute that had done research regarding ICBM warhead entry into the atmosphere from space. Now that peace had broken out between the two nations, Nemtchinov's group was applying its formidable computer programs to more purely scientific problems, such as the question of meteoroids exploding over Mars. Nemtchinov was a Russian of fearsome, bushy-eyebrowed visage, but bubbling with good humor and brimming with ideas about Mars. Our conversation soon led to an agreement to collaborate and an introduction to some of his staff members, including Olga Popova, a bright young Ph.D. who creates computer models of meteorites crashing into planetary atmospheres. Our idea was for the three of us to work together and try to understand what happens when meteoroids meet Mars.

I had come to this particular Moscow meeting from the International Space Science Institute (ISSI) in Bern, Switzerland, which sponsors international workshops. It occurred to me that a workshop at ISSI would be a special delight to the Russian scientists, who were in extreme financial straits after the collapse of state support for their institutes. (The director of a onetime flagship lab in Moscow told me that under the new free-market system, his government now provided only money to heat the building and to provide a base salary—*of around $35 a month!*—for senior researchers. They had to rent out some of the offices in their institute in order to provide funds for maintenance of their building.) Many Russian researchers were looking for a few months' work here and there as guests at foreign institutions, which in turn provided a fertile cross-pollination of ideas and was good for both sides.

So it came about that Ivan, Olga, and I met in Bern for two weeks at ISSI in the summers of 2001 and 2002. We were able to complete two papers, based on their calculations and the American MGS photos, showing that even the thin Martian atmosphere should break up meteorites smaller than a certain size, so that the smallest craters on Mars should range from the size of a dinner plate up to a few meters across. This cutoff was just too small to be seen in the MGS photos, so these predictions would have to be tested on some future mission. Olga and Ivan's computer models also showed that certain midsize stone meteoroids could explode in the Martian atmosphere, as they do in Earth's atmosphere, and create very small clusters of 10-meter craters, which we actually found in MGS images. Finally, we concluded that the big-crater clusters probably involved secondary fragments blown off Mars, as discussed in this chapter.

It was exciting to get some answers to the puzzles, but the most interesting thing for me was sitting with my Russian friends in pleasant, leaf-shaded café patios in beautiful Bern, comparing notes

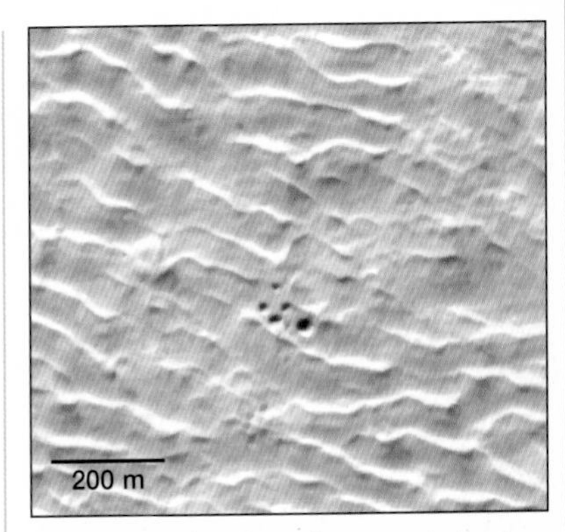

This cluster of 20-m-scale craters, isolated in a region of sand dunes, matches predictions of the Russian computer simulation for breakup of weak meteoroids in the atmosphere of Mars, producing a shower of fragments. (158W 45S, M19-00278.)

about growing up in systems where each of us was taught that the other was evil. During the second workshop, Olga had brought her eleven-year-old daughter, Julia, who was making her first trip from Russia to the heart of Europe. The streets of Bern were so safe that Julia went off each day on her own, armed with a map and her fledgling English and German, to see the medieval cathedrals, the zoo, or the apartment where Einstein discovered relativity. Julia was learning that the world was bigger than her own country. That was a good sign, I thought.

For me, there was also a certain surrealism to the whole experience. Amid these bustling streets full of happy, prosperous-looking people, I remembered that I would soon be returning to a land where sidewalks hold more derelicts than those in Bern, yet where talk-show hosts routinely assure listeners that Europeans, suffering under oppressive big-government taxation, all yearn to escape to America—a suggestion that made my Swiss friends laugh in disbelief.

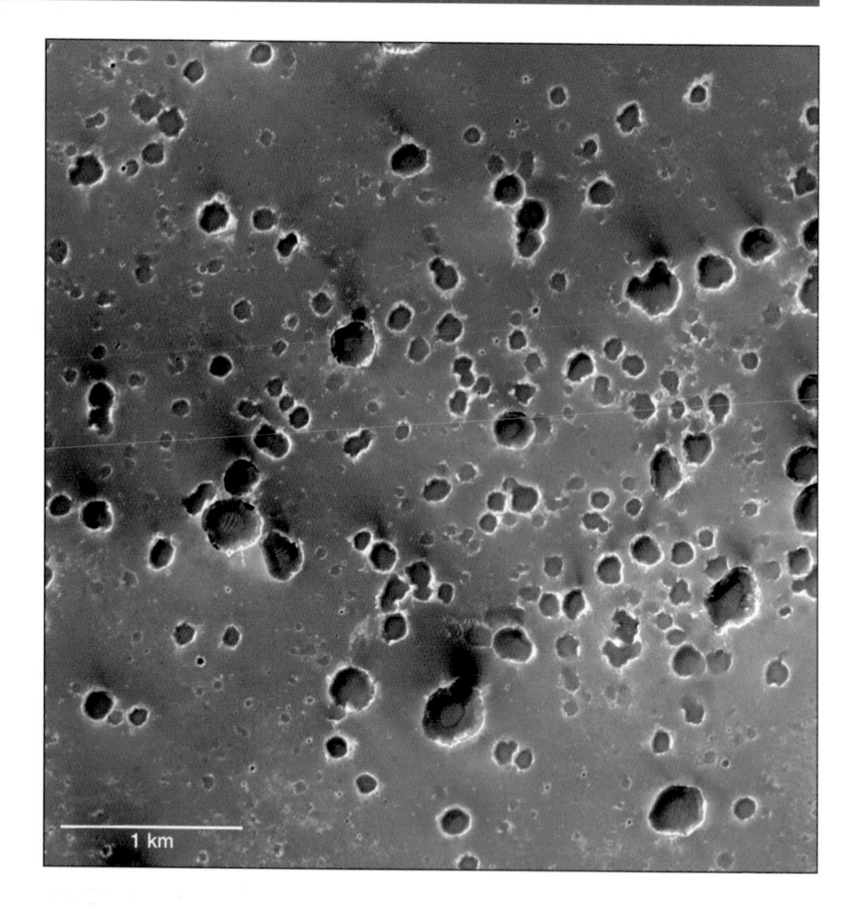

BACK TO THE DRAWING BOARD

Those ideas drove my Russian colleagues and me to a third theory, which seems to fit all the observations and combines aspects of the first two, fitting in with the existence of meteorites from Mars. To produce the Martian meteorites, our theory went, fast-moving rocks must have been blasted clear off Mars into deep space during impact explosions. Also, we noted that many large craters on Mars have nearby secondary craters due to slower-moving blocks of rubble up to a kilometer across, which had been lofted high into the atmosphere, only to fall back. Between those two extremes, some large masses of fractured rock must have been blasted high into space but not quite fast enough to escape from Mars. They would exit the Martian atmosphere, loop partway around Mars in about 20 to 45 minutes, and then fall back at ran-

A more dispersed crater cluster lies within the area of hematite deposits in Terra Meridiani. (4W 2S, MGS image mosaic by Daniel C. Berman.)

dom spots. These rocks would pass through the Martian atmosphere twice: first, they would be blasted up through the atmosphere; then, after the suborbital flight, they would crash back through the atmosphere. The speeds necessary to launch a rock off Mars would be of the order of 5 km per second, and the stresses generated on the upward flight through the atmosphere could start the fractured rock to begin separating into fragments. Suppose that during such a launch, a shattered, kilometer-scale debris crustal mass began to fall apart into 100-meter-size pieces, moving apart at speeds of around 10 meters per second during the upward ejection through the atmosphere. During a 30-minute flight (1,800 seconds), they would separate by 18 kilometers—a typical diameter for a crater cluster. A swarm of 100-meter-scale fragments could thus fall back and make the kinds of clusters we see on Mars.

Why don't we see these on the Moon? Lunar ejecta would not encounter an atmosphere on the upward or downward flight, so the shattered rock would not separate very much into individual fragments; and when the rock fell back to the Moon's surface, it would make a single, isolated crater, which would be hard to distinguish from the existing thousands of primary lunar craters of all sizes.

As exciting as it is to imagine a single interplanetary asteroid or comet slamming into Mars and making a huge explosion crater, somehow it is even more dramatic to imagine an ordinary afternoon in the desert of Mars when, without warning, in the space of a few seconds, a whole swarm of 100-meter fireballs crashes onto the landscape, making 20 or 30 near-simultaneous atom-bomb-scale explosions in a city-size piece of Martian real estate.

CHAPTER 37

LAND OF DUST DEVILS

At a Mars conference in Pasadena in 1999, Ohio Mars researcher Philip James gave a talk about the causes of global dust storms that periodically sweep across Mars. Noting that good theory must fit the data, he suggested (with tongue in cheek) that they are caused by the arrival of spacecraft from Earth. This theory fit the data, since spacecraft such as Mariner 9, the Russian Mars 2 and Mars 3 orbiters, and MGS arrived around the time of dust storms. Unfortunately, this theory lacks a theoretical underpinning.

A better hypothesis says that global dust storms are related to **dust devils** caused by strong summer heating of the atmosphere and surface. Earthbound telescopic observers discovered generations ago that Martian global dust storms arise most commonly during very warm summers in the southern hemisphere. The warmest Martian summers happen in the southern hemisphere because Mars is appreciably closer to the sun during southern summer than during northern summer. (This effect is negligible on Earth because our orbit is more circular and we remain at about the same distance from the sun during all seasons.) The strong southern summer sun warms the Martian ground, which in turn warms the air near the ground; this then creates pockets of hot air that start to rise. When the temperature difference between the near-ground air and the upper air is great enough, a pocket of warm, rising air

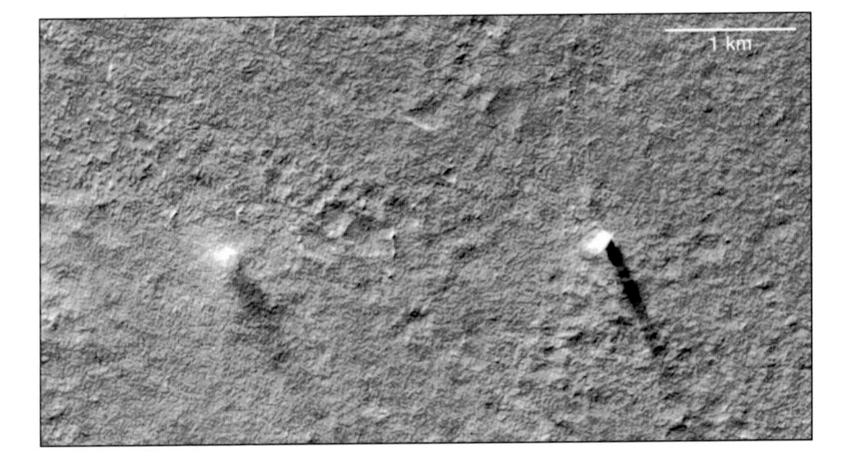

Two dust devils on the plains of Amazonis Planitia are revealed in this vertical downward view not only by the bright puffs of swirling dust, but also by their shadows (extending to lower right). The left example is diffuse and has a pale shadow; the right example is dense and casts a sharper, blacker shadow. Based on the shadow length, the dust devils rise hundreds of meters into the sky. (161W 35N, E05-02077.)

can keep expanding and rising. The eddying nature of air turns this into a whirlwind that picks up dust and debris, producing a tall, thin, spinning, wandering cloud of dust—a dust devil.

Dust devils on Earth can sometimes be spectacular, especially during the summer in western deserts. As wide as a car or a house, dust devils rise a hundred meters or more, wandering the landscape aimlessly like lost spirits. When they cross pavement or grassland, they may become nearly invisible, because there is little loose material to be sucked into the air. But when they cross loose soil, a plume of smoke-like dust spins up into the sky.

On Mars, where dust is everywhere, dust devils are even more spectacular. One of our pictures of the lost landing site of the first artifact on Mars, the Russian Mars 2 probe, showed a mysterious spiderweb network of dark straight, bent, and curved lines—tracks left by wandering dust devils.

Why do local dust devils trigger global dust storms on Mars? It's an example of feedback effects. The southern dust devils inject much dust into the atmosphere, creating the regional storms. The

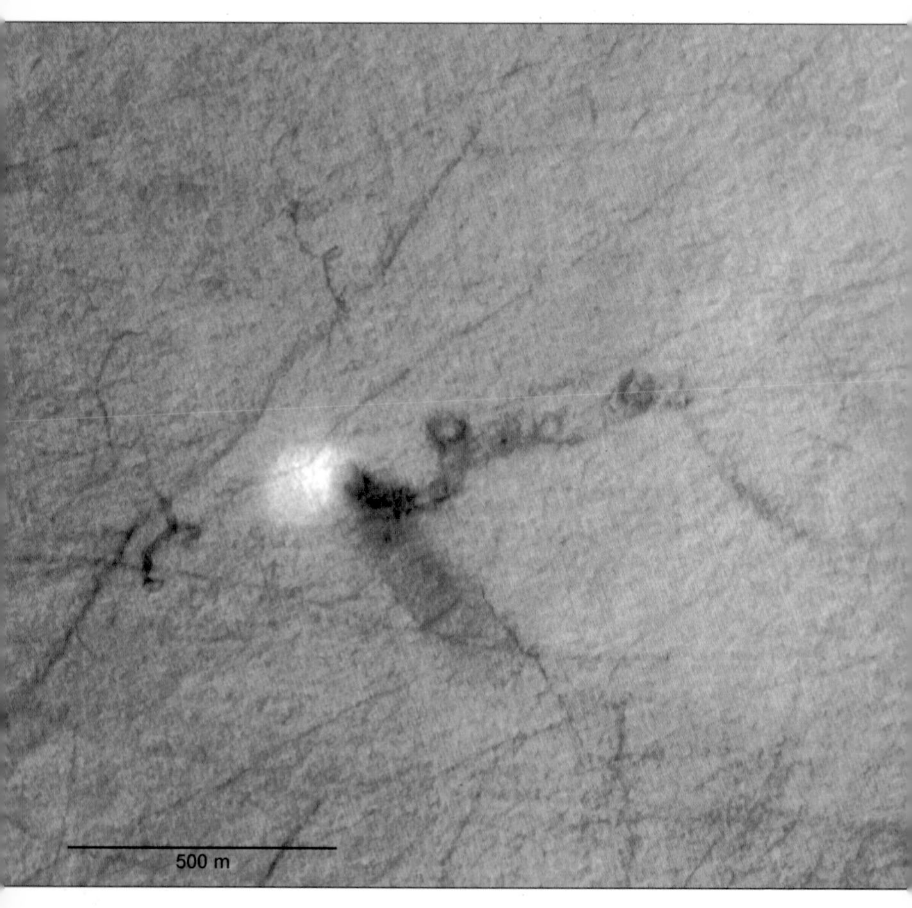

A dust devil and its footprints. The bright dust cloud casts a diffuse shadow toward the lower right. In addition, the wandering path of the dust devil is marked by a curious, looping trail, coming from the right. This area is a common spawning ground for Martian dust storms. Trails of other dust devils, both long and short, can be seen throughout the area. (243W 54S, M10-012567.)

dusty air absorbs sunlight, warming the upper atmosphere and changing the global wind patterns. The global winds shift the dust around the planet until the whole planet is covered by a pale, dusty pall and the normally contrasting markings can scarcely be seen at all through Earth-based telescopes. Once the dust is fairly uniformly spread, sunlight is cut off from warming the ground and the effect stops. Summer moves on toward autumn and the dust settles. Fall on Mars might refer to the fall of dust rather than leaves.

Dust devils may be common aspects of the Martian landscape in certain regions. Background and foreground include eroding buttes surrounded by talus slopes of loose debris. (Painting by author.)

Inset: *Dust devils can expand into large-scale dust storms, as seen in this example west of Las Cruces, New Mexico. Note typical Martian sky color in the wind-blown dust. (Photo by author.)*

One of the largest northern dust storms of 2002 (bright, yellowish cloud, upper left) pushes south from the north polar zone toward the Chryse region, where Viking I landed. Such regional dust storms may be triggered by local dust-devil activity. More bluish-white haze clouds can also be seen. (MGS wide-angle view, Malin Space Science Systems.)

CATCHING THEM IN THE ACT

We've already seen MGS photos (in chapter 15) of dust devil tracks that look like ghostly pencil lines crossing the Martian landscape. Usually the tracks imaged by MGS are darker than the background. They are presumably left as the whirlwind stirs the finest, bright dust off the ground, leaving the coarser, darker gravels behind. NASA's cameras have done more than make striking photographs of dust devil tracks; both the Viking and MGS cameras actually caught a few dust devils in action. Seen from orbit, an active dust devil appears as a blurry bright cloud in an area that was featureless in previous images. The giveaway is that the blurry cloud casts an extended shadow across the landscape, proving that it extends high into the air. Some have measured to be more than a kilometer in height.

Most people also may picture Martian landscapes as silent, static vistas. To me, it's a strange feeling to look up at the star-like red beacon of Mars and think of all the things that are actually happening there—dust devils swirling, water gushing out of slopes, volcanoes erupting—all without any witnesses. To paraphrase the riddle of the tree falling in an uninhabited forest, if a landslide crashes down a canyon wall on Mars and there's no one there to see it or hear it, does it really happen?

Yes.

DUST DEVILS AND ASTRONAUTS

Dust will be one of the great annoyances on Mars. It will find its way indoors. It may even become a serious enemy. Major dust storms may cause local brownout conditions in which landmarks or safe havens can't be seen. Dust devils move unpredictably and carry whirling, high-speed dust particles that will get into machinery, bearings, rover engines, air locks, and space-suit fittings. Someday one of them will cross the path of a Martian lander, rover, habitat, or explorer—with unhappy results. In my novel *Mars Underground,* about research bases on Mars, I imagined a Martian explorer caught by a vicious Martian dust devil. The high-velocity dust sandblasts his helmet faceplate to the extent that he can no longer see. He solves the problem by . . . well, read the book. It's fiction, of course; NASA personnel have told me that faceplate material is virtually indestructible—but then they haven't had to design yet for sandblasting hazards.

CHAPTER 38

THE MOST MAJESTIC DUNE FIELD

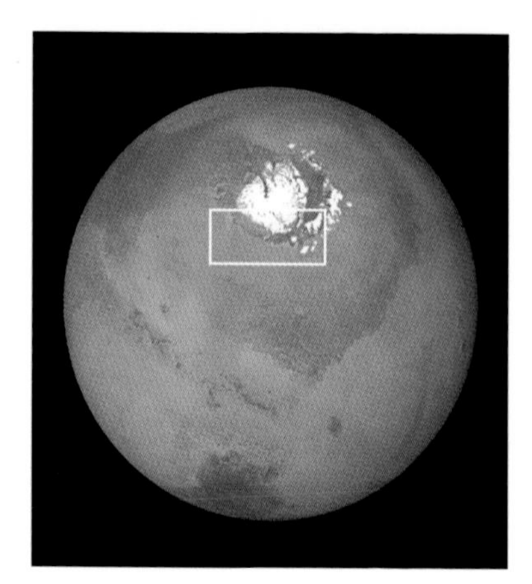

The desert landscapes of Mars range from dust-filled impact craters and desiccated river channels to classic vistas of endless *Lawrence of Arabia* sand dunes. The greatest concentration of such sand dunes girdles the north polar ice fields. It is the grandest, and perhaps the largest, tract of such dunes anywhere in the solar system.

This region was not particularly noteworthy to the telescopic observers from Earth in the nineteenth and early twentieth centuries. They mapped dark "collars" surrounding each polar cap, but there was no hint that these dark regions harbored massive sand-dune formations. These polar collars, like other dark patches of the planet, seemed to darken in the spring. To early scientists, this was one of the best proofs of vegetation. The melting of polar ice in the spring, they thought, moistened the polar soil, leading to a burgeoning of vegetation that lay dormant in winter. Today, researchers realize that the polar caps don't melt into liquid water but remain frozen year-round. The changes in the contrast of dark markings seem to be due partly to late-summer and winter hazes that reduce contrast, and partly to springtime winds that may sweep light-toned, fine dust off the surface, exposing darker gravels.

The north polar dune field was discovered in 1972 when Mariner 9 mapped the planet from orbit. The dunes are concentrated at very high latitudes, around 75° N to 82° N, and they extend

Trackless dunes of the north polar dune field. The sun never rises very high in the sky at polar latitudes; in this image at 1:27 P.M. local time, the sun is about 29 degrees above the horizon. (144W 80N, E02-00412.)

virtually the whole way around the polar ice deposits. In some places, this belt of sand reaches 500 km (300 miles) across. There is also some concentration of dunes around the south pole of Mars, though less well developed. Interestingly, there are fewer dunes in the Martian equatorial region; this is the opposite of Earth, where the equatorial region has more dunes because of the hot, dry conditions.

In some areas of the Martian north polar dune field, the dunes are patchy, breaking up into isolated crescents; but in other areas, they line up in rows of straight dunes, one after another, like waves marching across the sea. The spacing of the dunes, and the length of individual crescent-shaped segments, averages around 500 meters to 600 meters. All the forms and spacings are virtually identical to various classic dune-forms in Earthly deserts. The area involved on Mars is estimated at 1 million square kilometers—about the same as the area of active dunes in the Sahara Desert. It's a toss-up whether the Saharan or the Martian dunes constitute the largest dune field in the solar system, but I claim that the Martian dunes are the most majestic because of their pristine, alien, geometric beauty. It's as if our Arctic Ocean, along with the coastal regions of Asia and Canada, were replaced by an empty wasteland of shifting dunes, where few people could venture and where any research stations or oil rigs might be covered in a few years by shifting sands.

The motions of individual windblown sand grains are curious. Unlike dust motes, sand grains typically can't be blown high into the air or carried for miles. Rather, a sufficient wind or turbulence

THE SIZE OF SAND

Is it correct to use the term *sand* in discussing Martian dunes? Yes. Sand is defined by geologists according to particle size; technically, it is material in a size range of 0.06 millimeters to 2.0 millimeters—consistent with familiar descriptions of beach sand, et cetera. Geologists also use a series of technical terms for specific size ranges of smaller particles, but for ease of discussion and consistency with the general impression of Mars, I've referred to finer material simply as *dust*.

The bright patches are frost deposits concentrated in the low spots between these north polar dunes. (204W 81N, E05-00881.)

Inset: *Desert winds can produce strange erosion forms. In this formation, wind-driven saltating sand grains, about a foot off the ground, have eroded the lower part of a large rock, leaving a toadstool-like formation. Similar eroded forms may be found on Mars. (Death Valley, California; photo by author.)*

can dislodge a grain and propel it upward for a few feet. After being carried a few meters in the downwind direction, it promptly falls out again. In other words, an individual sand grain in a windstorm has a hopping motion, traveling in a short arc—a process known as saltation. If you're a beachgoer, you know the process of saltation: when a wind kicks up on a fine-sand beach, you feel grains striking your ankles and often see a mist of saltating sand grains that extends 10 centimeters (a few inches) above the surface. By contrast, smaller and light-dust grains get picked up by the air and carried for long distances.

Arizona geologist Ron Greeley and his colleagues have conducted wind-tunnel experiments simulating Martian conditions and found a similar situation on Mars (an example of their experiments was shown in chapter 2). Sand-size particles, especially

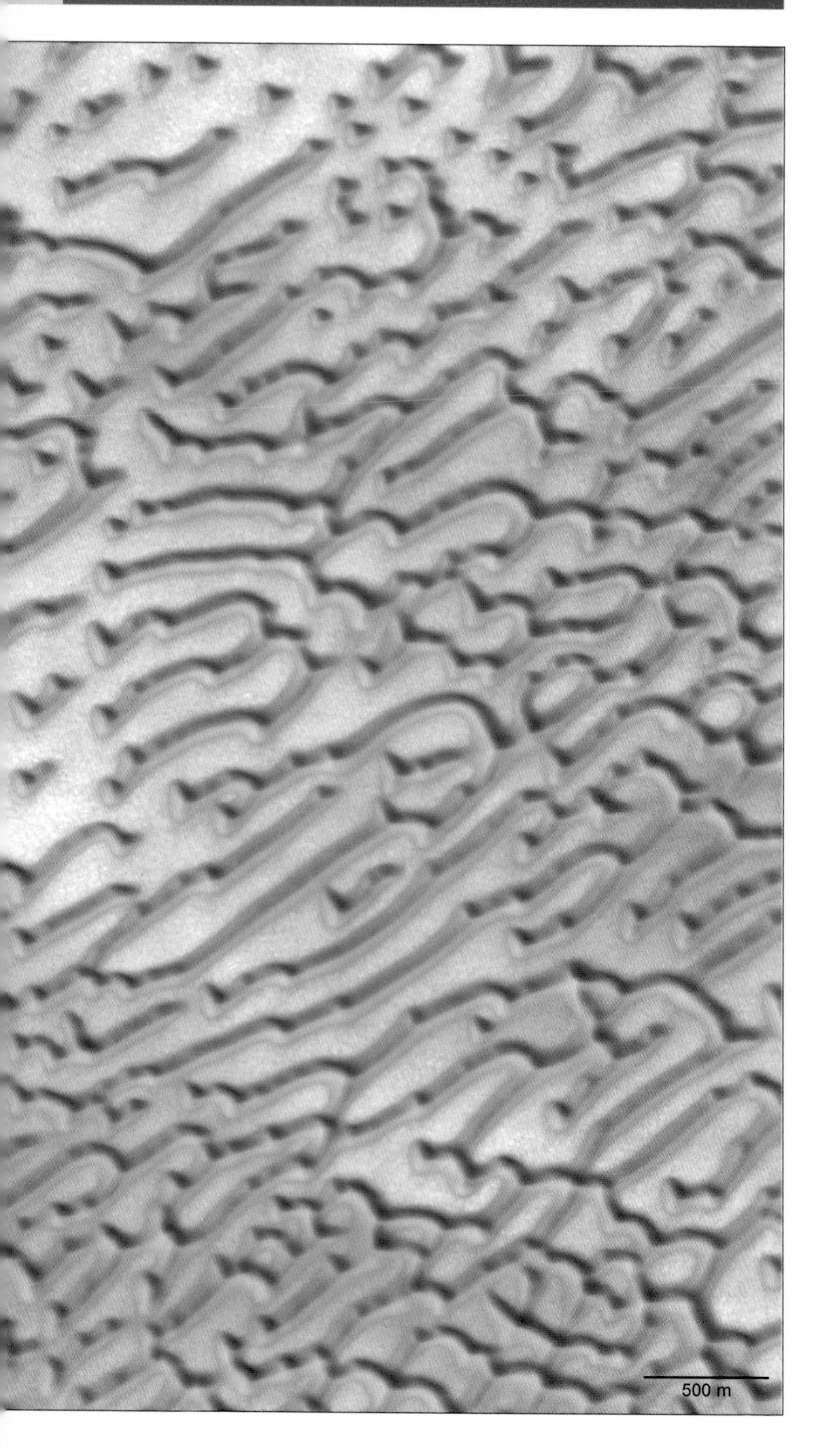
500 m

Wind patterns can create strange shapes among the north polar dunes (facing page). This area features a peculiar, reticulated appearance. (276W 78N, E02-00776.)

those around 1 millimeter across, are the ones lofted most easily, and typical Martian breezes of 12 miles per hour can cause grains to saltate up to a height of about a meter, then carry them 3 to 10 meters (9 to 30 feet) downwind in the low Martian gravity. The two Viking landers studied surface conditions over the course of a Martian year during the 1970s and found wind gusts up to about 50 to 60 miles per hour during periods of dust storms. These measurements, plus the direct observations of Martian dust devils and massive swirling dust storms that can cover much of the planet, make it obvious that both sand and dust are transported across Mars in massive amounts. Under Mars's low gravity, the grains and associated dust may create a mist as high as 6 or 10 feet, which could easily disorient an astronaut caught in the wrong conditions.

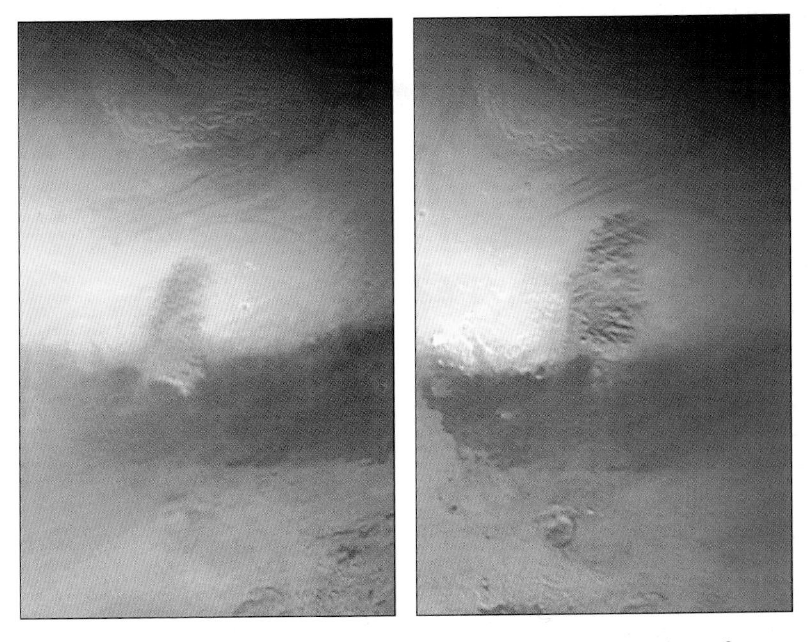

Two views, about two hours apart, show development of a dust storm, spreading from the north polar dune field north onto the polar ice cap at about 45 miles per hour. (MGS wide-angle view, Malin Space Science Systems.)

Dune formation depends on local or prevailing wind patterns. On Mars, dune deposition often results from wind blowing over crater rims or hills, leaving deposits of sand dunes or dust patches on crater floors or downwind on the outside of the crater or hill.

Because dunes are formed by winds, they change shape and move. On Earth, where shifts of dunes typically take years, highways and villages are often threatened by gradual encroachment of nearby dune fields. How fast do the dunes change on Mars? The gross shape of Martian dark markings have remained relatively constant for a century, but 300-km-scale changes in detailed outlines are not uncommon in a period of a few years. At a scale of hundreds of meters, comparison of MGS and Viking images showed changes in specific dune profiles in the 23 years between the two missions—but only among a minority of dunes. At a still smaller scale, as mentioned in chapter 17, the Viking 1 lander saw one small slippage of sand, a few inches across, on the face of a small dune nearby during the course of the yearlong mission. So, the Martian dunes probably evolve as fast as their terrestrial counterparts.

On Mars, as on Earth, the dunes would present a formidable challenge to human explorers. The dunes would be hard to traverse and present a threat in terms of dust clogging equipment. They would be hard to navigate. Scientifically, they are a hindrance because they cover the outcrops of intact rock and strata that tell the history of the planet. Yet, aesthetically, they are a wonder! The sensuous forms of slopes and curves will draw the eye onward, onward . . . to the hazy horizon and beyond. Deserts are places of sand and stars that have inspired prophets to new visions of humanity. It may be no accident that several major religions have come from the arid desert regions of Earth.

Someday, perhaps, the lonely Martian dune fields will inspire some new prophet.

CHAPTER 39

POLAR CAPS

A Tale of Two Ices

The brilliant white polar ice caps of Mars have long played a role in our understanding of the planet. Interpreted as early as the 1790s by William Herschel as arctic ice fields (like those of Earth), they inspired him to argue that Mars was Earth-like. For about a century, from the 1860s to the 1960s, scientists argued over what kind of ice was there. The first assumption was frozen water, as on

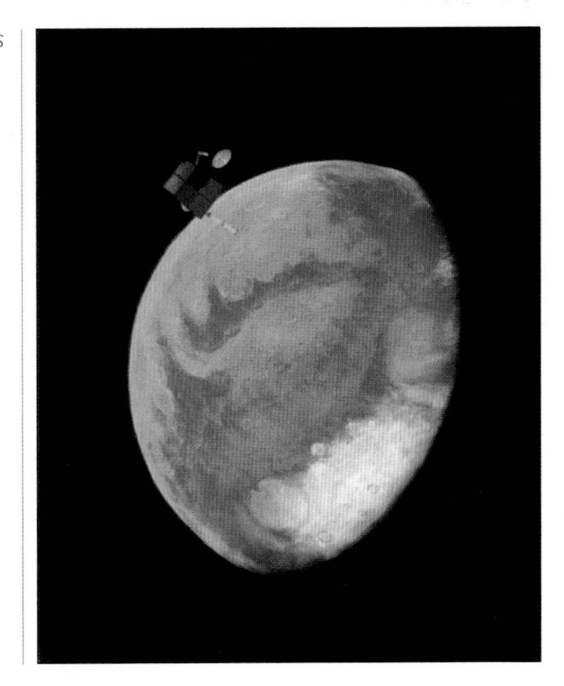

The Martian polar caps are never strongly tilted toward Earth, and thus it is difficult to study polar ice and other phenomena from Earth. Many modern space probes, however, have been put into polar orbits that pass over the polar ice fields and allow monitoring of daily, seasonal, and yearly changes. This permits better understanding of Martian climate cycles. (Painting by author.)

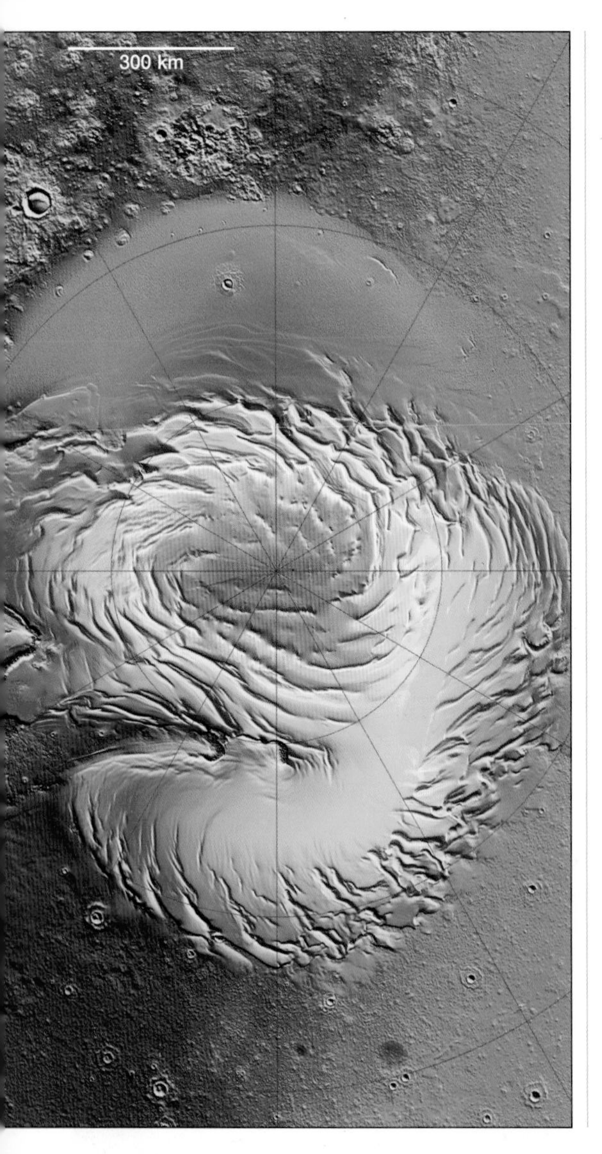

Comparison of topography in the north and south polar areas. Both poles are capped by a mound of layered deposits standing several kilometers above the surroundings, which are probably dust layers deposited by annual storms. The cause of the vaguely spiral pattern of valleys extending into the caps may be related to patterns of sublimation of the ice during the sun's motion around the sky as seen from the poles. Zero degrees longitude is at bottom in both views. Left: *North pole, showing the cap surrounded by the plains of Vastitas Borealis; the cap might have been a large island if there ever was a northern ocean.* Facing page: *South pole, showing the remnant structure of a huge impact basin (right). (MGS laser altimeter team.)*

Earth. Later researchers began to realize it might be frozen carbon dioxide—what we call dry ice—because CO_2 is the dominant gas in the Martian atmosphere and polar winters are cold enough to freeze it solid.

It turns out both arguments were correct! Each cap apparently has a relatively permanent central core of H_2O ice and a much larger seasonal deposit of CO_2 frost. There are two key facts to realize about the frosts and ices of the Martian polar regions: first,

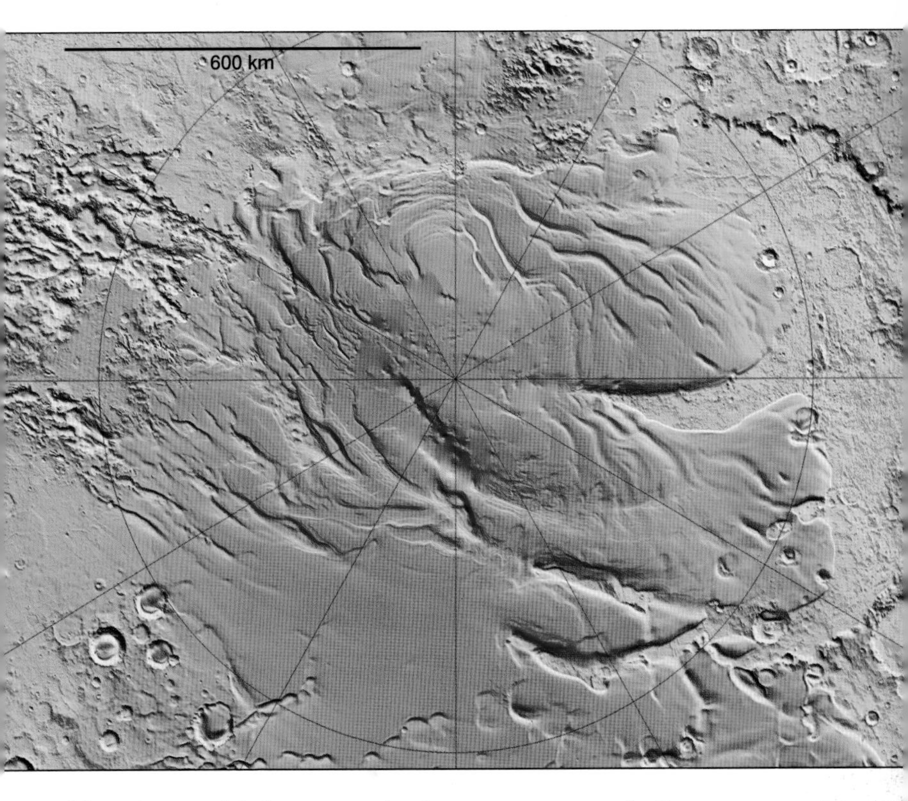

Mars is so cold that water is almost permanently frozen; second, CO_2 is to Mars as H_2O is to Earth. On Earth, during winter, it is water vapor in the atmosphere that condenses into frost or snowflakes, creating ice, ground-frost, or snow deposits. On Mars, during winter, it is carbon dioxide in the atmosphere that condenses into frost and frost deposits on the ground.

Vertical view downward onto the south polar ice cap, made by the Viking 2 orbiter in 1977. This view essentially shows the permanent summer ice cap of frozen water. The crude spiral pattern of rifts shows up in the ice-deposit pattern. (NASA, Viking 2.)

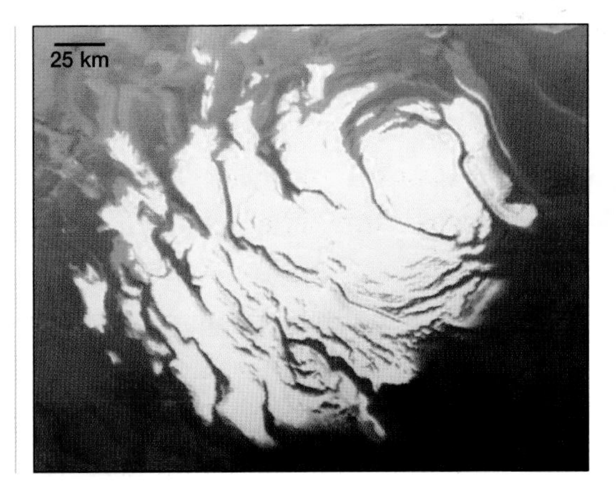

DOES IT SNOW ON MARS?

What early telescope observers, backyard amateur astronomers, and spacecraft see during the Martian year is that a small permanent cap persists during summer, more or less on the Martian pole. During fall and early winter, bluish white polar hazes and clouds form, extending from the pole down to moderate latitudes of 70°, 60°, and 50° like wintry cloud masses extending down from Canada over the northern United States. This oppressive cloud mass is called the winter **polar hood**, a term that could apply almost as well in northern winter latitudes from Pittsburgh to Paris. It's unclear from visual observations just what is happening under those winter cloud masses. Does CO_2 actually fall out of the atmosphere in Martian snowstorms? Or does Mars simply experience frost deposits, like hoarfrosts on chilly terrestrial mornings?

The MGS radar instrument has apparently solved the mystery by detecting clouds on Mars that form at night and reflect radar as if from solid grains. These clouds tend to disappear after sunrise, when the sun warms the scene. They occur in the areas where the broad winter CO_2 ice cap is forming. Judging from the radar properties, together with the temperatures, infrared spectra, and theoretical work, the MGS radar team concluded in 2002 that these clouds mark the first direct confirmation of CO_2 snow falling out of the Martian atmosphere. This conjures up the interesting but forbidding prospect of future high-latitude winter explorers being caught in snowstorms of dry ice during the Martian night.

As the clouds lift, earthbound observers see the bright white polar cap of frost and snow spreading to moderate latitudes, as far down as 60° to 65°, just as Earth's winter "polar cap" of snow spreads across Canada as far as Montana, Illinois, Pennsylvania, and Massachusetts. On Earth, the spreading "cap" is condensed water frost; but on Mars, with its lower temperatures, it is primarily carbon dioxide frost. Measurements from the MGS laser altimeter suggest that the total accumulation of CO_2 snow and frost is only a few meters. In spring, this Martian CO_2 frost burns off, partially revealing reddish soil at most latitudes and the stable, always-frozen H_2O ice deposit, solid as a rock, near the pole.

DIFFERENCES BETWEEN THE ICE CAPS

We've talked as if the north and south polar caps were exactly the same, but in frosty reality there are some differences between them. In northern summer, the cap becomes warm enough to burn off all the CO_2 frost and expose the H_2O ice cap. In fact, it's warm enough that water vapor starts to sublime off the water ice, and the atmospheric water vapor content increases. During summer in the southern hemisphere, something different happens. In spite of the fact that southern summer is warmer than northern summer—because Mars is closer to the sun—the southern cap never gets warm enough to burn off all the CO_2. This is one of the factors that confused scientists for a long time about whether the caps were carbon dioxide ice or water ice, but most scientists are convinced that under the southern cap is a permanent reservoir of frozen water, just as in the north.

The extended north polar ice cap. During winter, both the north and south caps extend outward in the form of carbon dioxide frost deposits to moderate latitudes. In this view, the central, permanent cap of frozen water is partially obscured by an orangish pall of airborne dust, which drifted over the cap from a regional dust storm in mid-2002. (MGS wide-angle view, Malin Space Science Systems.)

Why does the south cap stay cooler and retain its CO_2 frost? U.S. Geological Survey researcher Hugh Kieffer has studied this problem and concluded that it relates to the Martian dust storms. Remember, because southern summer is warm, dust devils form in the south and loft dust into the atmosphere. The dusty southern summer air blocks some of the midsummer southern summer sunlight, and the Martian south pole remains cooler than otherwise expected. This and other factors may explain why the southern cap doesn't lose all its CO_2 frost.

STRANGE LAYERS

Ever since the Mariner 9 mission, scientists have been struck not only by the polar ice but also by the geologic structure revealed in the polar regions of Mars. Mariner 9 images showed strange, layered sedimentary strata on both poles, stacked like pancakes. These are not pure ice layers. They have the color of soil layers, though they are probably mixed with ice or frost. Mariner and Viking scientists could see layers 10 meters to 30 meters thick, but MGS photographed some as little as 10 centimeters (a few inches) thick. The precise altitude measurements from the MGS laser altimeter indicate that the polar layers are stacked to a thick-

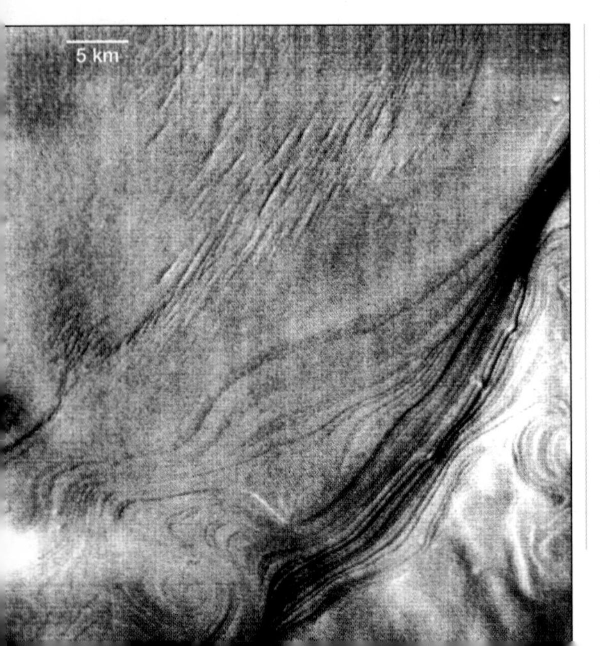

Mysterious layering of the polar deposits was discovered by the first close-up spacecraft mapping in 1972, as shown by this view of the S cap. (229W 75S, Mariner 9.)

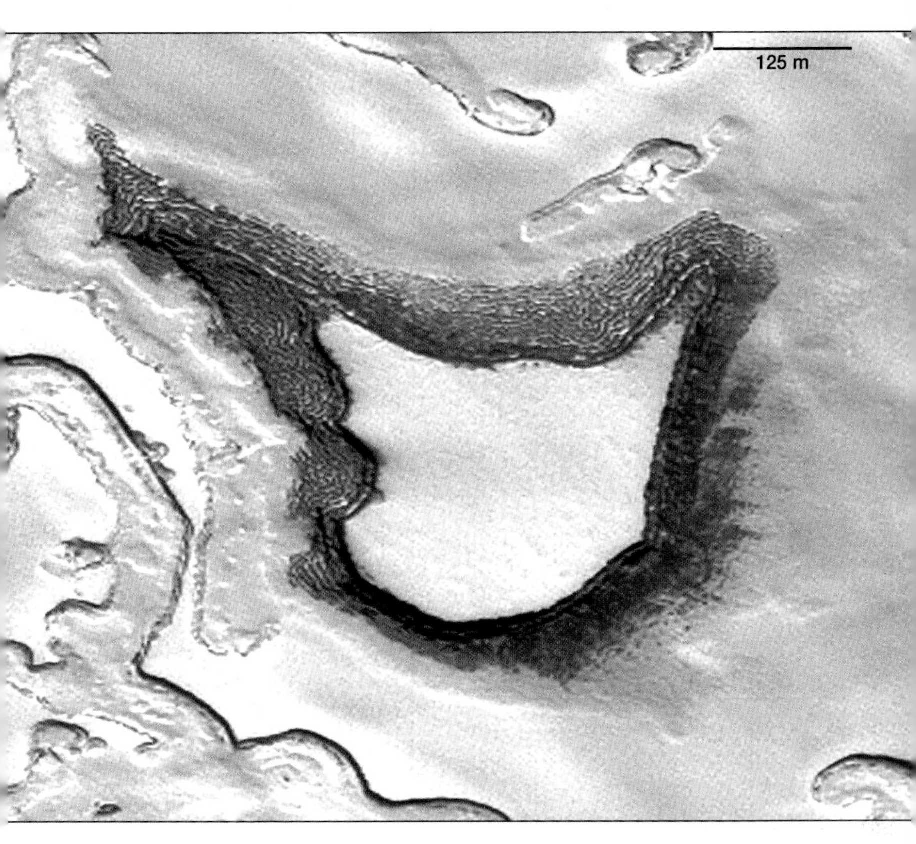

A mesa in the ice fields. Almost at the south pole, this mesa exposes layers of reddish brown strata in its walls. As seen from this area, the summer sun moves around the sky just above the horizon and burns frost off the mesa hillsides, leaving frost on flat surfaces. Hillside erosion and retreat at about 3 meters (10 feet) per Martian year have been observed by MGS in this area. (341W 87S, MGS image, NASA/JPL/MSSS release MOC 2-298; color processing by Malin Space Science Systems.)

ness of about 2.7 km (9,000 feet) above the surrounding plain in the north and 3.1 km (10,000 feet) in the south. The stacking of 3 km of thinly layered dusty sediments (along with the nearby dune fields visited in chapter 38) implies that the polar region is a vast repository for dust deposited in periodic episodes.

Why does dust collect in layers at the poles? It's probably a combined result of the cycles of global dust storms and axial tilt. Each global dust storm injects dust tens of kilometers into the Martian stratosphere, and this dust circulates over the poles. As winter comes, these dust grains serve as condensation nuclei for frost grains, in the same way that terrestrial dust grains can serve as nuclei for the condensation of hailstones. This frost precipitates onto the ground, depositing the dust grains down on the surface. The more dust that is blown into the polar regions, the more dust is dumped onto the polar cap. When spring comes, the frost can burn off, but the dust grains are left behind in a thin layer representing that one year, just as a tree ring represents a certain year's growth.

Slower cycles are superimposed on the annual cycles. We've already seen that the axial tilt swings through wide excursions on time periods of a few MY. On a shorter cycle of about 0.1 MY (100,000 years), the axial tilt goes through lesser excursions, and the planet's orbital eccentricity also changes (this being the ellipticity that brings Mars closer to the sun on one side of its orbit than the other, producing the warmer southern summer during the current era). So the variations on somewhat irregular overlapping cycles of 1 year, 0.1 MY, and a few MY probably produce variations in the intensity of dust storms and transport of grains to the poles. Therefore, there is a superlayering of the annual layers—centuries of intense dust deposition and "droughts" during which little dust is deposited.

FROST STORIES

Because frost is white and gravel is dark, the slightest deposits or meltings of ice can produce spectacular visual effects as a dark landscape gets frosty patches of blinding white or a uniform snowscape suddenly develops spotty dark "holes" where the first patches of dark soil are exposed. Photos show some terrains dotted by dark spots that look for all the world like aerial photos of New Mexico hillsides spotted by juniper bushes, which touched off an Internet flap that plants had been found on Mars. (An example is seen on p. 371.) But further Martian photography showed the Martian "bushes" widening into broad patches of barren dirt as the ice sublimed. Even solidly ice-covered terrain can develop its own strange topography from year to year as ice formations change. For example, a bizarre polar landform known as Swiss cheese terrain consists of a flat plane with many sharply rimmed circular-, oval-, or kidney-bean-shaped depressions, tens or hundreds of meters across. These have been seen to change shape within a single Martian year as the ice burns off.

All these sorts of landforms have received intense scrutiny as scientists tried to explain how they form and what they are telling

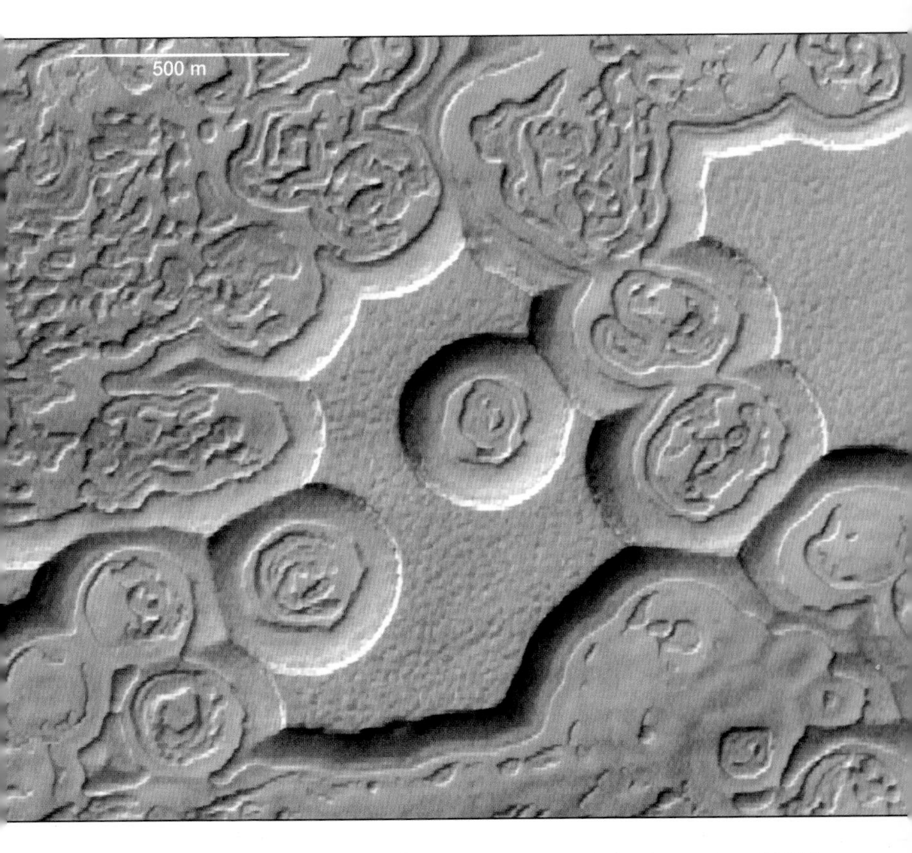

The polar ice fields are home to many odd formations that are probably caused by different rates of ice sublimation in different areas. In this so-called Swiss cheese terrain, circular depressions form by uncertain processes as ice sublimes. The flat surface stands about 4 m (13 ft) above the depressions. MGS has observed expansion of many depressions at a rate of a few meters per year. (76W 86S, M03-06646.)

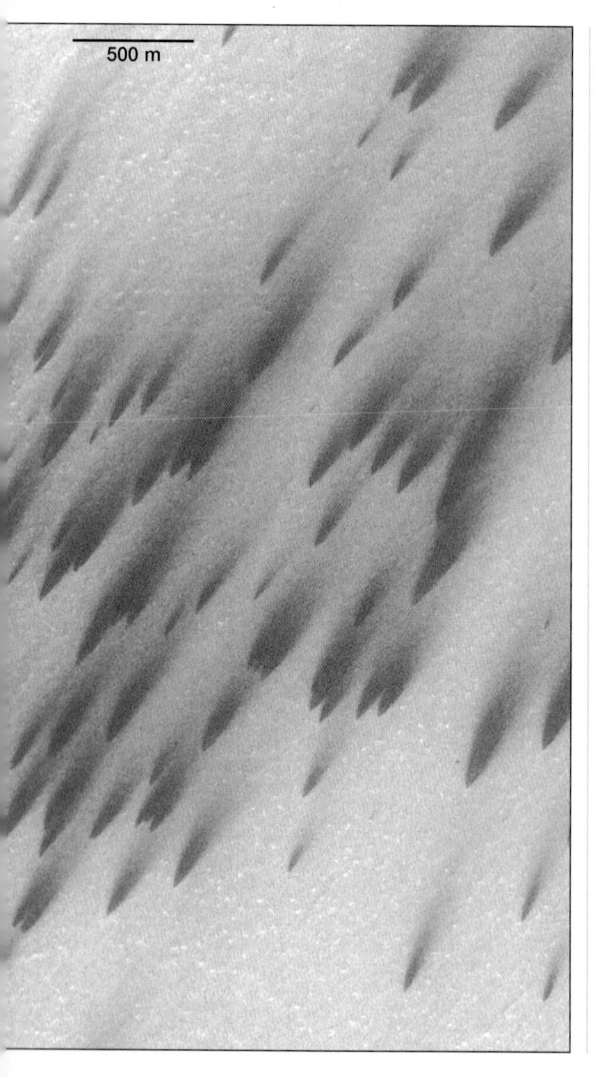

A field of dark "comets" on the south polar ice. These puzzling features may be caused by holes in the ice that expose surface soils. Strong winds from lower left to upper right could lift dust out of these holes and leave deposits on the ice. (363W 84S, E05-033620.)

An outlying part of the south polar deposits (facing page) *hints that the ice/soil mixture in the layered terrain (red) has flowed away from the pole, partly covering the rims of several craters (top, bottom, and several in the lower center). Scarcity of impact craters on the red ice-rich material suggests ongoing flow and/or constant resurfacing in modern geologic time. (264W 80S, MGS laser altimeter team.)*

us about the ice. For example, imagine you are standing in a frying-pan-shaped depression in the polar ice field. In summer, you see the sun moving all the way around the sky, just above the horizon, every day. That means all the walls of the frying pan will face directly toward the sun at one time or another, whereas the floor will always have diluted, slanting sunlight or shadows. The upshot is that the walls will get the most solar heating and will sublime and recede, meaning that the dimensions of the depression will expand each year. This explains the Swiss cheese terrain,

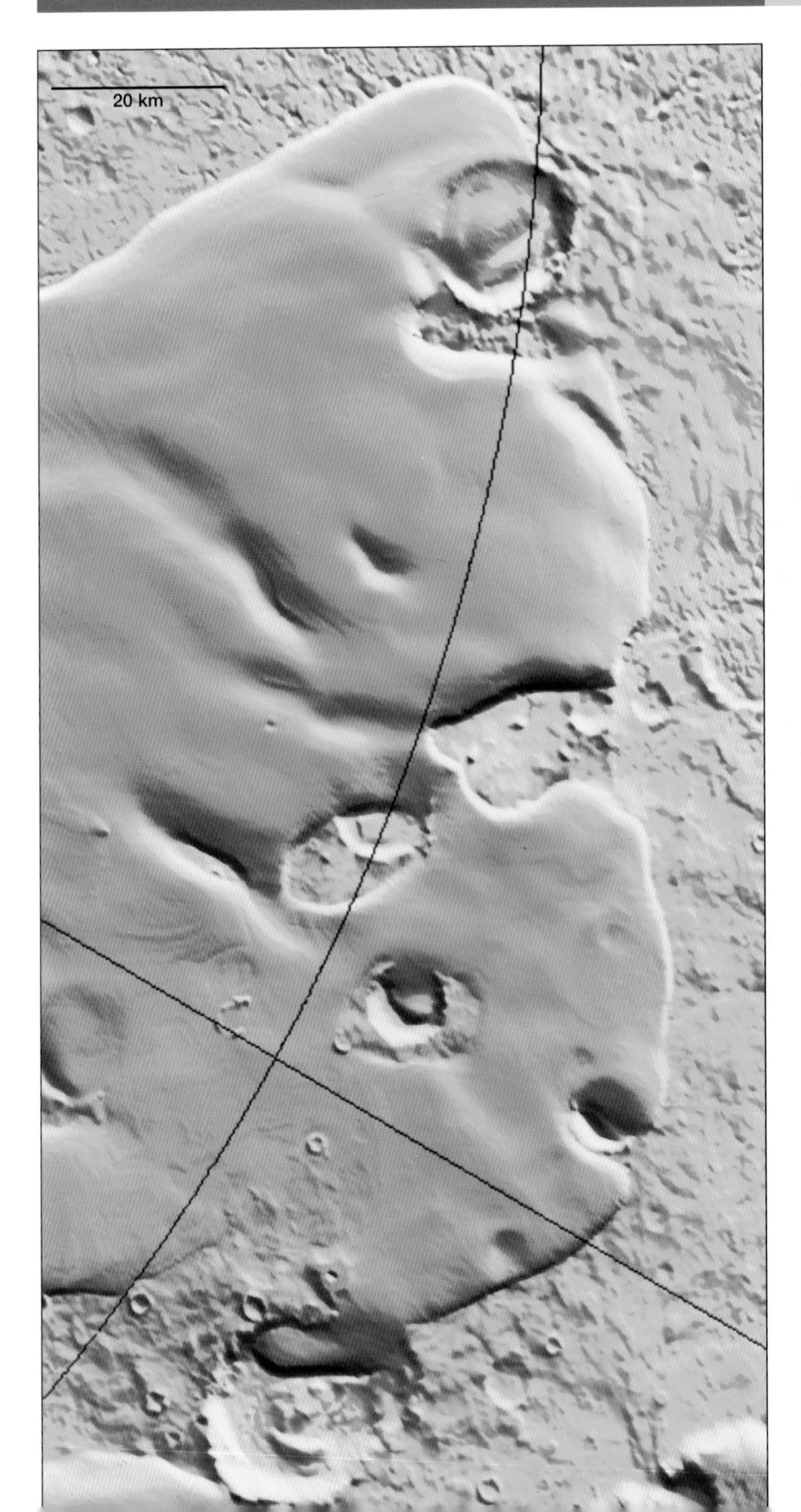
20 km

although it doesn't explain how the Swiss cheese holes get started in the first place.

Because the Swiss cheese holes have been seen to grow over several years, a number of researchers in 2003 have concluded that the polar ice is diminishing and that Mars must now be in a period of rapid climate change—an interesting possibility that might relate to the 100,000-year axial tilt cycles or the likelihood that gullies have formed from water on hillsides within the last few million years.

While the strange phenomena of the polar landscape pique our curiosity, we must remember that some of the features may fall more in the category of superficial local quirks rather than profound indicators of global history. You can get lost in trying to interpret every nuance of the interplay of ice and soil, the changes in ice patch shape, and the evolution of frost deposits banked up against this or that side of a polar ridge. Cornell researcher Peter Thomas, speaking at a science meeting in 2002, remarked ruefully that trying to understand the Martian polar processes and climate variations from studying the polar surface textures and features may be like trying to learn about forests by studying samples of tree bark.

IS THE ICE MOVING?

Most scientists have assumed that the stacked layers of soil and ice are relatively stable. However, there are hints that the whole mass can deform like a glacier. The laser altimeter maps, in particular, reveal craters near the edge of the layered terrain where lobes of smooth material seem to have flowed outward from the cap, half-engulfing craters (see photo on page 403). The images suggest that at least parts of the polar layered deposits are ice-rich and can flow out over nearby terrain. Whatever their properties, they are remarkably free of impact craters, which means that the polar surfaces are very young and continually renew themselves, either by deposition of new dust/ice layers or by flow, or both.

DECIPHERING THE PAST

Perhaps the most important aspect of the two polar caps is not the weird landforms of the icy surfaces or the details of snowfalls and hoarfrosts, but the secrets held by the deep layers. In Earth's polar areas, whether Greenland or Antarctica, scientists can drill deep into the ice and bring up cores that reveal the layers of deposits. A certain layer might mark ash ejected into Earth's high stratosphere by a volcano, settling later on the polar ice. Dating of the ash shows the time of the eruption. Isotopic variations in another layer of ice might reveal a period of unusual cold or warmth. In the same way, the Martian permanent ice cap and underlying layered deposits will offer a treasure trove of information about the history of the Martian climate. Do the layered deposits go all the way back to layers formed during the wet times of the Noachian era? Will the deepest layers reveal the nature of the Noachian and Hesperian climate changes? Will analysis of the layers themselves reveal the history of changes in axial tilt of Mars and clarify their role in climatic conditions? Time, and science, will tell.

Layers of history. MGS close-up shows complex polar layering in a gentle slope facing lower left. Sunlight is from lower left. (10W 87S, mosaic of MGS frames, by Malin Space Science Systems.)

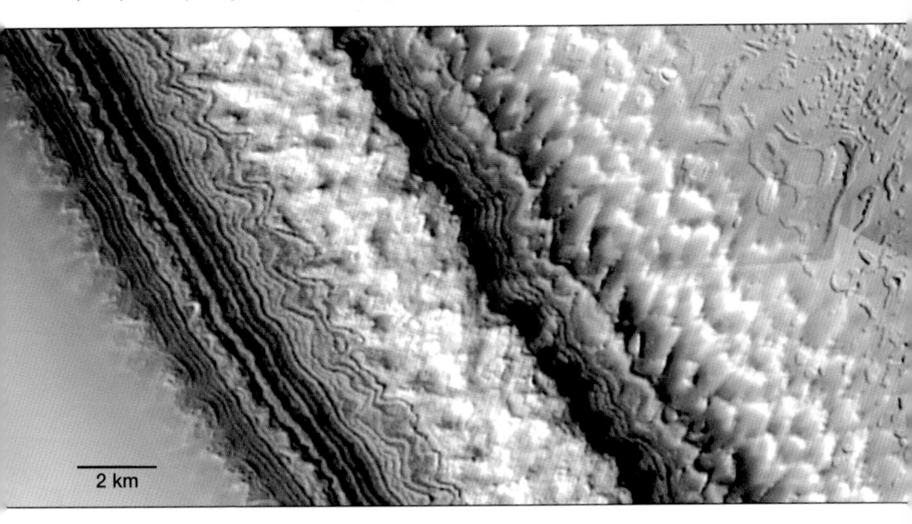

SHOULD WE LAND AT THE POLES?

Growing up on Earth, you might think the polar regions of Mars would be the worst place to land. Arctic regions are severe, and Mars is even colder than Earth. Paradoxically, the Martian poles have some real attractions. Martian night is perhaps the most fearsome Martian environment—the temperature drops to –100° C (–147° F)—but since Mars has about the same axial tilt as Earth, the summer pole is a land of midnight sun. In other words, at either Martian pole during the summer season, the sun never sets but simply circles in the sky about 10° to 25° above the horizon. This means there is always a chance to get some solar warmth, and if solar photovoltaic efficiency can be improved (as part of our drive to become more independent of oil?), then large solar panels might be an attractive option or backup to power a Martian polar base. Furthermore, water is easily accessible at the poles. Bright white water ice exists on the surface in the summer; during the winter, under a few meters of carbon dioxide—so it would be easy to harvest water there.

But the pole is a hostile place for a long-term base, with nearly an Earth-year of winter darkness (the entire Martian year runs 669 Martian days—or 687 Earth-days). On the other hand, a summertime base at the north or south pole is an attractive option because there would be 100 or 200 days of continuous, if modest, sunlight, which could power solar arrays. An early Martian expedition might very well set up shop for a few months on the polar ice!

Future lander probes may set down on the polar ice fields and allow close-up studies of ice formation and erosion processes. High-altitude ice crystals of either water or carbon dioxide may cause spectacular solar halo phenomena on Mars. (Painting by author.)

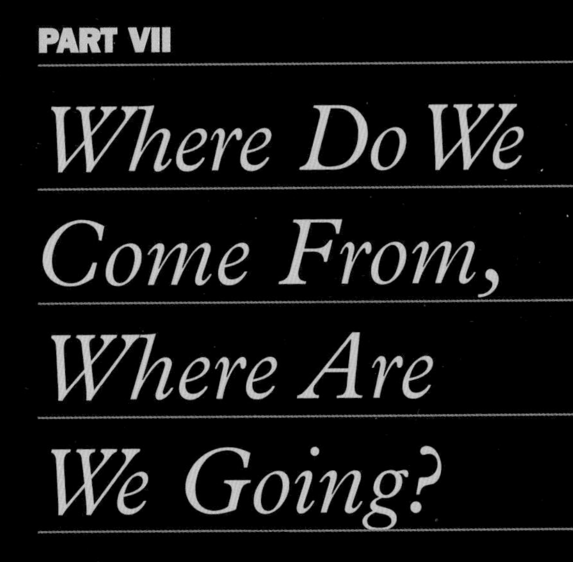

PART VII

Where Do We Come From, Where Are We Going?

CHAPTER 40

QUESTIONS THAT LEAD TO THE FUTURE

W*here Do We Come From, Where Are We Going?* is the mystical title that French painter Paul Gauguin gave to one of the enigmatic canvases he painted in 1897, during his sojourn in Tahiti. The phrase is relevant when we ask the big question of what Martian exploration means.

If you ask many scientists about this, their answers will reflect their own subdiscipline of research. If one of them were to tell you that Martian exploration is important because "we need to measure the argon isotope ratios in present-day Martian air in order to test theories of early atmospheric loss and volcanic degassing" or "we need to collect samples from a known Martian landing site so we can calibrate the cratering rate and measure the ages of the other stratigraphic units," you would not be entirely convinced it's worth spending $40 billion of our national treasury—or, more likely, of our "global treasury"—to send explorers to Mars.

Mars exploration's true value can be assessed only with more general answers. For example, such exploration challenges our national technological capabilities and leads to many economy-building spin-offs. We could pursue such a challenge merely by creating some sort of general technology-development administration, but long ago President Kennedy realized, after considerable thought, that a more effective way to stimulate our capabilities is to create a more dramatic, positive goal. If that goal is the establishment of our capability to work throughout the inner solar system, we'll learn to utilize many new resources. Once we've learned to use the 24 hours a day of "free" solar energy that streams through space, we can start to apply it to harvesting and processing materials such as asteroids made of pure nickel-iron alloy—whose fragments you can see in museums and planetariums. In my own view, learning to utilize the resources of our cosmic environment may be our planet's salvation. This may sound naively grandiose, but think about the big picture: our rapacious mining and processing of lower-grade ores and fossil fuels cause environmental pressures on all of us and reduce international politics to life-threatening squabbles over control of regions that have the last

Dunes upon dunes. In this area, crescent dunes of dark material have formed on top of an older surface of smaller dunes composed of brighter material. Such features testify to the continuing mobility of Martian sand and dust. (330W 48S, M03-03088.)

earthly resource reserves. National treasures, both monetary and creative, are soon devoted to bombs and bravado. How much better to strike out in a new direction that leapfrogs over these shortsighted wastes of talent, riches, and human life!

To wax still more philosophical, Mars will help us answer Gauguin's questions. It will tell us something about where we came from and where we are going. If humanity lasts another 1,000 years, those answers will be the life-affirming breakthroughs that will be most remembered from our generation.

WHERE DO WE COME FROM?

As we arrived in the twenty-first century, new discoveries were just confirming that planets—or at least bodies the size of our larger planets, such as Jupiter and Saturn—orbit around many other stars. Proof, in other words, of other planetary systems. We still don't know, however, if any of those worlds, or worlds closer to home in our own solar system, harbor life. Our deepest question—and the one that drives scientists toward Mars—remains unanswered after 10,000 years of human wonderment: Do alien creatures or even intelligent civilizations exist elsewhere in the universe? Or is life a unique creation that exists only on Earth?

The beauty of science and exploration is not just that it gives us answers but that it leads us by the hand onward to the next question that needs to be studied. It's not a one-way street like the older methods of seeking truth, such as appealing to great authorities of the past or hiring lawyers to get a decree from a royal or civic court. It's a direct dialogue with the universe around us, or, if you like, with God. As Albert Einstein argued in his 1931 book called *Cosmic Religion,* scientific curiosity and sense of religious wonder are the same basic impulse. "It seems to me," he said, "that the most important function of art and of science is to arouse and keep alive this feeling [and that the most] deeply religious people of our largely materialistic age are [the ones who pursue] research." We ask basic questions about the universe, and then we devise experiments and expeditions. If we are receptive, Nature gives us answers that propel us to the next level of understanding.

Whether or not life ever existed on Mars is a perfect scientific question because either answer leads to profound new conceptions. If we prove that life once evolved on Mars, then we are not alone in the universe and we may start investigating whether life follows the same course on all planets. On the other hand, if we prove that life never started on Mars despite the water-rich environment, it raises questions about whether our theories of life's origins are wrong and about whether we are really alone in the universe, after all.

A mysterious ring. This striking feature lies just NE of the hematite region, at 357W and 2N. It appears to be the exposed root structure of a 2.6-km (1.6-mile) impact crater and ejecta blanket, planed off by erosion and exposing the crater subsurface in cross section. Malin and Edgett have estimated that the whole structure was filled in and buried under more than 100 meters of hardened sediments before being planed off to this level. This picture testifies to the continuing exhumation of old features as overlying young strata are eroded away. (MGS 04-01289.)

Here, to summarize, is what we already know about Mars:

- The planet had abundant water, at least sporadically, on the surface during the first few hundred MY, probably until about 3,500 MY or even 3,000 MY ago.
- It almost certainly had a thicker atmosphere during that period, with perhaps one-third to two-thirds the surface atmospheric pressure as Earth has today.
- The initial atmosphere was mostly carbon dioxide, probably with some water vapor and other gases. That atmosphere was similar to Earth's atmosphere during the same period, because those are the main gases belched out by volcanoes.

Landslips. All over Mars, the MGS camera revealed dark and light streaks on hillsides, caused by dust or sand slipping downslope and being dislodged by recent disturbances such as wind, earthquakes, or meteorite hits. This image includes both dark (center) and light (upper right) examples. Among the dark landslips, the freshest are darkest (some joining dunes, left center). The older ones appear to fade into the background, probably as they become covered by thin films of windblown dust. These features testify to the ongoing erosion of Martian hills. (330W 9N, M03-07572.)

- Early Mars may have produced life, in the same way as Earth. Note that Earth's oxygen-rich atmosphere did not evolve until around 2,400 MY ago, when oceanic plant life (plankton) became abundant enough to consume substantial CO_2 and emit substantial oxygen (O_2). The lack of early O_2 molecules means that ozone (O_3) couldn't easily form on early Earth, which means that life formed on Earth when Earth probably did not have a protective ozone layer (O_3) to block solar ultraviolet radiation. So both early Earth and early Mars had CO_2 atmospheres and UV-blasted surfaces; life formed in shielded environments, for example, underwater or in moist soils.

- Early Mars may have been warmer than it is today and above freezing (due to the greenhouse effect of the CO_2 atmosphere?), or it may have been colder than it is today (due to the fainter early sun?).

- At least one Martian meteorite has been exposed to water about 670 MY ago or less, proving that limited amounts of water have been mobile on Mars within the past 15 percent of Martian time.

- Smaller, sporadic releases of water continue onto the surface in still more recent geological time, sometimes enough to carve long outflow channels such as Marte Vallis, and more often in small amounts sufficient to carve gullies on high-latitude cliff faces; this water probably comes from occasional melting of ice.

These findings suggest that early Mars had some kinds of conditions that allowed life to form on Earth. In particular, liquid water was present. For several decades, most scientists have believed that if conditions are right on any planet—particularly if there is liquid water as a medium to support concentrations of complex organic molecules—then life will eventually start at the molecular level, in tiny cell-like aggregations; and if these conditions last long enough, more complex organisms will evolve. At the moment, this is only a hypothesis. Mars may give us a chance to test it.

LIFE, CHEMISTRY, AND EVERYTHING

What, then, is the "right stuff"—the kinds of wondrous processes and conditions that led to the origins of life? Is it likely that complex life ever existed on Mars? Consider the following discoveries:

- In the 1950s, Chicago chemist Stanley Miller, working with Nobel laureate Harold Urey, showed that if a laboratory environment with water and carbon- and hydrogen-containing gases (for instance, methane and water vapor) is exposed to energy sources such as lightning (represented in the lab by sparks), then in periods as short as a few days, complex organic molecules such as amino acids are created by chemical reactions. Amino acids are building blocks of proteins, which are building blocks of living cells. This experiment has been repeated many times with various energy sources, including impacts, so we know that amino acids are easy to create throughout nature.
- As if to confirm the experiment, extraterrestrial amino acids have been found in carbon-rich types of meteorites from the asteroid belt—meteorites known as carbonaceous chondrites. This proves that the same process has occurred in other planetary environments on different kinds of worlds. In 1983, the geochemist Cyril Ponnamperuma announced finding a single carbonaceous chondrite that contained all five of the critical base molecules involved in carrying the genetic information in the giant RNA and DNA molecules that control living cells.
- Complex organic molecules, including formaldehyde (H_2CO), the amino acid glycine ($C_2H_5NO_2$), and even ethyl alcohol (C_2H_5OH), have been detected by astronomers in interstellar nebulae of gas and dust. This again proves that natural extraterrestrial processes easily produce at least the rudimentary materials from which cells are built.

MARS AND THE CREATIONISTS

An interesting hypothesis advanced by creationists is that supernatural processes created life, and that it was done only once—on Earth—with humans alone being given a kind of royal dominion over all other species. This view was originally tied to the Greek and medieval theories that Earth was the center of the solar system, the imperial Capital of the Universe, around which the sun, planets, and stars revolved. This Earth-centered solar system is now known to be wrong, but the creationist view is still tied to a fundamentalist religious framework in which human beings and other species were specially created as lords of the universe in one mighty act of creation, with no biological evolution and operating under direct supernatural franchise.

It's interesting to contemplate whether this Earthcentric view would predict life on Mars or elsewhere in the universe. Many of the "special creation" theorists deny not only biological evolution but the concept of cosmic evolution and the validity of radioisotopic dating. For information on the history of Earth, they rely on old manuscripts, written about 2,000 to 3,500 years ago (usually associated with one religion or another), from which they claim to be able to derive the age of our planet. Originally, this argument came from Renaissance scholars, who rationally and cleverly tabulated all the generations of humans described in early scriptures, back to Adam and Eve, and assumed that the total duration represented the age of Earth, since Adam was reportedly created only a few days after Earth and the universe. The most famous of these estimates came from Archbishop James Ussher in Ireland, who calculated in about 1650 that Earth was created on October 23, 4004 B.C.

In our modern age, when most educated people around the world accept radioisotopic dates more like 4,550,000,000 B.C., such estimates are often ridiculed, but I take a different view. The Renaissance scholars were measuring *something*—but it was not the age of Earth. Rather, it was the rough age of the oral and written traditions of humanity. We can now see that, in essence, they got a pretty good measurement that our cultural traditions stretch back 6,000 years.

Once, I found myself discussing an earlier book of mine, *The History of Earth* (Workman Publishing, 1991), on national radio with Larry King. After we mentioned the 4,500 MY age of Earth as measured by many labs, a number of people called in to say, "Your guest may be a well-meaning fellow, but I prefer to believe from scripture that Earth is only a few thousand years old." All I could do was repeat the evidence. On other occasions, I've also heard fundamentalist radio preachers trying to revive seventeenth-century views of the world and arguing that Satan put those nasty isotopes in the ground just to tempt scientists into the sin of pride. They insist that Earth is really only a few thousand years old, and in so doing they isolate their listeners from the real world of human dialogue with the cosmos.

We have to be careful about dismissing these sadly misguided souls, because fundamentalist creationists have gained control of various school boards in the United States, insisting that these outmoded ideas—special creation, no evolution, a 6,000-year-old Earth—be taught in public schools, alongside traditional science, as a legitimate alternative scientific hypothesis. My own Arizona state legislature passed such a measure a few years ago, but it was overturned after an outcry from the citizens. Fortunately, these attempts to rewrite geological evidence have been struck down as unconstitutional, because the founders of the United States wisely insisted on a separation of formalized religious views from governmental operations, including public schools. Historically, many countries have suffered when government and school authorities tried to impose preconceived ideology over scholarship, isolating young minds from international progress. For example, in the 1600s, Italy fell behind the rest of Europe in scientific prowess when Galileo's writings were suppressed by the Church, and in the twentieth century Russian biological research was retarded for a generation when Stalin and Soviet authorities insisted that learned traits, in addition to genetic ones, could be inherited. We learn about nature by studying nature, not by interpreting old writings. The best policy is to make your ideology consistent with what is known about the universe, rather than trying to force the universe to fit your ideology.

For me and most scientists, the radioisotopic evidence—from many labs in many countries—is overwhelming. Culture is 6,000 years old; the planets are 4,550,000,000 years old. We are all fortunate to live in a free country, where ideas can compete and people can believe what they want. But we have to guard the opportunity for our children to receive the clearest possible picture of what international science has pieced together after generations of struggle.

Halfway to Mars. This view in Iceland shows hillside gullies that match those of Mars, while reminding us of the tenacity of terrestrial plants that soften our arctic landscapes. Iceland also contains many Mars-like examples of volcanism, ground ice, and lava-ice interaction. Continuing research in arctic areas of Earth may shed much light on Martian features. (Photo by author, south side of Esja plateau.)

- Florida biologist S. W. Fox has shown that simple heating of dry amino acids can produce protein molecules, and when water is added, these can agglomerate into cell-like blobs called proteinoids. These resemble cells and can grow by attaching to each other, can take in the material from the surrounding medium, and can divide. They are not considered living, but they are so much like bacteria that even experts have trouble telling the difference while viewing them in microscopes.

- In the 1990s, studies of molecular biology, such as DNA structures, showed that all life-forms on Earth are descended from a class of simple microbial life-forms. These microbes are called extremophiles, because they tend to flourish in high temperature environments such as geothermal hot springs, as found in Yellowstone Park and elsewhere.

- Because Mars has both ancient and recent lava flows, Martian magmas are likely to have interacted with ground ice in many areas to produce geothermal hot springs at one time or another.

These findings about organic chemistry and Mars make it plausible that complex organic materials arose on Mars during

Noachian or Hesperian times. Amino acids and protein molecules could surely have been produced. Martian waters may have produced ponds full of proteinoids or similar agglomerates of organic materials. What we don't understand is how hard it may be to go from abundant organic molecules and protein-rich, cell-like blobs in ponds, lakes, and hot springs to full-fledged living cells. Does it happen on every planet with water and organic materials? Or did it take special conditions that are unique to Earth?

Microbial life seems to have begun on our planet in the interval of around 3,700 to 4,000 MY ago. Isotopic evidence of biology-produced carbon has been found in 3,700-MY-old rocks in Greenland, and evidence of fossil bacteria has been found in rocks at least as old as 3,400 MY. Evidence from the Moon shows that from the beginning, 4,500 MY ago, until about 3,900 MY ago, the bombardment rate of interplanetary debris crashing into Earth may have been too intense for life to form. Some calculations indicate that the larger impacts during that period may have been so energetic and frequent that they vaporized and blew away whole oceans! But as the impact rate dropped on Earth and conditions became tolerable, organic chemistry did its thing, and extremely primitive life developed "rapidly," within 100 MY or so. And if Mars had a similar, CO_2–rich, moist (at least sporadically) environment, why not there?

Mars and other planets also underwent intense meteorite bombardment until about 3,900 MY ago, and many researchers have assumed the impacts impeded Martian biology. However, in important work published in 2002, a team of researchers from England, USA, Canada, Germany, and New Zealand, headed by Charles Cockell of Cambridge and NASA, showed that the early impact facturing of Martian rock may actually have created enhanced habitats for microbes inside the rocks. In impact-damaged rocks from the eroded Haughton impact crater in a Mars-like environment in northern Canada, they found increased fracturing and porosity (giving microbes access to the rock

interior); increased translucency in the impact-shocked mineral crystals (allowing some sunlight to penetrate into the rock to foster photosynthesis); and increased protection from ultraviolet solar radiation inside the rock compared to the surface (giving microbes safer conditions inside than on the surface). Consistent with these findings, the researchers actually found a higher abundance of microbes inside the impact-fractured rocks of the crater than in the nearby rocks undamaged by the impact.

The crucial issue on Mars was probably whether life-forming conditions ever lasted long enough, at any one time, to allow

The holy grail of Martian exploration might be a deeply eroded gully that exposes ancient strata. Chemical measures of such ancient strata may reveal early Martian climate conditions, and such strata might even reveal fossils of early life-forms. Is that an example of a Martian fossil in the hillside at lower left? (Painting by author, based on natural features at Anza Borrego National Monument, California.)

microbial life to gain a foothold. This brings us to the key problem about the Noachian era. From the point of view of microbes, was it habitable or not? The most benign scenario is that Mars was water-rich or at least moist throughout that time, possibly much warmer than today due to the greenhouse effect, and with ground ice melted by higher early internal heat flow—creating oceans and rivers that didn't dry up until the Hesperian era. But a less fruitful scenario hasn't been ruled out. In the least benign view, Noachian Mars was colder than today (due to the lower solar luminosity at that time), or only marginally warmer (due to a barely compensating greenhouse effect), and water was frozen. Perpetually frozen water may prevent the mobility that organic materials have when dissolved in liquid water, and thus it would have prevented the chemical reactions that could have led to formation of living, reproducing cells. In this scenario, local volcanic events or climate anomalies produced local ice-melting episodes with rivers and lakes that lasted weeks or years, or even oceans that lasted a million years—but not long enough to allow the initiation of life.

DORMANT MARTIAN MICROBES?

However, an intermediate geological scenario is the most likely. In this scenario, Noachian Mars, due to its higher internal heat flow and some greenhouse effect, had more frequent melting of its ground ice than later Mars. Whereas the base of the ground ice today may be a few kilometers down, as discussed in chapter 8, the Noachian underground heat flow, produced by short-lived radioactive minerals present at the beginning, caused the base of the ground ice to be at, say, 1 km in depth. In that case, a large fraction of today's ice would have been liquid water, moving here and there in underground aquifers, sealed below a near-surface ground ice layer. Occasionally, perhaps due to local volcanism, groundwater gained access to the surface and burst out into rivers and lakes. High atmospheric water content may have caused rain and created the valley networks that we visited in chapter 9. As heat flow declined, and as loss of the atmosphere reduced the greenhouse effect and the surface pressure, the ground ice thickened and water on the surface became less stable. The last frequent outbursts of melted ground ice were in the Hesperian era, carving huge outflow channels such as Ares Vallis. Melting of ground ice and major outbursts grew less frequent through the Amazonian era, perhaps occurring only millions of years apart, or even hundreds of millions of years apart. This fits with the fact that the Martian meteorite Lafayette was exposed to water 670 MY ago, as we saw in chapter 25. Axial tilt processes might lead to much smaller and more localized high-latitude water release every few MY, explaining the high-latitude gullies of chapter 34. If water is mobilized only millions of years or hundreds of millions of years apart, would this scenario allow life?

A possible answer to this question came from an unlikely source. In 2000, Pennsylvania biologists Russell Vreeland and William Rosenzweig, working with Texas geologist Dennis Powers, published a paper unrelated to Mars but containing an amazing, relevant discovery: they reported finding viable bacteria in 250-MY-old terrestrial salt crystals. Two out of 53 sampled salt crystals yielded bacteria that were living and grew when "rescued" from the salt. The researchers found that these were highly salt-resistant bacteria, of a genus known as *Bacillus*, which probably were trapped in crystals as brines when an ancient playa evaporated and left salty deposits. A substantial number of reports have claimed to find living bacteria surviving in ancient deposits, but many have been questioned under the charge that the bacteria were from recent contamination. The report by Vreeland and his colleagues is considered one of the best documented because it is difficult for bacteria to get inside existing salt crystals—they must have been there when the salt formed 250 MY ago. This is one of the oldest claimed cases (another example involves bacteria found inside amber that is 25 to 30 MY old!).

Since these reports are still being debated, we need to maintain a bit of caution, but the idea of ancient dormant bacteria has been increasingly accepted. If bacteria lying dormant for 25 MY, or even 250 MY, can become active when nature simply adds water, we are led to a whole new prognosis for Martian biology! The new working hypothesis would be that clement conditions lasted long enough in the Noachian era for simple bacteria to evolve, which could have happened by 3,700 to 3,900 MY ago, as on Earth. Then, as water evaporated and salty ponds dried up, salt-loving bacteria evolved that could go dormant inside salt deposits or other havens until another ice melt produced liquid water. In this view, Mars would be like a very primitive version of a western desert dry water hole, which, after seeming lifeless for several years, bursts to life with frogs and other aquatic creatures that have evolved to lie hidden and dormant in the ground until big rainstorms turn the dry hole into a pond, and that then race through their reproductive cycle while the good times last. Surface landers might find lifeless soils sterilized by the solar ultraviolet light, but life might be hidden in crystals in salty soils, distributed around the planet. Local pockets of active organisms might arise in sporadic geothermal events tens or hundreds of MY apart.

LIFE UNDERGROUND

Dormancy helps when you live on Mars, just as it helps a hibernating polar bear. But you can't be dormant all the time, and the real question for a Martian microbe, as for a high school graduate, is How do you make a living?

Since we know the Martian soil is sterile and that microbes on the Martian surface would be blasted by sterilizing solar ultraviolet light, the question arises as to whether Martian microbes can prosper in sheltered environments, such as underground or in rocks. Martian research has created a new interest in microbial life underground on Earth. Biologists—especially those interested in Mars—are taking a new look at what's going on beneath our feet. Microbial communities have been found at hundreds of meters deep in basalt lava flows—environments not unlike those that could be found underground on Mars.

The problem is that we living creatures need energy sources to power the reactions in our cells. One camp asserts that all life on Earth ultimately depends on sunlight. We get our energy by burning oxygen, but oxygen comes from photosynthesis by plants, which in turn depend on CO_2 in Earth's atmosphere. Even creatures in the deepest parts of the sea depend on nutrition derived from life-forms that depend on life-forms that depend on surface plankton, which depend on sunlight. An alternate view is that Earth harbors a minority of life-forms that derive their energy from other sources, such as chemical reactions with sulfur compounds and other minerals coming from deep-seafloor geothermal vents, or minerals in underground rocks. At a 1997 NASA conference on possible Martian biology, one researcher noted that primitive bacteria can utilize many chemical reactions and can get "free meals" from sulfides and other chemicals in the hot fluids. Iron sulfide reactions, in particular, are a good source of energy. Tracing early life-forms to such energy sources, one scientist at the conference cited a bit of Kipling verse:

> *Gold is for the mistress, silver for the maid,*
> *Copper for the craftsman, cunning at his trade.*
> *"Good!" said the Baron, sitting in his hall,*
> *"But Iron—Cold Iron—is master of them all."*

At the same conference, biologist Todd Stevens talked about subsurface biology, noting that the biosphere on Earth extends several kilometers down, limited only by the increasing temperatures as we go downward, and the sources of energy. With eloquent simplicity, he posed the burning scientific questions about microbes: Are they down there? Are they doing anything down there? What are they doing? How fast are they doing it? How do we know? Do we care?

Studies of underground bacteria in basaltic lava formation showed that they get energy from chemical reactions by attacking the rock, even in low-oxygen environments. As one researcher in this field commented, "If our experiments are correct, microbial bacteria have some reason to erode rock, other than mere vandalism."

Clearly, the proposed Martian microbe fossils in meteorite ALH 84001 are crucial to the story! The characteristics of the proposed fossils in ALH 84001 were not unlike those described above—the possible microbes were concentrated in carbonate granules left by evaporating water inside rock. That controversy about whether these features really are fossil cells seems to be stalled. It may not be answered until more 4,500-MY-old Martian meteorites are found, which may provide more evidence of what was happening at that time on Mars—or, more likely, until rocks can be brought back from Mars itself to settle the arguments. All the while, we must keep open the chance that the reported microbes represent not full-fledged life, but some never-before-seen stepping-stone on the way to life.

LIFE BELOW THE ICE

Another detection of buried terrestrial bacteria in a Mars-like environment was announced in the journal *Science* in December 1999. Researchers in Antarctica drilled through more than 3 kilometers of ice overlying 200-km-long Lake Vostok, hidden below the Antarctic ice cap. This buried lake of liquid water is not unlike our earlier pictures of localized Martian aquifers several kilometers below the surface at the base of the Martian ground ice.

Biologists had long wondered what microorganisms, if any, might exist in the dark, frigid waters of the buried lake. They drilled to find out but stopped short of penetrating into the lake itself. The drill hole ended a hundred meters or so above the base

My Martian Chronicles

PART 14: WORKING WITH THE REAL MARTIAN CHRONICLER

Soon after Mariner 9's flight, the Mariner team asked for a volunteer to write an official NASA Special Publication, a popular-level book about what had been found.

I took on the project. It was coauthored by Jet Propulsion Lab engineer Odell Raper and called *The New Mars*. In the midst of the effort, I had the idea of contacting Ray Bradbury, the visionary author of the science-fiction classic *The Martian Chronicles*, and asking him to write a short preface about his response to the new discoveries about "his" planet.

With great trepidation, I dialed the number I'd been given for Mr. Bradbury, a childhood hero of mine. Ray came on the line, jovial and expansive, and he agreed to write the preface for a modest and reasonable fee. It was beautiful, poetic, and lyrical, with a nice line about how if we find even one tiny paramecium on Mars, it would be a priceless treasure for humanity. I was very happy. What a stroke of good fortune not only to talk to Ray but to have a true Martian pioneer write a preface to a NASA book about Mars!

A few months later, during the book's production, timid and nonlyrical NASA bureaucrats of the time banned the wonderful preface, under the theory that NASA couldn't appear to endorse any controversial opinion that Mars might somehow have produced even a single-celled life-form. My protests went unheeded. Two decades later, sociological polarity was reversed, and NASA was trumpeting its press conference about how one of the Martian meteorites might contain fossil microbes.

of the ice, but tapped water that was proven (by isotope chemistry) to come from the lake below the ice. In this water, researchers once again confirmed the incredible tenacity of life: they found substantial concentrations of living bacteria along with inorganic nutrients and dissolved carbon from organic sources. This finding does not prove that microbes could live in buried Martian water, because the Vostok bacteria may depend in part on nutrients filtering to those depths from other life-forms closer to the surface, but it does increase our awareness that life can persist in extreme environments—and perhaps in buried Martian aquifers as well.

The next step in studying Lake Vostok as a Martian analog would be to drill into the lake itself, but this step illustrates the profound problems in this type of work. If researchers drill farther through the ice into the lake, tons of aviation fuel and freon lubricants in the drill hole, not to mention possible bacterial contaminants from the surface, would drain into the lake, believed to be one of the last pristine water bodies on our planet. This, in turn, would raise questions about the possibility of any future studies of underground life in this forbidding polar habitat.

The same problems apply generally to Mars, where the more human activity occurs, the greater the chances of losing our ability to discriminate native Martian life-forms from terrestrial contamination. This is reminiscent of Heisenberg's uncertainty principle: It's hard to observe something without altering it.

But for now, perhaps the most interesting result is the way in which studies of terrestrial biology and potential Martian biology are converging into one awe-inspiring new field of cosmic biology.

THE TWO MYSTERIES

We are left with the two grand mysteries of the red planet. One is the planet's climate and environmental history. What was the role of water? Was the climate different? For how long? How did Mars change from a river-laced planet to a frozen desert? What does this tell us, if anything, about the possibilities of climate change on Earth? Are planetary climates less stable than we usually assume?

The second mystery is the question of life itself. Did microbial life appear in the ancient past on Mars, as on Earth? If life did evolve on Mars, what was it like? Was its RNA and DNA like ours? Is it underground? In the words of Todd Stevens, "Are they down there?" And if they're not down there, despite the water, why not? Is life harder to start than we currently think? Are we alone in the universe after all?

WHERE ARE WE GOING?

NASA's official strategic response to these questions about Mars is to "follow the water." This means designing missions that shed light on the history of water on Mars—measuring soil compositions, looking for carbonates and other evaporite deposits, identifying ancient lake beds, seeking ice deposits, et cetera.

About every two years, due to the configurations of Mars's and Earth's orbits, Mars and Earth come into positions where spacecraft can be launched from one planet to the other with minimum expenditure of fuel. The less weight used in fuel, the more can be allotted to scientific instruments (or, someday, humans). For this reason, missions to Mars tend to be spaced about two years apart. MGS and the Mars Pathfinder lander arrived in late 1997. During the 1999–2000 window of opportunity, NASA lost two missions, an orbiter (due to a stupid mistake in confusing metric with English units) and a polar lander (apparently due to a budget too skimpy to allow adequate testing of the landing procedure). In 2002, Mars Odyssey arrived in orbit and began more detailed studies of Mars mineralogy, as well as continued photomapping.

The year 2004 should be a banner year. Three landers are planned to arrive on the surface of Mars, with rovers that can study terrain around the landing sites. The launches, scheduled for 2003, have yet to happen at the time of this writing. The rover landings are planned for the first weeks of 2004. These missions are as follows:

- *Mars Exploration Rover (MER) missions organized by NASA.* Two separate launches will attempt to deliver two rovers to two different sites. A series of meetings over several years has been devoted to picking the landing sites to maximize the chances of finding interesting deposits and possible signs of ancient lake beds. One landing site is likely to be in the Terra Meridiani hematite-rich area discussed in chapter 12. The rovers will land with the air-bag system that was used successfully in Mars

Pathfinder when it landed in Ares Vallis in 1997. The packages zoom onto the surface at high speed; the air bags absorb the shock; the landers bounce and roll for a kilometer or so, and then come to rest. The bags deflate, and if all goes well, the so-called MER rovers emerge. Unlike the Pathfinder rovers, Mars Exploration rovers will be independent of the landing system and designed to roam for at least 90 days. They will carry normal cameras, a microcamera to photograph crystal structures of rocks, devices to scrape soil off rocks, and devices to measure compositions of the clean rock surfaces.

- *Mars Express and Beagle 2.* This is a mission from the European Space Agency with a strong role taken by the Italian Space Agency. An orbiting mother ship will carry instruments that include several derived from instruments on the gigantic, failed Russian Mars 96 spacecraft that ended up in the Pacific. American participation consists primarily of tracking and participation in some of the composition-measuring instruments. The greatest interest is in a small but daring instrument-laden rover that is being built in the United Kingdom. It has the amusing name Beagle 2, after the ship *Beagle* in which Charles Darwin sailed the Pacific, beginning in 1831, and gathered evidence that led to his theory of biological evolution. Beagle 2 will carry instruments similar to those on the MER rovers, designed to measure compositions of rocks and soils around the landing site. Many researchers consider Beagle 2 to be a long-shot mission, because it has an unusually high percentage of its total payload devoted to science instruments instead of engineering support, and it is being built by British teams without previous Martian experience. But for this very reason, it will be very exciting if the daring Brits can get their instruments onto Mars.

- *Nozomi.* This is a Japanese orbiter that will arrive at Mars in 2004, primarily studying charged atoms, or ions, in the upper atmosphere and solar gases near Mars.

Many other missions are in various stages of development in various countries. Some are being built, but others, slated farther in the future, are unfunded and merely in the planning stages. Some examples include:

- NASA plans a Mars Reconnaissance Orbiter, probably for arrival in 2006, with instruments to study the atmosphere. Advanced spectrometers and a high-resolution camera would continue the search for unusual surface materials and interesting landing sites.

- French researchers are designing four "Netlanders" to be placed on the surface. These are primarily seismic stations to search for Martian earthquakes. With a net of several seismometers, scientists can pinpoint quake location and depth, and study the interior structure of Mars. Each lander will probably carry a camera and geochemical instruments to study soils.

- Small, cheap, American "Scout" missions are being proposed to fill gaps left by data from other missions, or to possibly make up for failed missions. A host of different researchers have proposed missions under this program, and selections of winning proposals are still under way as of this writing.

- Italian and American designers have discussed a communications orbiter in order to relay data from other missions. Problems have arisen during past missions when data from working instruments could not be properly relayed to Earth. Landers operate best if they can send data up to an orbiting mother ship, whose more powerful equipment can relay data to Earth. A reliable long-term Martian communications satellite would free designers from worries about sending new communications equipment with each mission. The future of this proposed mission is uncertain as of this writing.

In a sample return mission, a rover would deposit soil and rock samples in a return rocket, which would then blast off for Earth. (Painting by author.)

ROCKS AND DREAMS: A SAMPLE RETURN MISSION?

Beyond the immediate series of scheduled Mars missions from various countries, a longer-term goal is a robotic mission to return rocks and soils from Mars. If we could pick an ancient lake-bed deposit, we could bring back bits of dried mud or rock and look for fossil microbes.

Prior to the two failed NASA missions in 1999, planners had expected such a mission near the end of the present decade. But the failures provided a sobering dose of reality; bringing back samples is much harder than successfully operating orbiters or rovers. The rovers have to go out and pick up rocks, but an additional launch mechanism must carry the rocks off Mars; they must be guided to Earth, and a tricky rendezvous must be made to recover the samples before they crash into the ocean or ground. On top of which, guidelines call for the sample to be maintained in sterile condition to avoid contamination and allow searches for Martian organic or biological materials. Facing these realities, program managers have pushed the sample return mission probably beyond 2010, and it has not been formally funded.

Nonetheless, such a mission is undergoing international design studies. Missions like this are so expensive that it makes sense to fund them by an international consortium, rather than having one country go it alone. Moreover, this approach has some nice benefits. Such cooperation helps the human race to invent the political mechanisms that will be needed as we address planet-scale technical problems, such as global pollution, climate change, and weapons inspections. Also, international cooperation is fun, because it means we all share in the human adventure of exploring nearby worlds, rather than letting it become a jingoistic ploy for one nation's power or prestige. It's like enjoying the Olympics as a whole event, in addition to rooting for your home team.

The Mars sample-return mission still faces many hurdles. In late 2002, for example, France, which had been participating to the tune of 400 to 500 million euros total (one euro roughly equals a dollar), decided to cut back—but Canada began contemplating a larger role. So the international constellation of participating nations keeps changing. We can only hope that within the lifetimes of readers of this book we will be bringing back Martian rocks from carefully chosen landing sites and using them to find the answer to the question: Are we alone in the universe?

HUMANS TO MARS?

Nature draws us forward, down the road of knowledge about our place in the cosmos. In the distance, Mars beckons. When it comes to human exploration of Mars, the challenge is more social and geopolitical than technical. We know how to do it. The spacecraft technology is available, and the human factor poses no insurmountable problem. When the Russian Mir space station was active, several Russian cosmonauts logged single continuous missions equivalent in duration to a flight to Mars. But the question is whether we humans have the will and resources to carry on, and whether we can agree on the benefits.

It's not unrealistic to say that there is a fork in the road leading to the future: either civilization will collapse, or humans will reach Mars! As far as the first option is concerned, the twenty-first cen-

First on Mars. In this conception of the first Martian landing, a robotic ship (far background) has placed supplies and a backup vehicle on Mars. It is followed by a piloted ship bringing the first explorers. (Painting by author; based on natural features at Pinacate Mountains Biosphere Reserve, Sonora, Mexico.)

tury may be critical. Earth is finite, and projections suggest that we may begin to exhaust our supplies of easily accessible resources within this century. (See my earlier book *Out of the Cradle,* Workman Publishing, 1984.) Some say we will always invent alternatives, but this is not guaranteed. In the meantime, we are witnessing a trend toward expending our energies and resources to blow up each other's cities. If many nuclear weapons are brought into play, the world economy would likely collapse, resources become scarce, and a second dark age could ensue. If this is the future we build for ourselves, society may never recover adequately to mount successful cosmic exploration. With all the "easy" coal, oil, and metal ores gone, it would be hard to have a second industrial revolution and climb out of the second dark age.

The other option is more attractive. If we can preserve a healthy world society, we will surely continue developing our capability to operate in space. And if that happens, we will certainly return to the Moon, learn the history of species-ending impacts on Earth, fly to asteroids to seek their metal resources, and visit Mars.

One silent threat to such deep-space operations is radiation. Atomic particles from the sun and from the rest of the universe zip through space, and when they zap human skin, they can break up DNA molecules and cause other damage—which is why the surface of Mars is sterilized. On Earth, we are shielded by our atmosphere. On the surface of Mars, the doses of this type of radiation (a misnomer, because it consists of tiny particles, not radiation like light or X rays) have an intensity about 1,000 times greater than those experienced on Earth. During solar storms, the radiation rates rise so high that an unshielded human being could be subjected to lethal doses during a solar storm of a few days and could die some days later. Fortunately, very modest amounts of material, such as a few inches of metal or other shielding material, block these atomic particles. Normal spacecraft design provides enough shielding for ordinary conditions, and planners for Martian voyages speak of creating "safe rooms" or strongly shielded parts of spaceships, to which astronauts could retreat for a day or so during a solar storm. In the same way, banking some Martian soil over the walls and roof of a Martian habitat would make it safe from solar storm exposure on the Martian surface.

If all goes well, some readers of this book may witness the first human footsteps on Mars—or perhaps even make them. It all depends on whether we can set aside our provincial earthbound squabbles long enough to consider our common future.

This quandary reminds me of a final joke, which comes from my long-suffering Russian colleagues: What is the difference between a realist and a dreamer? The realist thinks that someday a UFO will come down and hover over the UN building, and that the aliens will come out of the UFO and offer to share their technology and solve all our world's problems.

The dreamer thinks maybe we can get our act together and do it ourselves.

My Martian Chronicles

PART 15: MY FRIEND FROM MARS

I have a friend in Paris who, in essence, has already lived on Mars. His name is Charles Frankel, and he is a Franco-American geologist, science writer, and photographer.

He is also an all-around bon vivant of many talents, not the least of which is his Carl Sagan imitation and his ability to make fun of French and American foibles with equal enthusiasm. His 1996 book, *Volcanoes of the Solar System* (Cambridge University Press), is a notable popular introduction to that subject, and his own book about Mars is coming out soon.

Charles Frankel prepares to walk outside the "Mars habitat" in Canada.

Based on his varied background and interests, Chuck was able to join the crew of an experimental Mars habitat in arctic Canada near the Haughton impact crater, mentioned earlier. The habitat was a prototype self-contained Mars base, designed under the auspices of the Mars Society, an international group promoting prompt human exploration of Mars. Not content to wait for NASA or other national space agencies to begin Martian planning, they have created several modules, set them up in Mars-like locales, and staffed them with visiting crews of geologists, chemists, and engineers to test the technical and social requirements for Martian surface exploration.

On one of my trips to a scientific conference in Europe, I was able to catch up with him in a Parisian café near the bustling shopping area of Les Halles, a few blocks from Notre Dame. Chuck explained that this café was not only a good writing spot but also good for people-watching. Over drinks at our microscopic table in front of the café, we sat in the lazy afternoon sun and talked about his "Martian" experience. The base had been erected in a remote area of Canada called Devon Island, where the Mars-like terrain includes the ancient, eroded impact crater and rocky cliffs with erosion gullies. Wearing full space suits, expedition members made realistic Martian forays outside to test their ability to do meaningful field research under those conditions. I asked about the social and psychological aspects.

(continues on next page)

"We focused mainly on water consumption," he said. "We each managed to eat, drink, clean, and shower on an average of eight liters a day. It meant a brief shower every three days." Chuck explained that his particular crew of six, including Mars Society founder Robert Zubrin, got along well and was socially aided by a mix of four men and two women. He said that experience with crews having "too many alpha males in a cramped space" showed higher stress levels, more arguments in the habitat, and more conflicts with the "ground control" units operating out of Colorado or California.

"What about privacy?"

"The base was large enough to find quiet in the lab or at a console," Chuck said. "Privacy was not an issue. We all had our little cubbyholes but rarely retired to them except to sleep for four to six hours. If there were problems, going out on an EVA with a friend was a great way to let off steam!"

When I asked Chuck about wearing space suits, he said, "It gives you a wonderful sense of identity! You're *it*!—you're self-contained, on a mission. Radio communications can be a problem; you have to learn to convey only the essential—but include jokes, of course. Sometimes when we had radio problems, we reverted to sign language, like in scuba diving. And you learn to take advantage of a water pipe, a bolt, or a mike to scratch your nose! The difficulty of looking straight down at the ground is also a problem, especially

Geologists test concepts of Martian exploration outside the Devon Island Mars habitat in Canada. The boulder in the background was blasted out of the nearby Haughton impact crater and forms part of the eroded crater's ejecta blanket. (Photo by Charles Frankel.)

Experimental Martian habitat established by the Mars Society at the Haughton impact crater on Devon Island in northern Canada, as a site for Martian exploration concepts. (Photo by Charles Frankel.)

frustrating for a geologist. Also, fussing with the buttons and cameras. Moving around was not too hard for us, but the real Martian suits will probably be stiffer, because of the air pressure, and require more effort.

"But the main problem, as Apollo moon-walkers pointed out, is that time goes too fast. There's never enough of it to do the science, take the photos, and enjoy the experience!"

Chuck went on to tell me about a distinctly non-Martian problem. The main threat to the arctic Mars-walkers was polar bears! Wrapped in a space suit, with limited visibility and mobility, the walker was in danger of being taken by surprise. As a result, Inuit aides from a nearby village, armed with rifles, accompanied each foray out of the habitat.

The sun moved behind a nearby cathedral and the evening crowds came out into the promenade. Over the centuries, Paris cafés have witnessed many discussions about life, love, philosophy, art, science, and surrealism, but I wondered if there had ever been a stranger conversation in that café at Les Halles than our talk of polar bears, nose-scratching, and living in a base on Mars.

GLOSSARY

aquifer: A porous underground layer that can carry water. (The term does not apply to the water itself, which is called groundwater in an aquifer.)

basalt: The most common lava type on Earth and Mars. Generally dark gray, or weathered to reddish brown, basalts can be seen in most volcanic national parks, such as Craters of the Moon in Idaho, Sunset Crater in Arizona, and Hawaii Volcanoes National Park. Most of Earth's seafloor crustal rocks are basalt. The composition is a mixture primarily of minerals feldspar (plagioclase variety) and pyroxene.

caldera: A circular, oval, or irregular depression in a volcanic region, caused by collapse of surface layers into a magma chamber or underground volcanic cavity created by removal of magma. Sometimes called (less precisely) volcanic craters. Some calderas look superficially like impact craters. Calderas are often located at the summit of volcanic mountains. The term is usually applied to craters that are larger than 500 meters or 1 km in diameter; smaller features of similar origin are called collapse pits. See also CRATER.

canal: A term originally derived as a translation of Schiaparelli's term *canali*, for streaky markings on Mars, but adopted by Lowell specifically to mean hypothetical water-transferring, straight irrigation canals built by hypothetical intelligent Martians. Not to be confused with the modern term CHANNEL.

canali: An Italian term, meaning "channels" or "canals," used by Schiaparelli to refer to the streaky markings of Mars.

caprock: A layer of rock more resistant to erosion than other layers. As a landscape erodes, this layer forms a cap or shield against erosion on the tops of plateaus and mesas.

channel: A dry riverbed on Mars. Not to be confused with the term CANAL.

chaotic terrain: An area of jumbled hillocks and valleys, usually in a region of collapsed ground tens of kilometers wide.

continental drift: An early term for PLATE TECTONICS, referring to the slow drift of large blocks of Earth's crust.

convection currents: Patterns of circulation of semi-molten material inside a planet, due to concentration of heat in the center. Hot material rises from the center in MANTLE PLUMES in some areas, while material that has been cooled at the surface sinks in other areas. Note that most rock in the mantle is solid, but due to viscous-elastic properties, it can flow over long time periods, like a glacier. The whole process transfers heat from the center to the surface and causes a planet to cool off.

crater: A roughly circular depression. The term is usually applied to impact craters, but sometimes it is applied to volcanic calderas or collapse pits. See also IMPACT CRATER.

crust: The outer layer of lower density rock on a terrestrial planet. Under the oceans on Earth, it is a 5-km-thick layer of basalts. Under terrestrial continents, it averages about 30 km thick, but it is thicker under mountains. The crust on Mars is estimated to be on the order of 40 km to 50 km thick, but with large amounts of variation.

debris apron: A fan of debris that spreads out at a shallow slope from a Martian mountain. Often, such fans completely surround isolated peaks on Mars. They give an appearance as if material had melted and flowed or slumped out from the peak.

dendritic system: A branching, tree-like pattern of tributaries that connect to a main channel.

dike: A wall-like slab of rock formed underground (for example, when magma enters a vertical fracture and then solidifies). Often, if the surrounding fractured "host rock" is removed by erosion, dikes of resistant igneous rock are left standing on the surface, resembling artificial walls.

dipole magnetic field: A magnetic field with a well-defined north and south magnetic pole, as on Earth. A dipole magnetic field for a planet is believed to result from moving currents of material in a molten iron core of the planet; thus, a dipole field implies the existence of a hot, molten metal core in the planet.

duricrust: Weakly bonded Martian soils that break into clods and plates, similar to terrestrial dried muds and weakly cemented desert soils. The cementing probably comes from salts that are left by evaporating briny water.

dust devil: A swirling mass of dust raised 100 meters or more by tornado-like winds on Earth and Mars.

ejecta blanket: A layer of debris blown out of an impact crater surrounding the rim of the crater.

esker: A ridge of debris left by a retreating glacier or by water erosion under such a glacier.

exhumation: The process of removing overlying layers of soil or rock to reveal older surfaces underneath.

fault: A fracture along which the two sides have moved relative to each other.

ground ice: Deposits of underground ice.

groundwater (ground ice) sapping: A process believed to be involved in enlarging some Martian channels and other low areas bounded by cliffs or banks. If water (or ice) is exposed in the cliff face, the water may leak out or evaporate (and the ice may sublime), removing support from overlying layers and causing a local collapse. This enlarges the cliff, exposes the aquifer or ice layer, and the process starts over. In this way, channel systems may be enlarged as sapping eats back into the surrounding terrain. In some cases, it may produce tributary-like side channels. See also SAPPING.

igneous rock: A rock directly formed by solidification of molten magma, as opposed to other mechanisms, such as chemical precipitation of sediments or cementing of pre-existing granular materials.

impact crater: A circular, bowl-shaped cavity created by explosion during the high-speed impact of a meteorite (generally a fragment of an asteroid or comet).

lava: Molten rock material (magma) on the surface of a planet (as opposed to underground).

lava tube: A cavern-like underground conduit through which lava has flowed. A lava tube forms when a tongue-like flow of active lava starts to cool. The outside solidifies into rock, but the inside is still molten, slowing lava. If the lava breaks out of the lower end of the tube, the whole tube can be drained, leaving an empty cave.

lithosphere: The more rigid outer rock layer of a planet, distinguished from the more deformable, hotter underlying rocks. The lithosphere of Mars has been deformed in particular by the weight of lavas accumulated in the Tharsis dome.

magma: Molten rock material. Usually the term applies when the material is underground; once it has erupted on the surface, it is more commonly called lava.

mantle: The rocky zone outside the metal core of a planet, between the core and the thin crustal layer at the surface. The mantle includes most of the volume of the planet. Mantle rocks are hot and may be partly molten, and therefore can deform or flow slowly over long periods of time.

mantle plume: An ascending hot convection current in the slowly churning, plastically deforming mantle of a planet. (The mantle is hot rocky material that can deform slowly like a glacier.) Where the mantle plume rises to hit the underside of the crust, a hot spot is created, and this may be the site of volcanism. A long-term mantle plume may have created the Tharsis volcanic center.

meteorite: A rocky or nickel-iron fragment of an asteroid that reaches the surface of a planet.

meteoroid: A fragment of an asteroid or comet while it is still in space or is transiting through the atmosphere of a planet.

moraine: A ridge of rocky debris, originally carried in glacial ice and then dropped as the glacier melts.

orographic cloud: A cloud that forms above a mountain peak as air is carried upslope and cools. Orographic clouds are common over Olympus Mons and the other three high volcanoes of the Tharsis region.

outflow channel: A broad Martian riverbed with a typical width that exceeds 1 km and that often originates from a restricted area.

patterned ground: Surfaces with polygonal, fingerprint-like, or other patterns visible on a scale of tens to hundreds of meters. The patterns are reminiscent of those observed in terrestrial arctic tundra and are believed to be caused by effects of ground-ice deposits.

plate tectonics: The phenomenon in which the Earth's land and sub-ocean crust is fractured into plates, or continent-size blocks, which move relative to each other due to sluggish moving currents in the upper mantle, or zone beneath the crust.

polar hood: A cloud mass that forms over the Martian poles and high latitudes during Martian winter.

rampart crater: A crater with a thick ejecta blanket, sharply defined and often bounded by a raised lip. Such an ejecta blanket is believed to involve fluid mud, created by an impact into ice.

rays: Thin radial streamers of bright or dark material, extending out from an impact crater like a starburst pattern. They are deposits of fine dust that jetted out from the impact explosions.

rock glacier: A flowing glacial mass that comprises not just ice but a mixture of ice and soil. The surface of a rock glacier typically looks like a mixture of soil and rocks, but it masks a moving glacial system.

sapping: The erosion of hillsides or riverbanks by the seeping of water or loss of ice exposed in the face of the slope. This undermines the overlying material, causes local collapse, and eats into the hillside. See also GROUNDWATER SAPPING.

secondary impact craters: Craters that were formed not by asteroid or comet impacts directly from space, but rather by impacts of blocks of debris that are blasted out of large craters and then fall back to the surface, thus forming smaller craters.

sublime: To pass directly from the solid state into a gaseous state, without becoming a liquid. At the higher elevations on Mars, the pressure is so low that exposed ice would sublime into water vapor without creating a liquid form (just as frozen carbon dioxide, or dry ice, on Earth passes directly into a gaseous state). The process is called sublimation and may be important in causing the loss of ice from the upper few meters of soil at lower Martian latitudes.

tectonics: The study of fractures, faults, earthquakes, and motions that deform the crust of a planet.

terrain softening: Softening or "melting" of relief, such as crater rims and mountains, by an unknown process that possibly involves ice flow. Most terrain softening seems to have happened in early Martian history, during the Noachian era.

valley networks: Systems of shallow, modest-size drainage channels, with characteristic widths of 100 meters, usually in Noachian terrain and usually having dendritic tributary structure.

wind tail: A streak, either darker or lighter than the background, left on the leeward side of a crater, mountain, or other obstacle by dust dropped or distributed by prevailing winds.

SELECTED SOURCES

and Additional Reading About Mars

Baker, V. R. 1982. *The Channels of Mars* (Austin: University of Texas Press).

Batson, R.M., P.M. Bridges, and J. L. Inge. 1982. *Atlas of Mars* (Washington D.C.: NASA).

Binder, A. B., et al. 1977. "The Geology of the Viking 1 Lander Site." *J. Geophys. Res.*, 82, 4439–51.

Blunck, Jürgen 1982. *Mars and its Satellites: A Detailed Commentary on the Nomenclature* (Smithtown, N.Y.: Exposition Press).

Bridges, J. C., D. C. Catling, J. M. Saxton, T. D. Swindle, I. C. Lyon, and M. M. Grady. 2001. "Alteration Assemblages in Martian Meteorites: Implications for Near-Surface Processes," in *Chronology and Evolution of Mars*, eds. R. Kallenbach, J. Geiss, and W. K. Hartmann (Bern: International Space Science Institute, with Kluwer Academic Publishers).

Carr, M. 1981. *The Surface of Mars* (New Haven: Yale University Press).

Carr, Michael H. 1996. *Water on Mars* (New York: Oxford University Press).

Clifford, Stephen M. 1993. "A Model for the Hydrologic and Climatic Behavior of Water on Mars." *Journ. Geophys. Res.* 98, 10,973–11,016.

Clifford, Stephen M., and Timothy J. Parker. 2001. "The Evolution of the Martian Hydrosphere: Implications for the Fate of a Primordial Ocean and the Current State of the Northern Plains." *Icarus*, 154, 40–79.

Collins, Michael. 1990. *Mission to Mars* (New York: Grove Weidenfeld).

Collins, Stewart A. 1971. *The Mariner 6 and 7 Pictures of Mars* (Washington: NASA Special Publication 263).

Costard, F., F. Forget, N. Mangold, and J. P. Peulvast. 2001. "Formation of Recent Martian Debris Flows by Melting on Near-Surface Ground Ice at High Obliquity." *Science* 295, 110–113.

Croswell, Ken. 2003. *Magnificent Mars* (New York: The Free Press).

Gibson, E. K., Jr., D. S. McKay, K. Thomas-Keprta, and C. S. Romanek. 1997. "The Case for Relic Life on Mars." *Scientific American* 277, No. 6. 59–65.

Golombek, Matthew P. 1998. "The Mars Pathfinder Mission." *Science* 279, 40–49.

Grieve, R.A., and E. Shoemaker. 1994. "The Record of Past Impacts on Earth," in T. Gehrels (ed.), *Hazards Due to Comets and Asteroids* (Tucson: University of Arizona Press).

Hartmann, W. K. 1999. "Martian Cratering VI. Crater Count Isochrons and Evidence for Recent Volcanism from Mars Global Surveyor. Meteor." *Planet. Sci.* 34, 167–177.

——— and D. C. Berman. 2000. "Elysium Planitia Lava Flows: Crater Count Chronology and Geological Implications." *J. Geophys. Res.* 105, 15011–26.

——— and G. Neukum. 2001. "Cratering Chronology and Evolution of Mars," in *Chronology and Evolution of Mars,* eds. R. Kallenbach, J. Geiss, and W. K. Hartmann (Bern: International Space Science Institute); also *Space Sci. Rev.*, 96, 165–94.

——— and Odell Raper. 1974. *The New Mars: The Discoveries of Mariner 9* (Washington, D.C.: NASA Special Publication 337).

———, Ron Miller, and Pamela Lee. 1984. *Out of the Cradle: Exploring the Frontiers Beyond Earth* (New York: Workman Publishing).

———, T. Thorsteinsson, and F. Sigurdsson. (2003, in press). "Martian Hillside Gullies and Icelandic Analogs." *Icarus.*

Hoffman, N. 2000. "White Mars: A New Model for Mars' Surface and Atmosphere Based on CO_2." *Icarus* 146, 326–42.

Hoyt, William Graves. 1976. *Lowell and Mars* (Tucson: University of Arizona Press).

Jankowski, D. G., and S. W. Squyres. 1993. " 'Softened' Impact Craters on Mars: Implications for Ground Ice and the Structure of the Martian Megaregolith." *Icarus* 106, 365–79.

Kallenbach, R., J. Geiss, and W. K. Hartmann, eds. 2001. *Chronology and Evolution of Mars* (Bern: International Space Science Institute, in affiliation with Kluwer Academic Publishers).

Kargel, J.S., and R. G. Strom. 1996. "Global Cllimatic Change on Mars." *Scientific Amer.* 275, No.5, 80–88.

Kieffer, H. H., B. M. Jakosky, C.W. Snyder, M.S. Matthews, eds. 1992. *Mars* (Tucson: University of Arizona Press).

Malin, M., and K. Edgett. 2000. "Evidence for Recent Groundwater Seepage and Surface Runoff on Mars." *Science* 288, 233–35.

———. 2000. "Sedimentary Rocks of Early Mars." *Science* 290, 1927–37.

McKay, D.S., E. K. Gibson, K. L. Thoomas-Keprta, H. Vali, and C. S. Romanek. 1996. "Search for Life on Mars: Possible Relic Biogenic Activity in Martian Meteorite ALH84001." *Science* 273, 974–30.

Mellon, Michael T., and Roger J. Phillips. 2001. "Recent Gullies on Mars and the Source of Liquid Water." *Journ. Geophys. Res.* 106, 23, 165–73, 179.

Melosh, H. J. 1989. *Impact Cratering: A Geologic Process* (New York: Oxford University Press).

Moore, Patrick. 1977. *Guide to Mars* (New York: Norton and Co.).

Nyquist, K., D. Bogard, C.Y. Shih, A. Greshake, D. Stöffler, and O. Eugster. 2001. "Ages of Martian Meteorites," in *Chronology and Evolution of Mars*, eds. R. Kallenbach, J. Geiss, and W. K. Hartmann (Bern: International Space Science Institute); also *Space Sci. Rev.*, 96, 165–94.

Öpik, E. J. 1965. "Mariner V and Craters on Mars." *Irish Astron. Journ.* 7, 92.

Plescia, J. B. 1990. "Recent Flood Lavas in the Elysium Region of Mars." *Icarus* 88, 465–90.

Scott, D. H., K. L. Tanaka, R. Greeley, and J. E. Guest. 1987. *Geologic Maps of the Western Equatorial, Eastern Equatorial and Polar Regions of Mars, Maps. I-1802-A, B and C, Miscellaneous Investigation Series, 1986–1987.* (Flagstaff, Ariz.: U.S. Geological Survey).

Sheehan, William. 1996. *The Planet Mars: A History of Observation and Discovery* (Tucson: University of Arizona Press).

Sheehan, William, and S. J. O'Meara. 2001. *Mars: The Lure of the Red Planet* (New York: Prometheus Books).

Shoemaker, E. M., R. Hackman, and R. Eggleton. 1962. "Interplanetary Correlation of Geologic Time." *Adv. Astronaut. Sciences* 8: 70–89.

Spitzer, Cary. R., ed. 1980. *Viking Orbiter Views of Mars* (Washington, D.C.: NASA Special Publication 441).

Squyres, S. W., S. Clifford, R. Kuzmin, J. Zimbelman, and F. Costard. 1992. "Ice in the Martian Regolith," in *Mars*, eds. H. Caffer, B. M. Jakosky, C. Snyder, and M. S. Matthews (Tucson: University of Arizona Press).

Swindle, T. D., A. H. Thrombin, D. J. Lindstrom, M, K. Burkland, B. A. Cohen, J. A. Grier, B. Li, and E. K. Olson. 2000. "Noble Gases in Iddingsite from the Lafayette Meteorite: Evidence for Liquid Water on Mars in the Last Few Hundred Million Years." *Met. Planet. Sci.* 35, 107–15.

Tanaka, K. L. 1986. "The Stratigraphy of Mars." Proc. Lunar Planet. Sci. Conf. 17, Part 1. *J. Geophys. Res.* Supp. 91, E139–58.

Vreeland, R. H., W. D. Rosenzweig, and D. W. Powers. 2000. "Isolation of a 250 Million-Year-Old Halotolerant Bacterium from a Primary Salt Crystal." *Nature* 407, 897–900

Zubrin, Robert, with Richard Wagner. 1996. *The Case for Mars* (New York: The Free Press).

INDEX

(Page numbers *in italics* refer to illustrations.)

F

T